Water and Agricultural Sustainability Strategies

Water and Agricultural Sustainability Strategies

Editor

Manjit S. Kang
Punjab Agricultural University, Ludhiana, India

CRC Press is an imprint of the
Taylor & Francis Group, an **informa** business
A BALKEMA BOOK

CRC Press
Taylor & Francis Group
6000 Broken Sound Parkway NW, Suite 300
Boca Raton, FL 33487-2742

First issued in paperback 2019

CRC Press is an imprint of Taylor & Francis Group, an Informa business

Typeset by Vikatan Publishing Solutions (P) Ltd, Chennai, India.

ISBN-13: 978-0-415-57219-4 (hbk)
ISBN-13: 978-0-367-38438-8 (pbk)

Library of Congress Cataloging-in-Publication Data

Water and agricultural sustainability strategies / editor, Manjit S. Kang.
p. cm.
Includes bibliographical references and index.
ISBN 978-0-415-57219-4 (hardcover : alk. paper) -- ISBN 978-0-203-84792-3 (e-book) 1. Water-supply, Agricultural--India. 2. Water quality management--India. 3. Soil management--India. 4. Sustainable agriculture--India. I. Kang, Manjit S. II. Title.

S494.5.W3.W348 2010
631.5--dc22

2010002829

Visit the Taylor & Francis Web site at
http://www.taylorandfrancis.com

and the CRC Press Web site at
http://www.crcpress.com

Contents

Contributors vii
Foreword xiii
Preface xv
About the editor xix

Section I: Water and agricultural sustainability challenges

1. **Sustainable water systems for agriculture and 21st century challenges** 3
Ramesh Kanwar

2. **Challenges in securing India's water future** 21
Kapil K. Narula & Upmanu Lall

Section II: Sustainable water management strategies

3. **Soil water management in India** 29
R. Lal

4. **Conserving and optimizing limited water for crop production** 43
D.G. Westfall, G.A. Peterson & N.C. Hansen

5. **On watershed management** 57
Hemant Chowdhary & Vijay P. Singh

6. **Water management in northern states and the food security of India** 71
G.S. Hira

Section III: Sustainable soil management strategies

7. Sustainable management of Vertisols in central India 93
R.P. Rajput, D.L. Kauraw, R.K. Bhatnagar, Manish Bhavsar, M. Velayutham & R. Lal

8. Sustainable management of dryland Alfisols (red soils) in South India 109
S.P. Palaniappan, R. Balasubramanian, T. Ramesh, A. Chandrasekaran, K.G. Mani, M. Velayutham & R. Lal

Section IV: Soil and water quality management

9. Soil and water quality: Integral components of watershed management 133
K.R. Reddy & J.W. Jawitz

10. Water pollution related to agricultural, industrial, and urban activities, and its effects on the food chain: Case studies from Punjab 143
Milkha S. Aulakh, Mohinder Paul S. Khurana & Dhanwinder Singh

Section V: Soil and water management in dryland or rainfed agriculture

11. Manipulating tillage to increase stored soil water and manipulating plant geometry to increase water-use efficiency in dryland areas 169
B.A. Stewart

12. Integrated watershed management for increasing productivity and water-use efficiency in semi-arid tropical India 181
Piara Singh, P. Pathak, S.P. Wani & K.L. Sahrawat

13. Strategies for improving the productivity of rainfed farms in India with special emphasis on soil quality improvement 207
K.L. Sharma, Y.S. Ramakrishna, J.S. Samra, K.D. Sharma, U.K. Mandal, B. Venkateswarlu, G.R. Korwar & K. Srinivas

Section VI: Biotechnological and agronomic strategies

14. Biotechnology and drought tolerance 229
Satbir S. Gosal, Shabir H. Wani & Manjit S. Kang

15. Transgenic strategies for improved drought tolerance in legumes of semi-arid tropics 261
Pooja Bhatnagar-Mathur, J. Shridhar Rao, Vincent Vadez & Kiran K. Sharma

16. Viable alternatives to the rice-wheat cropping system in Punjab 279
B.S. Kang, Kuldeep Singh, Dhanwinder Singh, B.R. Garg, R. Lal & M. Velayutham

17. Weed management in aerobic rice in northwestern Indo-Gangetic plains 297
G. Mahajan, B.S. Chauhan & D.E. Johnson

Subject index 313

Contributors

AULAKH, MILKHA S.
Department of Soils
Punjab Agricultural University
Ludhiana 141 004, India

BALASUBRAMANIAN, R.
M.S. Swaminathan Research Foundation
Chennai 600 113, India

BHATNAGAR, R.K.
The Jawahar Lal Nehru Agricultural University
Jabalpur, India

BHATNAGAR-MATHUR, POOJA
Genetic Transformation Laboratory
ICRISAT
Patancheru 502 324, AP, India

BHAVSAR, MANISH
The Jawahar Lal Nehru Agricultural University
Jabalpur, India

CHANDRASEKARAN
M.S. Swaminathan Research Foundation
Chennai 600 113, India

CHAUHAN, B.S.
International Rice Research Institute (IRRI)
Los Banos, Philippines

CHOWDHARY, HEMANT
Department of Civil & Environmental Engineering
Louisiana State University
Baton Rouge, LA 70803, USA

GARG, B.R.
Punjab Agricultural University
Ludhiana 141 004, India

GOSAL, SATBIR S.
School of Agricultural Biotechnology
Punjab Agricultural University
Ludhiana 141 004, India

HANSEN, N.C.
Department of Soil and Crop Sciences
Colorado State University
Fort Collins, CO 80523, USA

HIRA, G.S.
Additional Director of Research, Agriculture (Retd.)
Punjab Agricultural University
Ludhiana 141 004, India

JAWITZ, J.W.
Soil and Water Science Department
University of Florida
Gainesville, FL 32611, USA

JOHNSON, D.E.
International Rice Research Institute (IRRI)
Los Banos, Philippines

KANG, B.S.
Punjab Agricultural University
Ludhiana 141 004, India

KANG, MANJIT S.
Office of Vice Chancellor
Punjab Agricultural University
Ludhiana 141 004, India

KANWAR, RAMESH
Department of Agricultural and Biosystems Engineering
Iowa State University
Ames, IA 50011, USA

KAURAW, D.L.
The Jawahar Lal Nehru Agricultural University
Jabalpur, India

KHURANA, MOHINDER PAUL S.
Department of Soils
Punjab Agricultural University
Ludhiana 141 004, India

KORWAR, G.R.
Central Research Institute for Dryland Agriculture
Santoshnagar, Hyderabad-500059, Andhra Pradesh, India

LAL, R.
Carbon Management and Sequestration Center
The Ohio State University
Columbus, OH 43210-1085, USA

LALL, UPMANU
Columbia Water Center
Earth and Environmental Engineering and
Civil Engineering and Engineering Mechanics
Columbia University
New York, NY 10025, USA

MAHAJAN, G.
Punjab Agricultural University
Ludhiana 141 004, India

MANDAL, U.K.
Central Research Institute for Dryland Agriculture
Santoshnagar, Hyderabad-500059, Andhra Pradesh, India

MANI, K.G.
M.S. Swaminathan Research Foundation
Chennai 600 113, India

NARULA, KAPIL K.
Columbia Water Center
The Earth Institute
Columbia University
New York, NY USA

Address in India:
Columbia Water Center
India Office (Regd.)
101, K-62
Hauz Khas Enclave
New Delhi 110 016, India

PALANIAPPAN, S.P.
M.S. Swaminathan Research Foundation
Chennai 600 113, India

PATHAK, P.
ICRISAT
Patancheru 502 324
Andhra Pradesh, India

PETERSON, G.A.
Department of Soil and Crop Sciences
Colorado State University
Fort Collins, CO 80523, USA

RAJPUT, R.P.
The Jawahar Lal Nehru Agricultural University
Jabalpur, India

RAMAKRISHNA, Y.S.
Central Research Institute for Dryland Agriculture
Santoshnagar, Hyderabad-500059, Andhra Pradesh, India

RAMESH, T.
Tamil Nadu Agricultural University
Coimbatore, India

RAO, SHRIDHAR J.
Genetic Transformation Laboratory
ICRISAT
Patancheru, Andhra Pradesh 502 324, India

REDDY, K.R.
Soil and Water Science Department
University of Florida
Gainesville, FL 32611 USA

SAHRAWAT, K.L.
ICRISAT,
Patancheru, 502 324
Andhra Pradesh, India

SAMRA, J.S.
National Rainfed Area Authority
Government of India
NASC Complex, DPS Marg, New Delhi-110012, India

SHARMA, K.D.
National Rainfed Area Authority
Government of India, NASC Complex, DPS Marg, PUSA,
New Delhi-110012, India

SHARMA, KIRAN K.
Genetic Transformation Laboratory
ICRISAT
Patancheru, Andhra Pradesh 502 324, India

SHARMA, K.L.
Central Research Institute for Dryland Agriculture
Santoshnagar, Hyderabad-500059
Andhra Pradesh, India

SINGH, DHANWINDER
Department of Soils
Punjab Agricultural University
Ludhiana 141 004, India

SINGH, KULDEEP
Punjab Agricultural University
Ludhiana 141 004, India

SINGH, PIARA
ICRISAT
Patancheru, 502 324
Andhra Pradesh, India

SINGH, VIJAY P.
Department of Biological and Agricultural Engineering
Texas A&M University
College Station, TX 77843, USA

SRINIVAS, K.
Central Research Institute for Dryland Agriculture
Santoshnagar, Hyderabad-500059, Andhra Pradesh, India

STEWART, B.A.
Dryland Agriculture Institute
West Texas A&M University
Canyon, TX 79016, USA

VALDEZ, VINCENT
Genetic Transformation Laboratory
ICRISAT
Patancheru, Andhra Pradesh 502 324, India

VELAYUTHAM, M.
M.S. Swaminathan Research Foundation
Chennai 600 113, India

VENKATESWARLU, B.
Central Research Institute for Dryland Agriculture
Santoshnagar, Hyderabad-500059, Andhra Pradesh, India

VELAYUTHAM, M.
M.S. Swaminathan Research Foundation
Chennai 600 113, India

WANI, SHABIR H.
School of Agricultural Biotechnology
Punjab Agricultural University
Ludhiana 141 004, India

WANI, S.P.
ICRISAT
Patancheru, 502 324
Andhra Pradesh, India

WESTFALL, D.G.
Department of Soil and Crop Sciences
Colorado State University
Fort Collins, CO 80523, USA

Foreword

This book, edited by Dr. Manjit S. Kang is a timely contribution, because there is now a global concern for the future of food and water security, in the context of climate change. Food and clean drinking water constitute the first among the hierarchical needs of a human being. Food security involves economic, environmental and social access to balanced diet and clean drinking water. An important requirement for food security is the enhancement of farm productivity in perpetuity without associated ecological harm, *i.e.*, achieving a paradigm shift from green to an evergreen revolution.

The heartland of the green revolution in India, Punjab, Haryana and Western Uttar Pradesh, is in a state of severe ecological distress. The water table is going down and soils are getting saline. It is clear that farmers will have to shift from exploitative practices to conservation farming. This volume presents authoritative papers on soil and water management, both in irrigated and dryland areas. Prof. Rattan Lal has written an authoritative contribution on soil and water management in India and all other chapters deal with important issues in mainstreaming ecology in technology development and dissemination.

Land-use decisions are also water-use decisions. Therefore, land and water management has to receive concurrent attention. In the area of water security, demand management should receive as much attention as supply augmentation. More crop and income per drop of water should become an international movement. In addition, in dry-farming areas, there should be a pond on every farm for harvesting rainwater.

The future of our agriculture depends on the sustainable management and equitable use of our natural resources. This book shows the way of achieving this goal. In my view, it should be widely read by scholars, students, extension workers as well as policy makers. We owe a deep debt of gratitude to Dr. Manjit Kang and all the authors for their effort towards introducing an era of evergreen revolution.

M.S. Swaminathan
Member of Parliament (Rajya Sabha)
Chairman, M.S. Swaminathan Research Foundation
Taramani Institutional Area
Chennai, India

Preface

"Rising oil prices have focused the world's attention on the depletion of oil reserves. But the depletion of underground water resources from overpumping is a far more serious issue. Excessive pumping for irrigation to satisfy food needs today almost guarantees a decline in food production tomorrow."

—Lester Brown

The world population is expected to increase by nearly three billion during the next 30 to 40 years, and food production will have to be doubled to feed this anticipated population. Agriculture is the biggest user of freshwater. Scarcity of water is pervasive across the world and the problem is expected to accentuate, as more pressure is exerted on the groundwater resources to produce additional food. Groundwater overdraft, if not controlled, is expected to exacerbate the current water-scarcity problem. Agricultural sustainability depends on the sustainability of water resources. Therefore, effective water-management strategies must be devised and instituted to ensure sustainability of agriculture. A 2000 World Water Council report estimated that during 2001–2020, the use of water by humans was expected to increase by about 40%, and that 17% more water than available would be needed to grow the world's food. The World Commission on Water for the 21st Century, constituted by the World Water Council to develop vision on Water, Life and the Environment has concluded, "*Only rapid and imaginative institutional and technological innovation can avoid the crisis.*"

According to the United Nations, 77 million people are expected to face water shortage by 2025, if people continue to use water at the current rate. As a result of scarcity of water, global annual food-production losses could mount to 350 million tons by 2025. The water scarcity and security issues need to be tackled at the local, regional, and global levels. According to World Water Vision (http://www.worldwatercouncil.org/), "*Water resources depletion generates tensions among water users and sometimes between states or countries. Every country wants to maintain sovereignty on its resources and to maximise its storage and withdrawals. Water scarcity already affects one third of the total world population.*"

Despite the expansion of irrigated agriculture, large tracts of the cropland area in many developing and developed countries are rainfed. It is estimated that 80% global agriculture is rainfed, which contributes about 60% to the global food production. In Asia, including India, the rainfed areas are dependent on monsoons, and are subject to drought stress and harsh tropical climate. Water scarcity and climate change are bound to affect rainfed regions that are vulnerable to drought. In most rainfed areas, water-use efficiency (WUE) varies between 15 and 25%. Enhancing WUE is a major strategy in irrigated agriculture, but in the rainfed areas, drought management is the main strategy to increase production. The challenge of enhancing WUE is, however, confounded by the severe problem of water quality either because of high concentration of naturally occurring salts in the aquifer or because of pollution and contamination by urban, industrial, and other anthropogenic activities.

India has about 17% of the world's population and about 16 million people are added to this total every year. Can India continue to feed itself and remain self-sufficient, say 40 years from now? The agricultural scenario in the most productive agricultural region in India, the Indo-Gangetic Plains, is bleak. Since the 1970s, this region, which includes the breadbasket states of Punjab, Haryana, and western U.P., has widely adopted the rice-wheat cropping system. The extensive use of this system has resulted in groundwater depletion and soil-health problems. The rice-wheat cropping system has severely stressed the groundwater resources, especially in areas irrigated with tubewell water. More groundwater withdrawal than recharge has lowered water tables in the Indo-Gangetic region. On the contrary, canal-irrigated areas have become prone to rising water tables, along with increased salinity.

In September 2007, we organized a three-day workshop on "*Water Management Strategies for Food Security and Environment Quality*" at Punjab Agricultural University, Ludhiana. The purpose of the workshop was to bring together expert agronomists, soil scientists, economists, agricultural engineers, and geneticists from the USA and India to discuss the issues related to water scarcity and to offer appropriate solutions to mitigate the effects of the groundwater depletion brought about by the use of unsustainable agricultural practices in India. Scientists from several U.S. universities and various agricultural organizations from India, such as agricultural universities, Indian Council of Agricultural Research and non-governmental organizations, participated in the workshop. The motivation for the workshop was to forestall the impending threat to food security in India, posed by the unsustainable use of water resources, especially in the Indo-Gangetic Plains. Nevertheless, the water scarcity and water security issue is also a major challenge of the 21st century in the rest of the world.

This book "*Water and Agricultural Sustainability Strategies*" mainly represents articles contributed by the workshop participants. A couple of additional authoritative articles are included to make the book more comprehensive and useful. The book has been divided into the following six sections, covering the various aspects related to water and agricultural sustainability issues:

1 Water and Agricultural Sustainability Challenges (two chapters)
2 Sustainable Water Management Strategies (four chapters)

3 Sustainable Soil Management Strategies (two chapters)
4 Soil and Water Quality Management (two chapters)
5 Soil and Water Management in Dryland or Rainfed Agriculture (three chapters)
6 Biotechnological and Agronomic Strategies (four chapters)

Section 1 relates to challenges to ensuring water and agricultural sustainability under the current scenario of water scarcity. Section 2 discusses sustainable strategies that can be used to manage water and soil resources efficiently. Groundwater recharge technologies are also enumerated in this section. Sections 3 and 4 cover soil quality issues as well as those related to water-quality decline, with special reference to groundwater pollution with arsenic and other elements. Soil and water management in different agroclimatic environments, with particular reference to dryland or rainfed agriculture, has been highlighted in Section 5. Biotechnological applications to develop crop varieties possessing drought tolerance and improved water-use efficiency are detailed in Section 6. The last section also discusses agronomic strategies to conserve water; for example, alternatives to the rice-wheat cropping system and weed control issues related to direct-seeded rice.

I am sure that the presented strategies to conserve soil and water resources to ensure water and agricultural sustainability will help to ensure both water security and food security, as well as to alleviate poverty, especially in developing countries. The knowledge contained in this book should be applicable not only to India but also to many other parts of the world.

I trust agricultural scientists, professionals and researchers, as well as graduate students in agronomy; soil science; agricultural engineering; agricultural economics; plant breeding, genetics, and biotechnology; water resources; hydrology; geography; and other agriculture-related fields will find this book useful in their research and educational pursuits.

I am grateful to Professor M.S. Swaminathan for graciously agreeing to write the Foreword. I am indebted to all the authors for supporting this extremely important cause. I trust this effort will prove useful in ensuring water and agricultural sustainability, as well as food security around the world.

Manjit S. Kang
Ludhiana
August 25, 2009

About the editor

Dr. Manjit S. Kang is currently the Vice Chancellor of the Punjab Agricultural University at Ludhiana. He previously served for more than twenty years as a Professor of Quantitative Genetics at the School of Plant, Environmental, and Soil Sciences, Louisiana State University, Baton Rouge. He obtained his Ph.D. degree in Crop Science (Genetics and Plant Breeding) in 1977 from the University of Missouri-Columbia. Earlier he earned his M.S. and M.A. degrees in Plant Genetics and Botany from the Southern Illinois University; and a B.Sc. (Agriculture & Animal Husbandry) degree from Punjab Agricultural University in Ludhiana, India.

In the course of his career, Dr. Kang has specialized in quantitative genetics applied in crop and plant improvement. His expertise is widely recognized and he received several prestigious awards and recognition from institutes all over the world. Dr. Kang is a Fellow of American Society of Agronomy and Crop Science Society of America. He served in 1999 as a U.S. Fulbright Senior Scholar to Malaysia. He serves as a Technical Editor of *Crop Science*, and as Editor-in-Chief of both *Journal of Crop Improvement* and *Journal of New Seeds*. He has also served as the Editor-in-Chief of *Communications in Biometry and Crop Science*. The Punjab Agricultural University-Ludhiana, India, recognized him for his significant contributions to plant breeding and genetics at its 36th Foundation Day in 1997. In 2007, he received from the US-based Association of Agricultural Scientists of Indian Origin (AASIO) the "Outstanding Agricultural Scientist Award". Dr. Kang served as a Sigma Xi (Scientific Research Society) Distinguished

Lecturer from July 2007 to June 2009. In February 2010, he was honored by Amity University, Noida, India, with an "Amity Academic Excellence Award".

Dr. Kang has edited or authored 10 books, including *Genotype-by-Environment Interaction* (1996; CRC Press), *Quantitative Genetics, Genomics and Plant Breeding* (2002; CABI Publishing), *Crop Improvement: Challenges in the Twenty-First Century* (2002; Haworth Press), *Handbook of Formulas for Plant Geneticists and Breeder* (2003; Haworth Press), *GGE Biplot Analysis* (2003; CRC Press), *Genetic and Production Innovations in Field Crop Technology* (2005; Haworth Press), *Agricultural and Environmental Sustainability: Considerations for the Future* (2007; Haworth Press), *and Breeding Major Food Staples* (2007; Blackwell Publishing). He has published more than 120 refereed journal articles in prestigious international journals, over thirty-five book chapters/essays, and more than a 120 other technical publications. Moreover, he has organized various international events in the fields of plant breeding, quantitative genetics and genomics and agricultural and environmental sustainability strategies.

Section I

Water and agricultural sustainability challenges

Chapter 1

Sustainable water systems for agriculture and 21st century challenges

Ramesh Kanwar
Department of Agricultural and Biosystems Engineering, Iowa State University, Ames, Iowa, USA

ABSTRACT

Availability and pollution of freshwater water supplies will be the dominant issues for the global society in the 21st century. With increasing population and climate change scenarios, the demand for water will continue to increase for agriculture, especially irrigation, and other economic uses to meet food, fiber and energy-security needs of the society. At the same time, there is good likelihood that the availability of freshwater will decrease with the fast melting of world's major glaciers. In addition, the water quality of rivers, lakes, oceans, and aquifers will further degrade from the discharge of domestic, municipal, industrial, and agricultural water. Currently, more than five million people die every year because of water-borne diseases and 1.5 billion get sick every year because of poor sanitation or use of poor quality drinking water. Therefore, water-resource management will continue to be of great importance, and global society needs to develop strategies on developing efficient methods for water use and management/reuse of poor quality water for agriculture. The objective of this chapter is to highlight some of the water challenges of the 21st century and identify some of the innovative water management strategies and agricultural practices that can be used in agricultural watersheds to improve water quality and water availability for all needs of society. This chapter also discusses issues related to global warming, melting of glaciers, rise of sea levels, and how these issues will affect water sustainability and food security issues of the 21st century.

Address correspondence to Ramesh Kanwar at Department of Agricultural and Biosystems Engineering, Iowa State University, Ames, Iowa 50011, USA. E-mail: rskanwar@iastate.edu

Reprinted from: *Journal of Crop Improvement*, Vol. 24, Issue 1 (January 2010), pp. 41–59,
DOI: 10.1080/15427520903307700

1.1 INTRODUCTION

Two of the several major challenges of the 21st century facing the global society are: 1) a sustainable food production system to meet food security needs of the increasing world population, and 2) impacts of climate change or global warming on water sustainability to meet all needs of the society to maintain a viable economic system for world peace. The world population is likely to increase to 9 billion people by the year 2050. This increase in population will demand more resources, including food and water. Several studies indicate that world food production must increase by 50% by 2030 and 100% by 2050 to feed the increasing population. Food and water are the two most important issues for world peace and social security in the 21st century. The key question facing the global society today is how to increase food production to feed the increasing population in the world. Another important question facing the global society is how much water would be needed for all the irrigation systems in the world in 2050 to produce enough food to feed 9 billion people on this planet. Answers to these questions are not very easy and will require innovative technical solutions, incentive-based policies, and huge investments in water projects worldwide.

1.2 GLOBAL WARMING AND WATER SUSTAINABILITY

1.2.1 Impact of climate change on glacier melting and sustainability of river water systems

Availability of water for all human needs on this planet is going to be affected severely by climate change. A summary of major sources of water availability on the earth is given in Table 1.1. Only 2.5% of the total volume of water available on earth is fresh water, and about 70% of this fresh water is in the form of glaciers or permanent ice locked up in Greenland, Antarctica and other major glaciers of the world, including Himalayan glaciers (Shiklomanov 1993). The Himalayan glaciers are the major sources of fresh water for the majority of the population in Asia. These glaciers are the source of water for

Table 1.1 Major water reservoir/sources on earth (Gleick 2000).

Water sources	*Volume (1000 km^3)*	*% total water*	*% total fresh water*
Salt water sources			
Oceans	1,338,000	96.54	–
Saline ground water	12,870	0.93	–
Saltwater lake	85	0.006	–
Total		97.48	
Fresh water sources			
Glaciers/ground ice	24,364	1.76	69.56
Ground water	10,530	0.76	30.06
Lakes	91	0.007	0.26
Rivers	2.12	0.0002	0.006
Marshes/wetlands/Soil moisture/Water vapor	40.9	0.003	0.12
Total		2.52	

some of the major rivers of the world (Indus, Ganges, Brahmaputra, Mekong, Yangtze, and Yellow River in Asia). The glacier-fed rivers originating from the Himalayan glaciers and surrounding the Tibetan Plateau are the largest source of river water from any single location in the world and support the daily livelihood of 1.5 billion people, which is about 25% of the total current population of the world. These glaciers are the major source of water for these rivers, especially during summer months, from the melting of ice and snowfall accumulated across centuries in Himalayan glaciers. These glaciers accumulate snow during winter months and, with slow melting during summer months, feed the rivers, which eventually become a major source of irrigation water for downstream agriculture and other economic activity in river basins. The Indus river basin has one of the largest irrigation networks in the world, consisting of about 16 million hectares. About 90% of Pakistan's crop is produced under irrigated conditions, and all of the water for Pakistan's irrigation systems comes from barrages along the Indus. The Ganges, Yangtze, and Yellow River also have some of the largest irrigated areas in the world – 17.9, 5.4, and 2.0 million hectares, respectively.

With global warming, Himalayan glaciers are melting much faster than earlier estimated and will affect the livelihood of 25% of the world's population, and the sustainability of the food production system is at risk. Some studies indicate that these glaciers will completely melt away by 2050 and will severely affect food security and agriculture in the river basins and water supplies for urban areas located along the riverbanks. Once these glaciers disappear, some of these rivers will have decreased water flow or even no flow in summer months, severely affecting people's livelihoods and posing a major challenge for human survival in these river basins. The only alternative for providing a stable water supply for agriculture in these river basins would be to develop mega water-harvesting plans to create additional water storage capacities in the watersheds by constructing a costly series of reservoirs and dams. At the same time, these reservoirs could have potential negative impacts on the environment and could cause the migration of millions of people in Asia to other locations. Therefore, innovative ideas and creative public policies would be needed to make sure these water-harvesting projects are sustainable.

Another consequence of global warming would be the rising sea-water levels by 5 to 8 meters, bringing many of the world's agricultural lands in the coastal areas under oceans, thus resulting in less area available for agriculture. Less land area available for agriculture and lesser availability of water for irrigation will make things extremely difficult to meet the demand of global food security in 2050. Between 1900 and 2000, irrigated area worldwide increased from about 50 million hectares to 250 million hectares (Gleick 2000), a remarkable progress in irrigated agriculture in the 20th century. In the year 2000, more than one billion ha of world area was cultivated, of which 26% of the area was irrigated, producing more than 40% of all food grown in the world (Gleick 2000). Global water use for three major categories (agricultural, industrial, and municipal use) is given in Table 1.2. Irrigation accounts for nearly 85% of all water consumed worldwide, which makes less water available for other uses (Table 1.2). India and China together have more than 36% of the world population to feed and more than 21% of the world population lives in South Asia. Although world food grain production has increased significantly, partly because of the gains made in irrigation systems, much improvement in feeding people has occurred in Asia (particularly India and Pakistan) because of green revolution and increased water use for irrigation. In spite of these gains, 830 million people remain undernourished – 45% of them in India and China alone. These data clearly indicate that food production alone cannot solve the local and regional food security needs and that things are going to be a lot more difficult in 2050 unless major efforts are made by the global society to address the challenges of climate change and water availability.

1.3 POTENTIAL SOLUTIONS FOR WATER AND FOOD SUSTAINABLE SYSTEMS IN RESPONSE TO 21ST CENTURY CHALLENGES OF CLIMATE CHANGE AND MELTING OF GLACIERS

Data on annual available renewable water resources for countries in South Asia, China and the United States are shown in Table 1.3. Agriculture continues to be the major user of renewable water withdrawals (Table 1.3). For some countries in South Asia (especially Nepal, Pakistan, and Sri Lanka), agricultural water use is more than 95% of the total withdrawal. This brings up more questions on the efficiency of water use for agricultural purposes by introducing innovative irrigation systems, such as drip irrigation in comparison with flood irrigation. Increased efficiency in water use in agriculture can save water for other uses. In addition, improved water-use efficiency in irrigation can result in more food production without increasing additional demands on fresh water.

Table 1.2 Global water use for agriculture, industry, and municipal (Gleick 2000).

Category	*1990*	*1950*	*1980*	*2000*	*2025*
Population (million)	–	2,542	4,410	6,181	7,877
Irrigated area (m. ha)	47.3	101	198	264	329
Agricultural use (km^3/yr)	406	849	1,688	1,970	2,331
Industrial use (km^3/yr)	3.4	14.4	59	85	133
Municipal use (km^3/yr)	4.2	14	42	64	84

Table 1.3 Annual renewable water resources and water withdrawal rates for South Asia, China, and the United States for the year 2000 (Gleick 2000).

Country	*Renewable water resource (km^3/yr)*	*Renewable withdrawal (km^3/yr)*	*Agricultural use (km^3/yr)*
Bhutan	95	0.02	0.01
Bangladesh	1210	14.6	12.6
India	1908	500	460
Nepal	210	29	28.6
Pakistan	429	156	151
Sri Lanka	50	9.8	9.4
China	2830	526	405
USA	2478	469	197

Maintaining a good standard of living will require renewable water resources capacity of 1000 m^3 per person per year in countries with a thriving economy (Bouwer 1993). China is developing future management plans on renewable water supplies of 500 m^3 per person per year to sustain its economy, whereas India's planners are using 250 m^3 per person per year. Many others will have less renewable water resources for their economic growth (Bouwer 1993). India is one of the few countries in the world that is blessed with rich water resources. The surface and groundwater resources totaling 231 Mham are plenty to meet India's growing irrigation and industrial development needs for the year 2050, although the distribution of water resources in India is highly variable, which is a huge challenge.

1.3.1 Improve infrastructure to store and distribute good quality water for drinking

One of the challenges India and other developing countries face is the lack of infrastructure for water storage in small to large reservoirs, drinking water treatment facilities, water distribution systems to carry good quality drinking water to homes, and lack of good infrastructure for good sanitation to prevent diseases and provide a higher standard of living for her people. One of the major problems India is facing is lack of capacity to harness the large renewable surface water resource from rains, in particular from monsoon rains, to meet its water needs for agriculture and increasing demand for its growing population and industry. Much of the surface water goes to the sea and outside India's borders. India is harvesting about 20% of total surface water through reservoirs, but capacity must be increased to meet the needs of current and future economic growth. Unless innovative and economically viable strategies are developed for water harvesting, groundwater recharge, and protecting the quality of existing water resources, food security and sustainable economic growth in India and similarly developing countries in the world are very much at risk to meet their future needs for water. Both developing and developed countries of the world need to create plans for making investments in water infrastructure (reservoirs, groundwater recharge, efficient water and waste treatment plants, efficient water delivery systems without leaky pipes, installing low-cost and low-flow toilets) and for investing in innovative irrigation systems because water is the backbone of agricultural and industrial economy, similar to investments in meeting energy needs.

1.3.2 Enhance water conservation by improving water-use efficiency and avoiding wasteful use of water

Agriculture (in particular irrigation) accounts for nearly 85% of all water consumed worldwide, and India's share for water use in agriculture is around 86%. Without irrigation, natural rainfed agricultural areas in the world would not be able to feed the world's current and future population in 2050. Currently, more than 500 million people live in countries with insufficient water to produce their own food and will depend on having to import food from other countries to meet their food needs. Expansion of irrigated areas is becoming more difficult because of lack of available land, limited availability of freshwater resources, huge cost of irrigation systems, and cost of bringing marginal lands under irrigation. When water is used effectively and

Table 1.4 Comparison in water savings between drip and flood irrigation.

	Water requirement (Ac-Inches)		
Crop	*Drip irrigation*	*Flow irrigation*	*Water savings with drip irrigation (%)*
Tomato	9.6	13.3	28
Capsicum	11.5	17.0	33
Potato	11.0	15.0	27
Brinjal	8.5	12.0	29

Source: Chandrakanth et al. 2004.

efficiently, it would result in water conservation by avoiding wasteful use of water. Water conservation can allow potentially more expansion, bringing more agricultural areas under irrigation. Current irrigation practices use open channels for water transport and flood methods of irrigation are highly wasteful and sometime result in waterlogging of productive agricultural land. Also, flood irrigation systems are highly inefficient because of very high evaporation and percolation losses. Many farmers use groundwater as a source for irrigation water, and inefficient irrigation systems are depleting groundwater aquifers at a fast rate. In some of Punjab and Karnataka in India, groundwater tables have declined more than 200 meters in the past 30 years because of over-pumping of aquifers, which has resulted in huge capital investments for installation of deeper wells with higher pumping costs. One of the best solutions is to encourage farmers to adopt a drip-irrigation system that will save tremendous amounts of water and will reduce pumping costs. Drip-irrigation technology is being adopted for widely spaced crops in addition to fruits and vegetables. In one of the recently conducted studies in Karnataka, it was demonstrated that use of drip irrigation could reduce water use by 27% to 33% (Table 1.4) and could give better yields in comparison with flood irrigation. In addition, because of effective and precise application of water to the root zone under drip irrigation, we found less pressure of weeds, and labor requirement was reduced by 20% to 30% under drip method compared with flow irrigation.

1.3.3 Develop strategies for using treated municipal and industrial wastewater for irrigation

In many countries, increased urbanization and industrialization is producing large quantities of wastewater being disposed of directly either to surface water or ponded on the land surface without any primary or secondary treatment. This wastewater is becoming a source of water contamination as well as spreader of diseases. In India, currently about 15% of total water is used for meeting domestic and industrial requirements, but this share will increase to 30% by 2050 because of a fast growth rate in these sectors (Anonymous 1999). Out of the 16,625 million liters of wastewater produced in India, 72% is collected and only 24% is treated before releasing to rivers and streams. Treating this wastewater will require an investment of $65 billion, but these treated wastewaters could be a very good source of irrigation water because of their high nutrient contents. At the same time, uncontrolled and indiscriminate disposal of these waters will contaminate natural water resources and will impact human health from environmental contamination. Using treated wastewater for irrigating high-value crops like vegetables and fruits is a good option and should be pursued in the future. Environmentally sound and cost-effective application of treated wastewaters for irrigating public lawns, forestry plantations, and non-edible crops having high transpiration rates will create sound economic systems.

1.4 WATER SUSTAINABILITY FOR GLOBAL FOOD SECURITY: PROTECTING WATER QUALITY FROM INTENSIVE IRRIGATION AND AGRICULTURAL PRODUCTION SYSTEMS

The availability of cropland for growing food is becoming another question for many of the world's fastest growing economies. Loss of prime agricultural land to urban and industrial development is a major concern in China, Indonesia, and the United States. Total cropland area per capita in the world has decreased from about 0.31 ha per person in 1983 to 0.25 ha person in 2000 (Gleick 2000). Because total area under cropland per person is decreasing, agricultural production systems are becoming intensive to grow more food per unit area of land. The intensification of agriculture, especially under irrigated conditions, has brought new environmental quality problems, including soil erosion, land degradation from salt accumulation, and decreased water quality caused by chemicals and animal manure. To provide global food and water security to the growing population, the final question would be: *what are the impacts of intensive irrigation and agricultural production systems on water sustainability and degradation of land and water resources?* Ecological sustainability of our planet and economic prosperity of global society are twin elements of global stability. About 30–40 years back, it was a popular belief that the goals of economic development for social prosperity of people and environmental quality were mutually exclusive. Today, we need a better understanding between the industrial and agricultural development and environment to bring about long-term sustainability on the planet. The first and foremost component of a comprehensive environmental assessment policy for countries should be that economic developments through industrial and intensive agriculture models are environmentally sound and ecologically sustainable.

If we look at the global availability of water resources and available productive land areas to sustain agriculture to grow enough food to feed a population of about 9 billion in 2050, we see a frightening outlook. The impact of this increased population will be severe on the environmental quality of land and water resources and overall sustainability of the global economic system. The increased population will put enormous stress on the available land and water resources that would be used even more intensively for industrial and agricultural production systems, while trying to maintain environmental quality. An increased population will require that we must grow more food with less water, using more intensive agricultural production systems, possibly requiring more use of pesticides and inorganic fertilizers. Intensive agricultural production systems were introduced in the 1960s with the advances in improved crop varieties, mechanization, sprinkler and drip-irrigation systems, and increased availability of pesticides and fertilizers. More recent experiences from developed countries, especially in Europe and the United States, have shown that modern and intensive agricultural production systems have increased land degradation and water contamination and are adding enormous stress on finite sources of available water. The increased use of chemicals and animal manure as fertilizer has increased pesticide and nitrate concentrations in surface and groundwater sources in agricultural watersheds (Council for Agriculture Science and Technology 1985; Kanwar, Baker & Baker 1988; Kanwar, Colvin & Karlen 1997; Hallberg 1989; Kanwar & Baker 1993; Kanwar 2006; Bakhsh & Kanwar 2007; Bakhsh et al. 2009). Higher concentration of nitrate in a groundwater well used for drinking water was first recognized as a health problem in 1945 when two cases of infant methemoglobinemia (blue baby syndrome) were reported in Iowa (Comly 1945) and in South Dakota (Johnson, Bonrud & Dosch 1987).

Many countries in the world, especially the United States, Brazil, Argentina, India and the rest of South and Southeast Asia are blessed with productive land and availability of water for their agriculture and economic development. For example, India's 30% cropland area is currently irrigated, but this irrigated area produces 56% of the country's cereal grain. India's rainfed agriculture occupies 53% of cultivated area and produces only 44% of food grains. Although food production in India and the rest of South Asia has increased significantly, India and Pakistan, in particular, have seen a sharp degradation of its natural resources (soil, water, and air). The demand for productive land and water resources will continue to grow worldwide to provide food security to the burgeoning population. In the global context, India is feeding 16% of the world population with only 2.4% of the world's geographical area (Kanwar 2003). The per capita availability of land in India has decreased from 0.9 ha in 1951 to about 0.25 ha in 2000. It is quite possible to increase the intensity of Indian agriculture to another 300% as India is blessed with good quality land, abundant water resources, plenty of sunshine hours annually, skilled labor, and an excellent network of research and extension institutions in the country. India's grain production increased from 50 million tons in 1947 to more than 210 million tons in 2000–2001. This increase in grain production has been higher than population growth rate in the 20th century, and India is a successful model in the world community to provide food security to its massive population.

Following are some of the examples of degradation of soil and water quality because of intensification of agriculture worldwide:

1.4.1 Nitrate and pesticide pollution of groundwater

Several studies conducted in the United States have shown that many of the herbicides, such as atrazine, alachlor, and cyanazine, are lost mainly in surface runoff (Baker & Johnson 1983) but have also been found in the shallow groundwater sources (Kanwar 1991; Kanwar et al. 1993: Kalita et al. 1997). In one of the recently completed studies in Iowa, the flow-weighed average NO_3-N concentrations in subsurface drainage are presented in Table 1.5 for four corn-soybean cropping patterns under fertilizer and manure management systems (Hoang et al. 2009). For six year average, flow-weighed NO_3-N concentrations in shallow groundwater ranged from

16.7 to 40.6 mg/l (Table 1.5), much higher than required for drinking water quality standard of 10 mg/l in the United States set by U.S. Environmental Protection Agency (1992). Significant differences were also found among plots receiving manure each year in comparison with plots receiving manure to corn phase only. Corn and soybean plots receiving manure each year showed the highest level of NO_3-N concentrations (40.6 mg/l) in groundwater in comparison with treatments receiving manure every other year to corn plots only. These data indicate that significant amounts of nitrate leaching to shallow groundwater systems can occur if a large amount of nitrogen is applied to corn as well as to soybean plots.

In another study conducted in the United States, Kanwar, Colvin, and Karlen (1997) reported NO_3-N and pesticide losses to shallow groundwater systems under intensive corn and soybean production rotation for conventional and conservation tillage systems. Data from this study on NO_3-N lossses with subsurface drain water, which ranged from 4.8 kg/ha in 1992 to 107.2 kg/ha in 1990 from plots receiving N-fertilizer at rates of 168 kg-N/ha to 200 kg-N/ha, are given in Table 1.6. Three year average (1990–1992) NO_3-N losses with subsurface drain water were much higher under continuous corn

Table 1.5 Treatment means for annual subsurface drainage NO_3-N concentration (mg/l) (Hoang et al. 2009).

	Years[a]						
Treatments	*2001*	*2002*	*2003*	*2004*	*2005*	*2006*	*Average (2001–2006)*
CNMS	24.9a	16.9b	26.8b	36.6c	25.7b	22.5c	25.5b
CMTS	25.9a	31.8a	29.4b	70.4a	43.2a	16.1d	40.6a
SNM	15.8c	19.3b	16.1c	18.6d	14.4c	43.2a	16.7c
SWFM	31.5a	20.7ab	44.6a	50.1b	43.5a	34.0b	37.4a

[a] Means within years and on average followed by the same letter are not significantly different at p = 0.05.
CNMS = corn after soybean – fall manure applied to corn only.
CMTS = corn after soybean – fall manure applied to corn and soybean every year.
SNM = soybean after corn – no manure applied to soybean.
SWFM = soybean after corn – fall manure applied to soybean.

Table 1.6 Average yearly NO_3-N losses to groundwater as a function of tillage and crop rotation for three years (1990–92), kg/ha (Kanwar, Colvin & Karlen 1997).

Year	*Crop rotation*	*Chisel plow*	*Moldbd plow*	*Ridge-till*	*No-till*
1990	Cont. corn	100.0	58.1	83.4	107.2
1991	-do-	76.0	62.7	58.2	61.7
1992	-do-	17.0	16.6	10.2	14.9
Average (1990–92)	-do-	64.3a	45.8a	50.6a	61.2a
1990	Corn-soybean	52.4	38.0	30.3	36.5
1991	-do-	36.3	35.5	29.4	30.3
1992	-do-	15.3	9.1	11.2	4.8
Average (1990–92)	-do-	32.1a	27.5a	23.7a	23.9a

Table 1.7 Average herbicide loss to groundwater as a function of tillage and crop rotation for three years (1990–92), g/ha (Kanwar et al. 1993).

		Herbicide loss with subsurface drain water g/ha			
Crop rotation	*Herbicide*	*Chisel plow*	*MB plow*	*Ridge-till*	*No-till*
Continuous corn	Atrazine	4.4	2.17	5.9	7.3
	Alachlor	0.36	0.06	0.34	0.31
Corn-soybean	Cyanazine	0.10	0.25	0.19	0.16
	Alachlor	0.05	0.62	0.39	0.16
Soybean-corn	Metribuzin	1.7	1.70	3.4	2.5
	Alachlor	0.79	0.06	0.11	0.16

in comparison with the corn-soybean rotation. Although NO_3-N concentrations were greater under conventional tillage system (moldboard plow + disking) than under a no-till system, total NO_3-N losses with subsurface drain flow were higher under the no-till and chisel plow systems because of greater volume of water moving through the soil. This shows that all tillage practices used for corn-soybean production systems will contaminate the groundwater system under intensive agriculture.

Data on herbicide leaching with subsurface drain water as a function of tillage and crop rotation are presented in Table 1.7 (Kanwar et al. 1993). These data clearly indicate that atrazine losses to groundwater were greatest in comparison with other herbicides. In addition, no-till and ridge-till systems resulted in even much greater losses of atrazine, cyanazine, and metribuzin to groundwater because of the preferential movement of these herbicides through macropores (worm holes and natural fractures). The results of this study showed that total yearly average losses for atrazine and alachlor ranged from 2.2 to 7.3 g/ha and 0.06 to 0.62 g/ha, respectively. This study also clearly showed that intensive crop-production systems resulted in surface and groundwater contamination of Iowa soils.

In highly intensive agricultural areas of India and the rest of the world (such as Punjab, Haryana, and Uttar Pradesh), intensive grain-production systems have used fertilizer and pesticide application rates similar to those used in the United States. In these areas, nitrate and pesticide pollution of groundwater sources have been reported. In Punjab, the average NO_3-N concentrations of 2.25 mg/l to 10 mg/l in shallow groundwater have been reported (Gupta et al. 2000). However, Bajwa (2001) reported that nitrate concentrations of more than 100 mg/l had been observed at times in selected tubewell waters during the time of excessive irrigation in Punjab. Several studies have indicated that 11% to 48% of applied nitrogen in maize-wheat production systems leached to groundwater systems. Gupta, Jain, and Arora (1999) detected nitrate concentrations of 12 to 16 mg/l in the Talkatora Lake near Jaipur in the state of Rajastan in India as a result of urban pollution.

1.4.2 Loss of forest lands to make room for agriculture

About 200 years back, Europe was predominantly a forestland, but today a majority of the forests in Europe have been cleared to create new land for agriculture and a growing urban population. Nepal was prominently a forestland country about 200 years back. Today, much of forestland in Nepal has been cleared for food, fiber, and fuel, causing large-scale floods and enormous erosion of agricultural and forest lands. India has 15% of the world's population, but today she is left with only 2% of the world's forestland. India had been losing forests at a rate of 1.5 million hectares per year to agriculture between 1972 and 1982. Grasslands and forestlands are very important for the sustainability of ecosystems on our planet. Deforestation and exploitation of grasslands have increased soil erosion extensively and caused flooding of lowland areas and overflowing of rivers. Unless global society puts good agricultural and forest-management practices in place to protect land and water resources, further environmental degradation of our ecosystem is inevitable in developed as well as in developing countries.

1.4.3 Soil erosion

Soil erosion is the number one source of water pollution worldwide. In the Midwestern part of the United States, where intensive agricultural production systems were introduced in 1960s, soil erosion rates varied from 5 to 30 tons/ha in 1970s, but recently farmers have brought soil erosion rates down to sustainable limits by implementing no-till and conservation-tillage systems on their farms. In India, it has been reported that nearly one-third of the cropland area of 161.5 Mha suffers from water and wind erosion. Intensive agriculture on hills and steeper soils; grazing on steep hills by sheep, goat, and cattle; lack of conservation practices in watersheds; and heavy rainfall are the main causes of severe soil erosion and siltation of rivers and reservoirs. On many agricultural lands in south Asia, China, Africa, and the Middle East, soil erosion rates vary from 20 to 100 tons/ha/yr depending on their geographical location and production system. Soil erosion is a serious land-degradation problem in the rainfed agricultural areas of Asia and needs immediate attention to improve soil's productive capacity (Velayutham & Bhattacharyya 2000) and maintain sustainability of marginal lands used for agriculture.

1.4.3.1 Controlling soil erosion

Research and common-sense experience have shown that no-tillage and conservation-tillage systems must be practiced on highly erodible soils to control erosion. These practices will help increase organic matter and reduce soil erosion. Other practices include contouring, terracing, grass waterways, strip cropping, and innovative crop rotations. In addition, planting trees and shrubs on hill slopes and promoting rotational grazing along stream banks and forest areas will reduce erosion significantly. Another important program in Asia will require significant public investments to construct a series of reservoirs for water-harvesting projects to minimize the effect of floods caused by heavy monsoon rains. These floods are the major cause of large-scale soil erosion and siltation of reservoirs and river bottoms.

1.4.4 Salinization of soils

Excessive irrigation with low irrigation efficiency, inadequate drainage for canal-irrigation systems, and overexploitation of groundwater for irrigation has resulted in the accumulation of salts in soils. Abrol and Bhumbla (1971) estimated that about 7 million ha soils are affected by salinity and alkalinity in Gangetic Plain alone, and nearly 50% canal-irrigated soils are suffering from salinization and alkalization. Excessive flood irrigations are causing salinity problems in many areas of India, Pakistan, and China. The salinity problems are increasing at such a fast rate that these areas may become totally unfit for producing any vegetation and will bring social and economic discomfort to the society.

1.4.5 Decreasing supplies of groundwater

Groundwater is the major source of irrigation water worldwide. About 60% of the cropland area worldwide is irrigated with groundwater. In North America and several countries of Asia (in particular China, India, Pakistan, Nepal, and Sri Lanka), groundwater is the major source of irrigation water. In Punjab, Haryana, Karnataka, and western Uttar Pradesh, shallow groundwater has been used for decades to meet extensive demand for irrigation water, causing significant water withdrawal from good quality aquifers. In Punjab, nearly 70% of the area is irrigated with tubewells (shallow-water table wells); the number of tubewells increased from 192,000 in 1971–1972 to 800,000 in 1993, resulting in an average lowering of water tables by 0.2 m to 0.8 m per year. In the semi-arid regions of the southern Indian state of Karnataka, groundwater tables have fallen by as much as 200 meters during the past 30 years. This has resulted in more discharge from aquifers compared with their annual recharge from rainwater, which is a highly unsustainable irrigation system. Without sound policies on groundwater use for irrigation and groundwater-recharge projects in watersheds, society is moving toward unavoidable disasters.

1.4.6 Increased waterlogging and salinity in productive soils

Poor planning and mismanagement of irrigation systems, particularly in Uzebekistan, India, and Pakistan, have resulted in rising of water tables in canal-irrigated areas, causing waterlogging, contamination of groundwater with agricultural chemicals, and salinization of some of the most productive agricultural lands in the world. The problem of rise in water table is more severe in arid and semi-arid regions of India and certain other areas of central Asia. Uzebekistan and Kazakistan are two other countries where excessive irrigation has raised water tables by as much as 29 m. Lack of subsurface drainage to collect excessive percolation water from surface-irrigation methods has resulted in the rise of water tables. One of the best examples can be cited in India, where Indira Gandhi Canal was brought to irrigate the driest or desert areas of Rajasthan in 1961. Water tables rose at a rate of 1 m per year after the beginning of intensive irrigation practices in Rajasthan (Hooja, Niwas & Sharma 1994). In other areas of Rajasthan, excessive canal irrigations have made areas unfit for cultivation during the monsoon season (Rao 1997). Rise in water tables has caused waterlogging on more than 8.5 Mha and have added salinity problems to additional 3.9 Mha (Singh & Bandhopadhyay 1996). The state of Uttar Pradesh in India has seen an annual addition of 50,000 ha area to salinity buildup caused by mismanagement of irrigation practices (Yadav 1996).

1.5 BMPS FOR WATER SUSTAINABILITY AND TO MINIMIZE CONTAMINATION OF SURFACE AND GROUNDWATER

Best management practices (BMPs) are those that can be used for protecting water quality and enhancing water sustainability as well as controlling soil erosion. The following BMPs could possibly be used to control environmental degradation of soil and water resources.

1.5.1 Chemical rates and methods of applications

Several studies have shown that reductions in chemical application rates could reduce the amount of chemicals available for leaching to groundwater or surface runoff. Hoang et al. (2009), Bakhsh and Kanwar (2007), Kanwar, Baker, and Baker (1988), and Baker and Johnson (1983) have summarized the results of several field studies indicating higher NO_3-N concentrations in groundwater were related to higher N applications. Therefore, the best practice to control surface and groundwater pollutions would be to apply fertilizer and other chemicals at reduced rates to meet the crop N and P uptake requirements.

1.5.2 Placement of chemicals

One of the approaches to reduce the leaching of chemicals to groundwater and/or surface water under rainfed and irrigated conditions would be to incorporate the chemical with soil. If a chemical is broadcast applied, it will mix with the incoming rainfall or irrigation in a thin mixing zone (about 10 to 20 mm thick) and will either leach to subsoil layers or become part of the runoff water. If a chemical is incorporated, it is less susceptible to runoff losses. With banding practice, the rate of chemical application can be reduced to 50% or more (Kanwar 2003). Kanwar and Baker (1993) have shown that banding of herbicides had a significant effect on water-quality improvement.

1.5.3 Timing of chemical application

Appropriate timing of fertilizer and chemical applications under irrigated or rainfed agriculture can enhance the efficient use of chemicals (especially N) and result in decreased leaching losses to groundwater. Kanwar and Baker (1993) have shown that split-N or multiple applications at different times during the growing season resulted in lower residual soil N and NO_3-N concentrations in shallow groundwater.

1.5.4 Water conservation by improving irrigation and drainage systems

One of the best practices for water sustainability is to improve the efficiency of irrigation systems, which is very low for surface irrigation systems. Although irrigation efficiencies have been improving across years, in some countries one of the major problems is low irrigation efficiency. Majority of the farmers in developing countries need to be provided with the latest knowledge and training to improve their irrigation methods. In addition, introduction of new methods, such as sprinkler and micro or drip methods, can save water and will significantly improve irrigation efficiencies. New methods of irrigation will help in increasing nutrient-uptake efficiencies and reduce water contamination from the use of agricultural chemicals. Irrigation and drainage practices are typically considered production practices rather than BMPs for water-quality enhancement. The method of irrigation and rate, amount, and timings of irrigation applications are important considerations for irrigation management. Several new irrigation systems that deserve consideration are low-energy precision application (LEPA) method, sprinkler, drip and sub-irrigation methods, and use of chemigation. Local agricultural production systems, scale of economy, and hydrologic and geologic factors must be considered before selecting the best-suited irrigation-management practice for local farmers in different countries. Kalita and colleagues (1997) and Kalita and Kanwar (1993) reported that shallow groundwater table/sub-irrigation management practices in certain areas of the Midwest United States could be used to reduce the risk of groundwater contamination and to increase crop yields by using subsurface drainage water for irrigation.

1.5.5 BMPs for minimizing salinity and protecting groundwater

One of the best-known BMPs to correct the salinity/alkalinity problem of salt-affected soils in irrigated areas is to provide adequate drainage to flush out salts from the cropland with surface or subsurface drainage. This will do two things. First, it will reduce the buildup of salts in the soils and, second, it will reduce the leaching of salts and agricultural chemicals to shallow groundwater. Without adequate drainage systems, accumulated salts cannot be flushed out. Another good practice that has been used is the reclamation of sodic and saline soils through chemical and biological amelioration. Sodic soils can be easily reclaimed using gypsum and solubilizing calcium and sodium salts and flushing them out of the active root zone. These methods have been found to be extremely successful in India and Pakistan. Other methods are biological controls, including increasing vegetation and proper management of irrigation and drainage practices. Leaching and drainage are the essential parts of reclamation for chemical and biological methods.

1.5.6 Conservation tillage systems

Any conservation-tillage system that leaves at least 30% of the soil surface covered with crop residue is defined as a conservation-tillage system. Several conservation-tillage systems (namely no-till, ridge-till, and chisel plow) are being used to reduce soil erosion and energy-input costs, but these systems may require more pesticide use to control weeds in cropland areas. In recent studies conducted at Iowa State University (Kanwar et al. 1993; Kanwar, Colvin & Karlen 1997; Kanwar & Baker 1993), it has been concluded that conservation-tillage systems increased infiltration, organic carbon, adsorption of chemicals, and microbial activity, and could potentially reduce the chemical leaching to groundwater. Conservation-tillage practice could be an effective BMP for reducing energy requirements, controlling surface and groundwater pollution, and reducing soil erosion under highly intensive agricultural production systems.

1.5.7 Cropping systems

Different cropping systems have been found to reduce leaching of agricultural chemicals to surface and groundwater systems because crop rotations affect the use of chemicals. For example, corn-soybean rotation will not use nitrogen fertilizer in the soybean years, whereas continuous corn practice will use nitrogen year after year. In addition, crop rotations offer a greater diversity of pesticide use within a watershed to control nonpoint-source pollution. Kanwar, Colvin, and Karlen (1997) reported that continuous corn practice is not a sustainable production system because this system requires N application rates for corn every year and results in higher NO_3-N losses to the groundwater. At the same time, Kanwar, Colvin, and Karlen (1997) found corn-soybean rotation system to be a sustainable system for Midwest farmers if fertilizer applications rates could be matched with crop-uptake requirements.

1.5.8 Buffer strips around surface water bodies and well buffer zones around groundwater wells

Several studies have reported that vegetative filter strips and grassed strips of certain widths at the end of agricultural fields and along river banks or open ditches can reduce nitrate and pesticide losses to surface water ranging from 19% to 22%. Therefore, vegetative filter strips and waterways have the potential to serve as BMPs to control water quality and sedimentation problems in streams and rivers.

1.6 SUMMARY AND CONCLUSIONS

Much of the world has been blessed with good land and water resources, but increased population pressure, human greed, disregard for maintaining healthy ecosystem, poor public environmental policies, and mismanagement of natural resources in the latter part of the 20th century have created significant problems for the long-term sustainability of global society on this planet. A couple of major impacts of the human population on the environment quality have been the rapid depletion and pollution of water resources and increased carbon emission into the atmosphere to cause global warming. For example, South Asia has been blessed with two major natural resources, relatively productive land and a good reservoir of water resources. At the same time, South Asia has 21% of world population and one of the highest population densities and population growth rates. Increased population pressure has already reduced the land area per person, which is expected to shrink per capita in cultivable land further in the years to come. Demand for finite water resources is increasing and has already increased five times in the last 10–15 years. With continuously increasing population, contamination of water resources is on the rise. Also, an increase in population in South Asia and the rest of the world means further intensification of agricultural production systems to double food production by 2050 to feed a population of 9 billion people. This means demand for irrigation water and agricultural chemicals will increase to produce more food, resulting in the pollution of soil, water, air, and our fragile ecosystems on the planet. Ever-increasing demands on our natural resources have added enormous stress on the available land and water resources. Unless individual countries develop good land- and water-management policies and state-of-the-art, highly efficient irrigation and crop-production technologies for water conservation, reduced carbon emission, and environmental enhancement, the core issues of food security and water sustainability for the global society are very much at risk in the 21st century.

REFERENCES

Abrol, I.P. and D.R. Bhumbla. 1971. Saline and alkali soils in India: Their occurrence and management. Paper presented at FAO/UNDP seminar on soil fertility research. *FAO World Soil Research Rep.* 41: 42–51.

Anonymous, 1999. *Integrated water resource development-A plan for action.* Ministry of Water Sources, Government of India, New Delhi, India.

Bajwa, M.S. 2001. *Oral communication: Nitrate pollution of groundwater in Punjab.* Ludhiana, India: Punjab Agricultural University.

Baker, J.L. and H.L. Johnson. 1983. Evaluating effectiveness of BMPs from field studies. In *Agricultural management and water quality*, pp. 281–304. eds. F.W. Schaller and G.W. Bailey. Ames, Iowa: Iowa Sate University Press.

Bakhsh, A. and R.S. Kanwar. 2007. Tillage and N application rates effect on corn and soybean yields and NO_3-N leaching losses. *Trans. ASABE* 50(4): 1189–1198.

Bakhsh, A., R.S. Kanwar, J.L. Baker, J. Sawyer and A. Mallarino. 2009. Annual swine manure applications to soybean under corn-soybean rotation. *Trans. Am. Soc. Agri. Biol. Eng.* 52(3): 751–757.

Bouwer, H. 1993. Sustainable irrigated agriculture: Water resources management in the future. *Irrigation J.* 43(6): 16–23.

Chandrakanth, M.G., A.C. Hemalatha, B.S. Chaitra and N. Nagaraj. 2004. Economic efficiency in ground water use in Karnataka. Presented at the Chennai Water Seminar, Chennai, India, 23 September, 2004.

Comly, H.H. 1945. Cyanosis in infants caused by nitrate in well water. *J. Am. Med. Assoc.* 129: 112–117.

Council for Agriculture Science and Technology (CAST). 1985. *Agriculture and ground water quality. CAST report no. 103.* Ames, Iowa: CAST.

Fraser, P., C. Chilvers, V. Beral and M.J. Hill. 1980. Nitrate and human cancer: A review of the evidence. *In. J. Epidemiol.* 9: 3–9.

Gleick, P.H. 2000. *The world's water 2000–2001. Biennial report on freshwater resources.* Washington, DC: Island Press.

Gupta, A.B., R. Jain and K. Arora. 1999. Water quality management for the Talkatora Lake, Jaipur – A case study. *Water Sci. Tech.* 40(2): 29–33.

Gupta, S.K., P.S. Minhas, S.K. Sondhi, N.K. Tyagi and J.S.P. Yadav. 2000. Water resources management. In *Natural resource management for agricultural productivity in India*, eds. J.S.P Yadav and G.B. Singh, pp. 137–244. New Delhi, India: Indian Society of Soil Science.

Hallberg, G.R. 1989. Pesticide pollution of ground water in humid United States. *Agric. Ecosyst. Environ.* 26: 299–367.

Hooja, R.V., S. Niwas and G. Sharma. 1994. Waterlogging and possible remedial measures in Indira Gandhi canal command area development project. In *National seminar on reclamation and management of waterlogged soils, Central Soil Salinity Research Institute, Indian Council of Agricultural Research, Karnal, India.*

Hoang, C.K., R.S. Kanwar, J.L. Baker, M. Helmers, C.H. Pederson, A. Mallarino, J. Sawyer and T. Bailey. 2009. Impact of swine manure on soybean yield and water quality. Paper published in proceedings of the World Soybean Research Conference VIII held in Beijing, China on August 10–15, 2009.

Johnson, C.J., P.A. Bonrud and T.L. Dosch. 1987. Fatal outcome of methemoglobinemia in an infant. *J. Am. Med. Assoc.* 257: 27296–2797.

Kalita, P.K. and R.S. Kanwar. 1993. Effect of water table management practices on the transport of Nitrate-N to shallow groundwater. *Trans. ASAE* 36(2): 413–422.

Kalita, P.K., R.S. Kanwar, J.L. Baker and S.W. Melvin. 1997. Groundwater residues of atrazine and alachlor under water table management practices. *Trans. ASAE* 40(3): 605–614.

Kanwar, R.S. 1991. Preferential movement of nitrate and herbicides to shallow groundwater as affected by tillage and crop rotation. In *Proceeding of the national symposium on preferential flow*, eds. T.J. Gish and A. Shirmohammadi, pp. 328–337. Am. Soc. Of Ag. Engr. St. Joseph, MI, USA.

Kanwar, R.S. 2003. Water quality and chemicals. In *Food security and environmental quality in developing world*, eds. R. Lal, D. Hansen, N. Uphoff and S. Slack, pp. 169–192. Washington, DC: Lewis Publishers.

Kanwar, R.S. 2006. Effects of cropping systems on NO_3-N losses to tile drain systems. *J. Am. Water Res. Assoc.* 42(6): 1493–1502.

Kanwar, R.S. and J.L. Baker. 1993. Tillage and chemical management effects on groundwater quality. In *Proceedings of national conference on agricultural research to protect water quality*, pp. 490–493. Ankeny, IA: SCS.

Kanwar, R.S., J.L. Baker and D.G. Baker. 1988. Tillage and split N-fertilization effects on subsurface drainage water quality and corn yield. *Trans. ASAE* 31(2): 453–460.

Kanwar, R.S., T.S. Colvin and D.L. Karlen. 1997. Effect of ridge till and three tillage systems and crop rotation on subsurface drain water quality. *J. Prod. Agric.* 10: 227–234.

Kanwar, R.S., D.E. Stolenberg, R. Pfiffer, D.L. Karlen, T.S. Colvin and W.W. Simpkins. 1993. Transport of nitrate and pesticides to shallow groundwater systems as affected by tillage and crop rotation practices. In *Proceedings of national conference on agricultural research to protect water quality*, pp. 270–273. Ankeny, IA: SCS.

Rao, K.V.G.K. 1997. Man's interference with environment in water-use problems of waterlogging and salinity. In *National water policy-Agricultural scientists' perception*, pp. 68–80. Proceedings of the Round Table Conference, August 12–14, 1994. National Academy of Agricultural Sciences, New Delhi, India.

Shiklomanov, I.A. 1993. World fresh water resources. In *Water in crisis*, ed. P.H. Glick, pp. 13–24. Oxford: Oxford University Press.

Singh, N.T. and A.K. Bandyopathyay. 1996. Chemical degradation leading to salt affected soils and their management for agricultural and alternative uses. In *Soil management in relation to land degradation and environment*, pp. 89–101. Bulletin Indian Society of Soil Science, New Delhi, India.

U.S. Environmental Protection Agency (EPA). 1992. Drinking water regulation and health advisories. Washington, DC: Office of Water, U.S. EPA.

Velayuthum, M. and T. Bhattacharyya. 2000. Soil resource management. In: *Natural Resource Management for Agricultural Productivity in India*, eds. J.S.P. Yadav and G.B. Singh. Indian Society of Soil Science, New Delhi, India.

Yadav, J.S.P. 1996. Extent, nature, intensity and causes of land degradation in India. In *Soil management in relation to land degradation and environment*, pp. 1–26. Bulletin Indian Society of Soil Science, New Delhi, India.

Chapter 2

Challenges in securing India's water future

Kapil K. Narula
Columbia Water Center, India Office, New Delhi, India
The Earth Institute, Columbia University, New York, New York, USA

Upmanu Lall
Columbia Water Center
Department of Earth and Environmental Engineering and Civil Engineering and Engineering Mechanics, Columbia University, New York, New York, USA

ABSTRACT

The agriculture sector in India accounts for more than 85% of the total water use for irrigation. Within irrigation, the share of groundwater in the net irrigated area is around 50% to 55%, which is responsible for two-thirds of the total agricultural production. Agriculture also accounts for non-point-source pollution that arises from excessive use of fertilizers and pesticides. The Green Revolution was partly based on increased fertilizer use and intensive irrigation. During the past three decades, food production was tripled and poverty witnessed a decline. However, agriculture intensification led to overexploitation and pollution of groundwater resources, causing degradation of the environment and depletion of natural resources. Over and above, absence of significant public investment in surface water-based irrigation infrastructure, free or highly subsidized electricity, unregulated groundwater pumping, and irrational pricing policies have promoted cropping choices independent of resource endowments of water and energy. Over the long term, the current trends toward resource depletion and degradation, combined with a lack of proper reforms and policies that ensure agricultural sustainability and threats from climate change, pose a huge risk to farmers' income and can force poorer farmers out of the agricultural sector. In turn, these problems would lead to increased migration and poverty and decreased crop production, which are serious threats to food security. In this paper, we have discussed various challenges to water sustainability and food security of India.

Keywords: security, depletion, degradation, policy, pricing, subsidy, sustainability, climate change

Address correspondence to Kapil K. Narula at Columbia Water Center, India Office (Regd.) Address: 101, K-62, Hauz Khas Enclave, New Delhi 110 016, India. E-mail: kkn2104@columbia.edu

Reprinted from: *Journal of Crop Improvement*, Vol. 24, Issue 1 (January 2010), pp. 85–91,
DOI: 10.1080/15427520903310621

2.1 INTRODUCTION

India supports 1/6th of world's population, 1/25th of the world's water resources, and 1/50th of the world's land. India also supports about 20% of the world's total livestock population, of which more than half are cattle. Agriculture is the major (85%) user of the available fresh water. On the basis of population projections, the current average per capita availability of water, which is around 1,600 m^3/capita/year, is expected to drop to around 1,000 m^3/capita/year by 2050 (see Figure 2.1). Climate-change impacts on the availability of future water resources are uncertain, but the frequency of extremes (floods and droughts) is expected to increase.

Climatic variability has always been a source of water stress in India. High spatial variability in rainfall and high inter-annual variability exacerbate aridity and its impacts on agriculture. Figures 2.2 and 2.3 depict spatial variability in rainfall and occurrence of rainfall (measured in standard deviation of normal rainfall). Monsoon failures and floods have significant social impacts. Both food security and rural livelihoods are adversely impacted because of reduced grain production. A former finance minister of India once quipped that his budget was a gamble on the monsoon. Surface reservoirs are an infrastructure response to mitigation of flood and drought impacts. However, per capita water storage (200 m^3/capita) is the lowest in India relative to comparable countries and world average (e.g., 1,960 m^3/capita for the United States and 1,100 m^3/capita for China; and world average = 900 m^3/capita). Increasing storage is thus a critical infrastructure need, and the Government of India has proposed a massive River Inter-Linking project to store and convey water across the country. However, this project faces much opposition, and in the absence of critical analysis and access to information, its future is rather uncertain.

In the absence of significant public investment in surface water-based irrigation infrastructure during the last 30–40 years, the need for reliable water supplies has translated into extensive and essentially unregulated groundwater pumping by individuals across the country. This trend has been facilitated by free or highly subsidized electricity or other energy sources for agricultural pumping that is sanctioned by states in which a large population is engaged in agriculture. Groundwater irrigation now

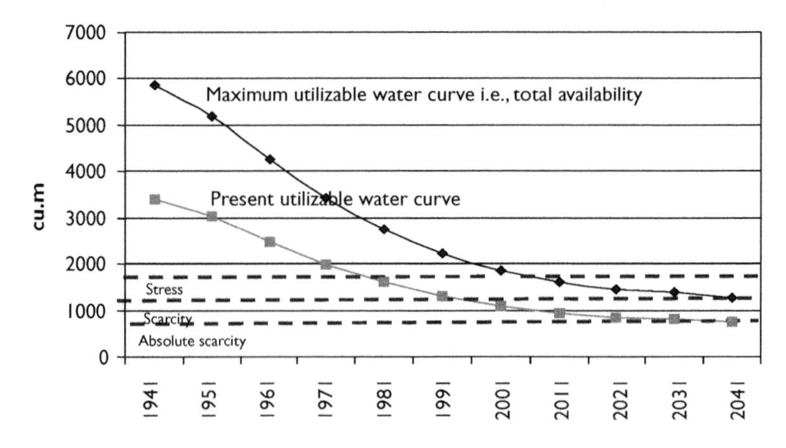

Figure 2.1 Declining availability of water per capita.

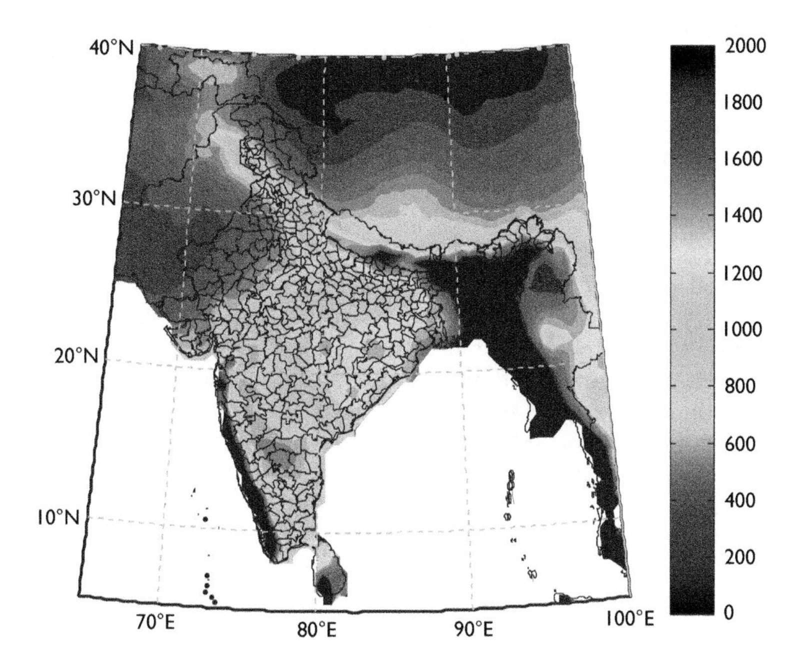

Figure 2.2 Mean annual rainfall spatial distribution.
Source: Maps prepared at Columbia University.

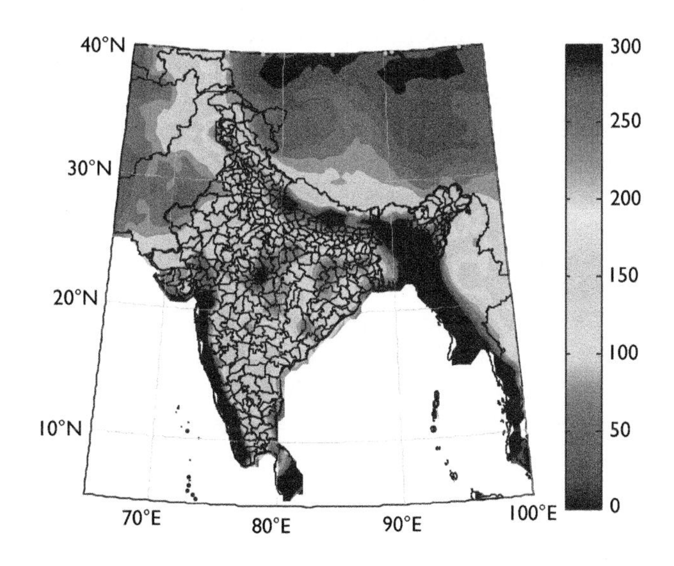

Figure 2.3 Mean annual rainfall standard deviation spatial distribution.

contributes to almost 50% to 55% of total irrigation water use and is responsible for two-thirds of the total agricultural production or approximately 10% of the total GDP (Ministry of Agriculture (MOA) 2006–2008). It is also responsible for almost 30% of the total electricity consumption. Both groundwater and electricity consumption are higher in agricultural states like Andhra Pradesh, Tamil Nadu, and Karnataka in southern India; Gujarat in the west; and Uttar Pradesh, Punjab, and Haryana in the northwest than in other states. There, agricultural electricity use is between 35% to 45% of the total generated electric use and groundwater use is between 70% to > 100% of estimated annual recharge. In such places, aquifers are being depleted (mined) at average rates of 0.2–0.5 m per year, with higher rates prevailing locally. Farm water-use efficiencies are typically very low (5% to 15%), given the free provision of energy and water (Narula & Lall 2009).

The Central government's policies for food security are a contributor to this situation. It offers minimum support prices above international and local prices through a food grain procurement program. These "incentives" and subsidies for energy, fertilizer, and water have promoted food self-sufficiency, with annual food grain production increasing from 51 million tons in the early 1950s to 206 million tons at the turn of the century. However, food grain production may need to double to meet the requirement of about 380–420 million tons in 2050. Dramatic increases in irrigated area and crop yields per unit water used and per unit land are needed if such a target is to be achieved within the country. These, in turn, may lead to very high rates of water-resource depletion, degradation and pollution, leading to a resource and environmental catastrophe, and subsequent food insecurity. For example, Punjab, the rice bowl of India, has witnessed a sharp fall in groundwater table; the average rate of groundwater table fall has been 25–30 cm/year (in certain areas, it is more than 1 meter per year). The fall in the share of canals for irrigation is nearly matched by the increase in the share of groundwater extraction systems. The region has also seen a cropping-pattern shift from a rainfed crop (maize) to a water-intensive crop (paddy). Punjab now has 97% of its cropped area under assured irrigation, which largely comes from groundwater sources. This discernable shift in cropping pattern has been mainly attributed to access to subsidized inputs and assured income through minimum support price for rice. As a result, area of rice as a percentage of net-cropped area has increased nearly 2.5 times during last two decades. It is worthwhile to note that the perverse pricing policy has not only encouraged a shift toward a more water-intensive crop like paddy, but has also killed all incentives to conserve water by abolishing canal water and the lift irrigation water rates in the 1990s, and not charging for power for irrigation use. Groundwater development, a ratio of groundwater use to natural recharge, in the state is around 150%, well above the annual replenishable groundwater resource. Overexploitation of water resources has aggravated water-quality problems, which now affect almost half of the state. Public and private action and investment are urgently needed to achieve a more sustainable trajectory for Indian agriculture, water, and energy futures.

Another example is that of North Gujarat. Agriculture is central to Gujarat's economy, and the state is an important producer of a range of cash crops. However, because of unsustainable water-use patterns, there is a serious concern that the region may soon face significant water problems with devastating consequences. Water tables have already declined more than 80 meters in the last 30 years in North Gujarat (e.g., Mehsana district), and further declines could eventually cause irreversible salinization. Livelihoods are also negatively impacted; many farmers in such areas are no longer able to generate net incomes that exceed the cost of the subsidized electricity supplied to them. In other words, the net economic impact of their farming is negative to the state. The energy subsidy is currently structured so that farmers can pump water from any depth without incurring any marginal costs. While this type of subsidy provides for the immediate livelihood of the farmers, it encourages overexploitation of water resources. Farmers are not motivated to do anything other than to continue to deepen their wells and pump groundwater to the greatest extent possible. In addition, the subsidy leads to long-term negative results for the farmers because, by overusing groundwater, they are creating a dangerous situation that would ultimately result in the collapse of groundwater supply. Farmers are now no longer able to generate net incomes that exceed the cost, to the state, of the subsidized electricity supplied to them. In other words, while the state exchequer extends, after recovering per horsepower charges, subsidized power of nearly Rs. 30,000 per hectare of irrigated land in North Gujarat, farmers' net returns from agriculture are at best two-thirds of this amount from the same hectare.

The needed policy reform to promote more efficient water use is difficult in a populist democratic environment, where politicians cater to the perceived desire of the rural masses through an ever-growing web of subsidies and support mechanisms in the short run. Irrational pricing of agricultural inputs and products has promoted cropping choices independent of resource endowments of water and energy. However, the increasing competition for water and energy resources between the relatively affluent urban and industrial users and the rural poor involved in agriculture is creating new tensions and dynamics that may turn out to be an agent of change (Narula & Lall 2009).

The private sector has so far not engaged in a serious assessment of the needs for or development of water infrastructure at any large scale. Recognition of the role the private sector could play in either developing infrastructure or improving agricultural water productivity or of the associated investment needs and opportunities is only just emerging. Addressing cost recovery for electricity supplied to the agricultural sector for pumping groundwater and the large water losses from urban water-supply systems are emerging areas of concern where limited private sector engagement is now evident.

The absence of a concerted public and private sector effort to improve the water-supply reliability and access has led to increasing conflict between rural and urban/industrial users. Protesting local impacts of water use by bottling companies, such as Coca-Cola and Pepsico, is one example. Interstate and international disputes on the shared use of river waters have also emerged without a clear, long-term resolution in the face of growing populations and climate variability. States typically have not had the capital to invest in or properly maintain water infrastructure, and multiple states are engaged in disputes where such developments have taken place, impacting ability to store or access water. Even small-scale rainwater-harvesting systems have proliferated to such an extent in some areas that downstream flows in rivers are negatively impacted (Narula & Lall 2009).

The final area of concern is the widespread pollution of rivers, lakes, and groundwater because of the release of untreated municipal and industrial wastewater, and because of the excessive application of fertilizers and pesticides in agriculture (Central Pollution Control Board 2003; Central Groundwater Board 2004). Virtually all streams in the country now have shown the presence of chemical and biological contaminants at concentrations well in excess of international health standards.

Water quality observed in rivers during the 1992–2006 period, measured as BOD (biochemical oxygen demand), showed that out of the total riverine length of around 45,000 km, close to 40% had high (BOD > 6 mg/l) and moderate (BOD = 3 to 6 mg/l) pollution. More than 25% of the riverine lengths of the Ganga river (The Ganges) basin (comprising Yamuna river basin, too) and Godavari and Krishna river basins had BOD greater than the standard of 3 mg/l. Similarly, non-point sources of pollution, arising because of excessive use of inorganic fertilizers and pesticides, are of immense importance in agricultural areas in India. The chemical behavior of nitrates suggests that nitrates are not attracted to soil particles; they easily leach down and are displaced by percolation water to the groundwater systems.

The overall impact of deteriorating quality and diminishing availability of water translates into water insecurity. A freshwater crisis is gradually unfolding in India. Solving the water-use problem in agricultural areas will ultimately require a range of solutions. These include restructuring of the entire supply chain – not just the product-supply chain but also the supply chain of knowledge – supporting infrastructure and any specialized inputs that might be needed; a shift in cropping patterns; and the creation of incentives for capital investments in devices that improve water-use efficiency. Most importantly, water conservation must make economic sense to the farmer.

REFERENCES

Central Groundwater Board (CGWB) 2004. *Nitrates in ground water of India.* New Delhi, India: Central Groundwater Board: Ministry of Water Resources.

Central Pollution Control Board (CPCB) 2003. *Water quality statistics of India.* Monitoring of Indian National Aquatic Resources Series (MINARS) No. MINARS/23/2004–05. New Delhi, India: Central Pollution Control Board.

Ministry of Agriculture, Government of India 2006. *Agriculture Statistics at a glance 2006.* New Delhi, India: Directorate of Economics and Statistics, Department of Agriculture and Cooperation, Ministry of Agriculture, Government of India. Accessed in December 2008 at http://agricoop.nic.in/Agristatistics.htm

Ministry of Agriculture, Government of India 2007. *Agriculture Statistics at a glance 2007.* New Delhi, India: Directorate of Economics and Statistics, Department of Agriculture and Cooperation, Ministry of Agriculture, Government of India. Accessed in March 2009 at http://agricoop.nic.in/Agristatistics.htm

Ministry of Agriculture, Government of India 2008. *Agriculture statistics at a glance 2008.* New Delhi, India: Directorate of Economics and Statistics, Department of Agriculture and Cooperation, Ministry of Agriculture, Government of India. Accessed in March 2009 at http://agricoop.nic.in/Agristatistics.htm

Narula, K.K. and U. Lall 2009. *Water security challenges in India.* Asia's next challenge – Securing the region's water future. A report by leadership group on water security in Asia. Asia Society, April 2009.

Section II

Sustainable water management strategies

Chapter 3

Soil water management in India

R. Lal
Carbon Management and Sequestration Center, The Ohio State University, Columbus, Ohio, USA

ABSTRACT

The growing demand for food in India, due to rising population and rapid industrialization, necessitates judicious and sustainable management of soil and water resources. While the increase in agronomic production of irrigated rice-wheat system (RWS) has been impressive since the 1970s, the low water- and fertilizer-use efficiency has caused degradation of soil quality and pollution of surface and ground waters. With increasing competition for soil and water for non-agricultural uses (e.g., urbanization, industrial use), the strategy is to increase production per unit input of scarce resources through agricultural intensification, and by reducing losses and optimizing the input-use efficiency. Strategies of soil-water management in India's dry farming regions include conserving water in the root zone, water harvesting, and recycling, and integrated nutrient management (INM) involving use of crop residue mulch and biosolids. Similarly, soil-water management in India's irrigated regions (e.g., RWS in the Indo-Gangetic Basin) include INM, crop-residue management, efficient irrigation methods (furrow, drip), irrigation with saline water, alternative cropping systems, and no-till farming. Soil water, the precious and scarce resource, must be used prudently. Undervaluing a scarce resource can lead to its misuse, overexploitation, depletion, pollution, and contamination and eutrophication.

Keywords: water-use efficiency, irrigation, integrated nutrient management, no-till farming, residue management

---

Address correspondence to R. Lal at the Carbon Management and Sequestration Center, The Ohio State University, Columbus, OH 43210 USA. E-mail: lal.1@osu.edu

Reprinted from: *Journal of Crop Improvement*, Vol. 23, Issue 1 (January 2009), pp. 55–70,
DOI: 10.1080/15427520802418293

3.1 INTRODUCTION

India's arable land resources are excellent, and are second only to those of the United States. India is also endowed with a range of climates that permit production of diverse crops throughout the year in one region or another. Renewable fresh-water resources of India are adequate to vastly expand the irrigable land area from 60 Mha now to the largest in the world. Indian farmers, with a track record of quadrupling grain production across four decades (1965 to 2005), are hardworking and innovative. With these natural and human resources, India has the capacity to feed the world. Yet, Indian agricultural production has stagnated since the 1990s, farmers are demoralized and some committing suicide, the country's grain reserves are frequently depleted, and India has imported wheat several times during the 2000s. Yields of upland crops in India are low and comparable to those of Sub-Saharan Africa (Table 3.1). Yield of crops in 1995 in India as a ratio of those in China were 64% for irrigated crops and 33% for rainfed crops. Projected yield of crops in 2025 in India, as a ratio of those in China, will be 62% for irrigated and 36% for rainfed crops. In contrast, water use in India in 1995, as a ratio of that in China, was 183% for rice and other crops. Water use in 2025 will be 178% for rice and 188% for other grains (Table 3.1). Therefore, understanding the cause-effect relationship of these complex questions is important to enhancing productivity and use efficiency. Identifying the causes of agronomic stagnation in India is important not only to India, but to global food security, and peace and stability.

Food security has four complex and interrelated components: production, access, consumption, and retention or utilization. Food production, the principal focus of this review, depends on a range of biophysical (soil, water, climate) and social/economic factors. India's endowment of soils, diverse and developed in a wide range of ecoregions/biomes, is impressive. Yet, soils have been taken for granted, misused, neglected, and degraded. Similar to its soils, India's water resources are also highly diverse and abundant. However, water resources have also been undervalued and treated as a common resource. Consequently, both surface and ground waters are highly polluted and contaminated.

Table 3.1 Yield of cereals (Mg/ha) and water use (cm) in India compared with other countries/regions (Adapted from Rosegrant Cai & Cline, 2002).

	Irrigated		*Rainfed*		*Total*		*Water consumption*			
							Rice		*Other grains*	
Region/Country	*1995*	*2025*	*1995*	*2025*	*1995*	*2025*	*1995*	*2025*	*1995*	*2025*
Asia	3.2	4.65	1.7	2.5	2.5	3.7	83	83	58	55
China	4.2	6.0	3.6	4.7	4.0	5.6	60	59	41	38
India	2.7	3.7	1.2	1.7	1.8	2.7	110	105	75	72
South Asia	2.2	3.3	1.2	1.9	2.0	3.1	89	87	73	70
Sub-Saharan Africa	2.2	3.2	0.9	1.2	0.9	1.3	112	108	78	80
Latin America	4.1	5.7	2.1	3.0	2.4	3.4	95	85	81	78
World	3.5	4.9	2.2	2.8	2.6	3.5	85	84	52	50

The serious problem of degradation of the quality of soil and water resources has a strong adverse impact on agricultural productivity and sustainability. Therefore, the objective of this manuscript is to outline the conceptual basis of innovative technological options for sustainable management of soil and water resources, with specific reference to soil-water management in India (SWAMI).

3.2 FOOD DEMAND AND CROP YIELDS

India's population of 1.1 billion in 2007 may stabilize at 1.65 billion by 2050, making it by then the most populous country in the world. The increase in population, along with an increase in standard of living, is rapidly increasing the demand for cereals. The cereal grain consumption (e.g., wheat, rice, maize, sorghum, millet) will increase from 220 million tons (Mt) in 2007 to 260 to 300 Mt by 2020 (Kumar, 1998; Bhalla, Hazel & Kerr, 1999), necessitating increases in yields of rice and wheat by 56% and 62%, respectively (Paroda, Woodhead & Singh, 1994). Yet, the grain yield of rice (*Oryza sativa*) and wheat (*Triticum aestivum*) is declining or stagnant. The yield of wheat declined and that of rice stagnated between 1999 and 2006 (Figure 3.1). Grain yields of maize (*Zea mays*), sorghum (*Sorghum bicolor* L.) and millet (*Panicum miliare* L.) are low and have been stagnant since 1998–99 (Figure 3.2). Yields of soybean (*Glycine max*), chickpea (*Cicer arietinum*), safflower (*Carathamus tinctorius*), rapeseed (*Brassica campestris*), and groundnut (*Arachic hypogea*) are among the lowest in the world (Figure 3.3). Indeed, yields of most crops in India have stagnated or declined during the 2000s (Figures 3.2, 3.3, 3.4) because of a range of interactive biophysical and social/economic factors. Important among biophysical factors are low use efficiency of water, nitrogen (N), and energy-based inputs because of depletion of soil organic carbon (SOC) pool and the overall decline in soil quality. The rice-wheat system (RWS) is the dominant production system in the Indo-Gangetic Basin (IGB) (Timsina & Connor, 2001). Decline in its productivity is occurring in conjunction

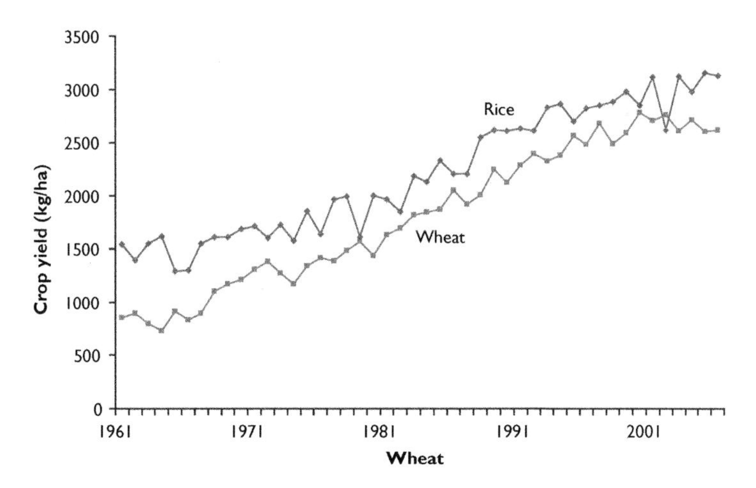

Figure 3.1 Temporal changes in grain yields of rice and wheat in India between 1961 and 2006 (redrawn from FAO stat).

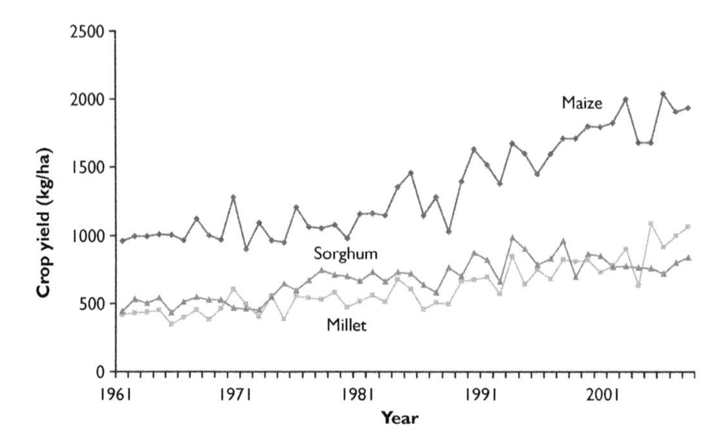

Figure 3.2 Temporal changes in grain yields of maize, sorghum, and millet in India between 1961 and 2006 (redrawn from FAO stat).

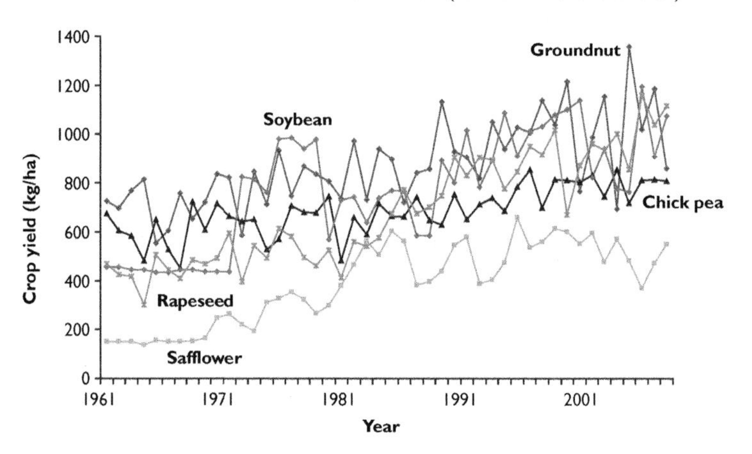

Figure 3.3 Temporal changes in grain yields of groundnut, soybean, chick pea, rapeseed, and safflower in India between 1961 and 2006 (redrawn from FAO stat).

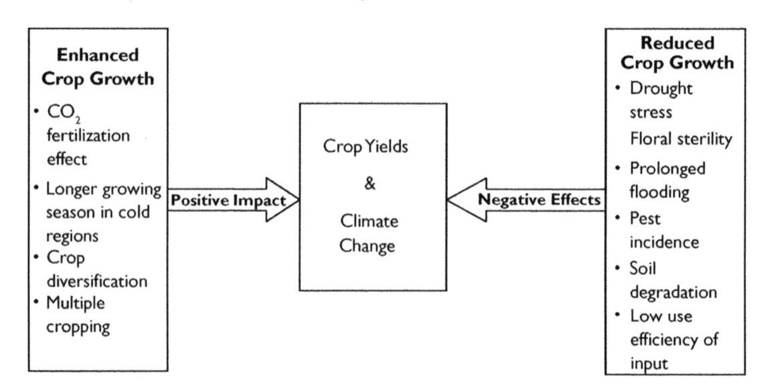

Figure 3.4 Positive and negative effects of climate change on crop yields.

with either decline in ground water (tube-well irrigation; Table 3.2) or rise in water table and soil salinity (canal irrigation) because of the widespread use of improper irrigation practices (e.g., flood irrigation, pre-monsoon planting of flooded rice). Soil degradation, exacerbated by depletion of SOC pool and the nutrient reserves (N, P, K, and micronutrients), is another factor responsible for declining productivity (Bhandari et al., 2002; Singh et al., 2004). Depletion of SOC and nutrient reserves is partly due to removal of wheat straw for fodder (Samra, Bijay-Singh & Kumar, 2003), burning of rice straw, and use of animal dung for cooking. Yet, sustainability of agricultural production in the IGB (northwest region of Punjab and Haryana) is important because the region contributes about 52% of India's food production (Abrol, 1999).

Overexploitation of water resources, unequal access to water, and adverse impacts on ecosystem services are among the major issues impacting sustainable use of water resources throughout the developing countries (Vincent, 2003). The overall water-use efficiency of irrigated crop production in India is about 30% (Sharma & Rao, 1997). Groundwater, supporting more than 50% of the total irrigated area in India, is being depleted due to overdraft, and consequently resulting in falling water table and dry wells.

3.3 WATER BUDGET

Water, nutrient, and energy management are strongly interrelated and are the principal factors affecting crop production in tropical and subtropical regions of India. Low water productivity (WP), agronomic/biomass production per unit input of water in an agroecosystem, is the principal constraint to enhancing production in arid and semiarid regions of India. Enhancing WP necessitates understanding components of the soil hydrologic budget for the root zone to establish a cause-effect relationship and interaction among different components (Equation 3.1):

$$ET = P + I - (R + D + E + W) + \Delta S \tag{3.1}$$

where ET is evapotranspiration or crop water use, P is precipitation, I is irrigation, R is runoff, D is deep drainage, E is direct soil evaporation, W is water use by weeds, and ΔS is change in soil-water storage. The goal of soil and water management is to optimize ET and ΔS by minimizing components listed within the parenthesis in Equation 3.1. Within the ET, the goal is to optimize the T component, which constitutes the direct water use by the crop for photosynthesis. Alleviation of drought

Table 3.2 Changes in water table depth in Punjab, India (Adapted from Sidhu, 2005).

	Water table depth (m)	
Eco-regions	1993	2003
Sub-montane zone	17.6	18.5
Central zone (Sangrur dist.)	13.8	20.6
Central zone (Patiale dist.)	11.7	18.4
Central zone (Mage dist.)	9.4	18.6
Southwest zone	6.8	8.9

stress implies optimizing ET and ΔS components by minimizing losses caused by the components of the hydrologic cycle listed in the parenthesis in Equation 3.1.

For sandy soils of the IGB, under hot arid and semiarid climates, both D and E components are very high, especially in flooded rice cultivation. Loses caused by D (percolation) may be as much as 71% of the total water applied (Dastane et al., 1970). Furthermore, 82% of the total energy input is used in lifting groundwater for tube-well irrigation (Singh, Singh & Bakshi, 1990). Delaying rice cultivation until the onset of the monsoon season may reduce evaporation losses (Singh, Gogri & Arora, 2001). Losses caused by soil evaporation (E) can be reduced by eliminating or reducing tillage, using crop-residue mulch and adopting efficient methods of irrigation (e.g., furrow, drip, sprinkle).

3.4 IMPACT OF CLIMATE CHANGE

Increase in global temperature is likely to strongly impact crop growth and agronomic yields (Figure 3.4). The CO_2 fertilization effect can increase yield of wheat by 20% to 30% (Amthor, 2001). There may also be some increase in growing-season duration in cold regions at higher altitudes (e.g., middle Himalayas). In general, however, climate change may have severe adverse effects on crop growth and agronomic production in India. Factors responsible for decline in crop yields include increase in frequency and severity of drought stress, prolonged flooding due to extreme events, soil degradation, and the attendant decline in use efficiency of input (e.g., fertilizer). Increase in incidence of diseases and insects may also reduce crop yield. An important factor with severe adverse impact on grain yields may be high temperatures, especially during the reproductive stage of crop growth (e.g., flowering, grain filling, or pegging; Wheeler et al., 2000; Porter, 2005). The rate of photosynthesis peaks between 20°C and 30°C for most crops. However, the rate of respiration increases exponentially with increases in temperature, with the highest respiration rate being around 35°C. Rapid increase in temperatures in March in the IGP can cause shriveling of wheat grains (Porter & Gawith, 1999), as is frequently observed in Punjab region of India and Pakistan. High temperatures also lead to floral sterility in rice. Both growth and development exhibit temperature optima. For C_3 crops (e.g., rice and wheat), the optimal range is 20°–30°C, followed by a rapid decline beyond 40°C (Porter & Semenov, 2005). Vara Prasad et al., (2000) observed that the yield of groundnut was adversely affected by temperatures >40°C. For a temperature range of 32° to 42°C, fruit-set declined from 50% to 0%. Ray and Sinclair (1997) reported stomatal closure of maize in response to drought by soil drying. With adverse impacts on crop yields, it is important to identify adaptive strategies (e.g., time of sowing, early-maturing varieties, new cropping systems, irrigation methods, etc.).

3.5 STRATEGIES FOR ENHANCING AGRONOMIC PRODUCTIVITY IN INDIA

Optimizing SWAMI is crucial to enhancing agronomic productivity and meeting the food needs of the growing population with rising standards of living. An important strategy is to adapt to climate change (Figure 3.5) by: 1) improving soil quality, 2) increasing water availability, 3) adjusting time of sowing, and 4) growing improved varieties and GM/biotech crops in appropriate crop combination and sequences. Improving soil quality involves restoring degraded soils by: 1) increasing SOC pool; 2) enhancing soil structure; 3) balancing and increasing nutrient reserves; and 4) improving activity and species diversity of soil biota (Figure 3.5).

Similarly, alleviating drought stress involves: 1) improving water-infiltration rate; 2) water harvesting by safe disposal and storage of excess runoff for use as supplemental irrigation; 3) using an efficient irrigation system; and 4) conserving soil in the rest zone by minimizing losses.

3.5.1 Rainfed agriculture

Crop yields in rainfed agriculture (predominantly in southern, central, and western India) are low because of severe drought stress and high temperatures. Several experiments have shown that conserving water in the root zone, through practices that enhance water infiltration and decrease evaporation losses, can increase crop yield by 33% to 173% (Table 3.3). In addition, water harvesting for supplemental irrigation can be extremely important in these drought-prone regions. Increasing recharge involves installation of check dams, percolation tanks, recharge tube-wells, and rainwater conservation. Water harvesting and storage are key options in drylands of peninsular India (Gunnell & Krishnamurthy, 2003). Remote sensing (Ines et al., 2006), field-water management (Bouwan & Tuong, 2001), crop planning (Reddy and Kumar, 2008), and modeling to plan watershed management (Kaur et al., 2004) are some of the modern innovations needed to improve water-use efficiency (WUE). One

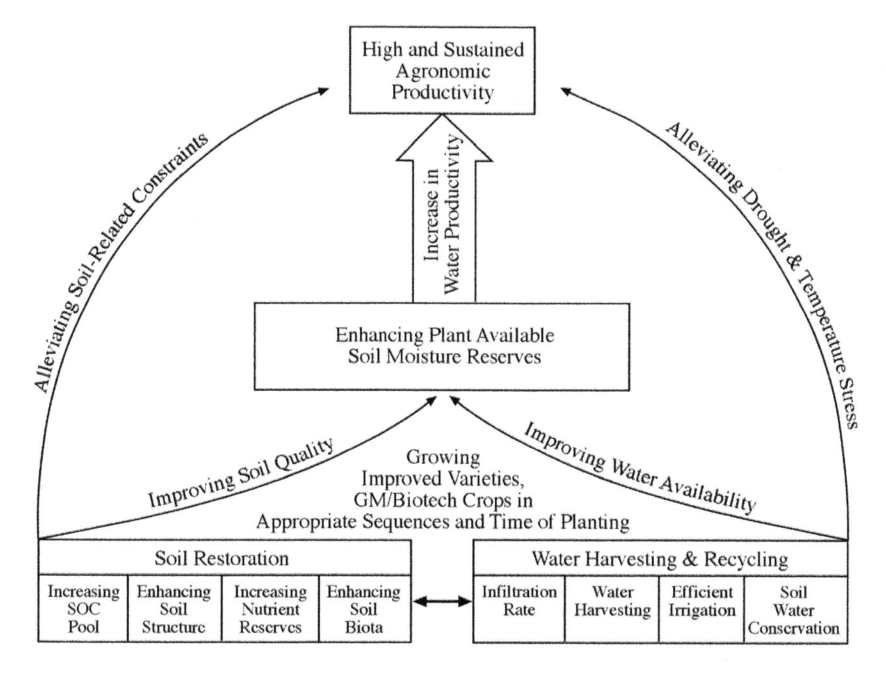

Figure 3.5 Synergism between soil quality, water management and improved germplasm to enhance and sustain agronomic yield and water productivity.

such supplemental irrigation, applied at the critical stage of crop growth, can increase agronomic yield by 23% to 250% (Table 3.4). Effectiveness of these soil-water management techniques can be enhanced when used in conjunction with improved varieties and INM strategies (Lal, 2006).

3.5.2 Irrigated farming

Enhancing irrigation WUE by reducing losses and improving plant water use is important to sustainable management of limited water resources.

3.5.2.1 *Residue management*

Low irrigation and nutrient-use efficiency (NUE) are partly attributed to low SOC pool. Therefore, enhancing SOC pool through crop residue management and application of biosolids is an important strategy for long-term improvement in soil quality. The data in Table 3.5 show that residue retention, along with the use of recommended rates of fertilizers and farmyard manure (FYM), produced the highest yield. In addition to enhancing soil quality by recycling nutrients, crop-residue retention is also important to improving soil structure and hydrological practices.

3.5.2.2 *No-till farming*

No-till (NT) sowing of wheat after rice is gaining popularity among farmers in the IGB. One of the advantages of NT sowing of wheat is the saving in time of two to three weeks for seedbed preparation that leads to increase in yield of wheat (Singh et al., 2002; Hobbs & Gupta, 2004). With the use of an appropriate seed drill, equal or higher yield of wheat is obtained with NT compared with conventional tillage (Sidhu et al., 2007). However, surface retention of residues of high C:N ratio and broadcasting fertilizer leads to N deficiency in wheat (Srivastava, Panwar & Garg, 2000; Timsina & Connor, 2001; Thuy et al., 2008) and to stunting of seedling growth.

Table 3.3 Impact of soil and water conservation practices on crop yield in India (Adapted from Sivanappen, 1995).

Crop	*Grain yield (kg/ha)* *Control*	*% Increase with soil and water conservation*
Castor	700	100.0
Chickpea	700	100.0
Cotton	800	80.0
Groundnut	600	173.3
Pearl millet	710	109.9
Pulses	450	33.3
Sorghum	600	33.3
Wheat	1800	77.8

Table 3.4 Impact of one supplemental irrigation on crop yield in different regions of India (Recalculated from Channappa, 1986; Sivanappan, 1995).

Crop	*Site/State*	*Grain yield (Mg/ha)* Control	*% Increase with one 5-cm irrigation*
Wheat	Agra, Uttar Pradesh	2.2	23
Maize	Banglore, Karnatka	5.4	80
Cowpea	Banglore, Karnatka	0.2	200
Sorghum	Bellari, Kartnatka	0.4	250
Sorghum	Bijapur, Karnatka	1.7	41
Wheat	Dehradun, Uttranchal	2.1	71
Wheat	Ludhiaana, Punjab	1.9	116
Upland rice	Rewa, Madhya Pradesh	1.6	75
Wheat	Rewa, Madhya Pradesh	0.6	217
Sorghum	Solapur, Maharashtra	1.0	80
Barley	Varanasi, Uttar Pradesh	2.6	31

Table 3.5 Crop yields with recommended management practices in the Indo-Gangetic basin (Recalculated from Gregory, Ingram & Brklacich, 2005).

Crop	*No fertilizer*	*Recommended fertilizer*	*Recommended fertilizer + FYM*
Rice (Mg/ha)	3.74 (100)	5.67 (151.6)	6.41 (171.4)
Wheat (Mg/ha)	1.71 (100)	3.97 (232.2)	4.60 (269.0)

Numbers in parenthesis are relative yields compared with control as 100.

3.5.2.3 *Irrigation management*

Water management is a major issue that must be addressed. The decline in productivity by 15% of the RWS in the northern region (Ambast, Tyagi & Raul, 2006) is attributed to decline in the yield of both wheat and rice. The tube-well density in the region is >15/km^2, which is responsible for the rapidly receding groundwater [5 m–15 m recession across 20 years from 1986 to 2006]. The remedial measures include reduction in groundwater withdrawal and increase in groundwater recharge. The reduction in groundwater withdrawal can be achieved through judicious management of irrigation. Alternatives to flooded rice, delayed transplanting until onset of monsoons, aerobic-rice production, and maize-wheat and other systems must be adopted.

3.5.2.4 *Nutrient management*

Adopting INM is crucial to increasing and sustaining crop yield in the RWS in IGB (Singh et al., 2004; 2005). Application of FYM, in conjunction with chemical fertilizers, can increase yield of rice by 70% and that of wheat by 170% (Tables 3.5–3.7). There is an important synergistic effect when chemical fertilizers are used in combination

Table 3.6 Average grain yield (kg/ha) of wheat and rice at three sites in the IGB under improved nutrient management (Adapted from IISS, 2006–07).

Site	Crop	Control	50% NPK	Recommended NPK	NPK + FYM
Barrackpur	Rice	983	1325	1688	1871
	Wheat	723	1340	1533	2097
Pantnagar	Rice	1935	4195	4655	5540
	Wheat	1315	3120	3600	4480
Ludhiana	Rice	2079	3219	3966	4840
	Wheat	1676	2842	3859	4765
	Maize	–	2410	3770	–

Table 3.7 Average grain yield (kg/ha) of crops in central and southern India under improved nutrient management (Adapted from IISS, 2006–07).

Site	Crop	Control	50% NPK	Recommended NPK	NPK + FYM
Bangalore	Millet	466	2039	4072	4484
	Maize	237	1457	2628	3123
Udaipur	Maize	1550	–	2960	3480
	Wheat	1780	–	3770	4340
Jabalpur	Soybean	619	1293	1533	1749
	Wheat	1448	3670	4602	5144
Ranchi	Soybean	617	1232	1496	1832
	Wheat	684	1969	2707	3235
Raipur	Rice	2873	5052	6392	6896
	Wheat	677	1232	1842	2042

with organic amendments, especially in the light-textured soil of the IGB in which the SOC pool has been severely depleted because of residue removal for fodder and other purposes. Long-term experimental data are needed to model crop growth under a range of fertility management options (Bhattacharyya et al., 2007).

3.5.2.5 *Irrigation with saline water and urban water*

A large proportion of groundwater in India is brackish with a high salt content (Minhas, 1996). Thus, a prudent strategy is to use saline water for irrigation. While so doing, the goal is to prevent salt buildup in the soil. Strategies of using saline water must be based on: 1) assessment of the leaching-requirement level; 2) frequency and amount of irrigation; 3) mixing of water of different salt concentrations; 4) appropriate method of irrigation; 5) tillage method and crop management; 6) soil-surface management (e.g., structure, crusting); and 7) nutrient management, along with soil amendments. It is important to develop guidelines for using saline water for different soils, cropping systems, and agro-ecosystems. Management of declining ground water is an important issue (Ambast, Tyagi & Raul, 2006) that needs to be objectively and urgently addressed. Some of these strategies may involve use of saline-drainage water for irrigation (Sharma & Rao, 1998) and on-farm water management to minimize losses (Tyagi et al., 2005). Conjunctive use of canal water with saline water is one such option (Mandare et al., 2008). Urban water (grey water) is another major resource that needs to be used for irrigation. However, health-related issues need to be adequately addressed.

3.5.3 Soil organic carbon management

The SOC pool in most cultivated soils of India is low (0.1% to 0.3%) and below the threshold level needed for the soil's ecosystem functions. In addition to nutrient retention and uptake, managing SOC pool is essential for judicious use and conservation of soil water. The strategy is to maintain a positive C budget in the soil by application/recycling of biosolids (e.g., crop residues, manure, compost, sludge, animal waste). It is in this context that management of crop residues needs a careful appraisal. The data (Table 3.8) show that even with application of the recommended rates of NPK and FYM, the rate of SOC sequestration ranges from 0 to <400 kg/ha/yr. Furthermore, CO_2 emissions are lower with NT used in conjunction with residue retention than with plow-till and residue removal or burning (Table 3.9). It is important to identify and adopt agricultural practices that minimize C-footprint and farming operations.

3.6 CONCLUSION

Stagnation or decline in agricultural productivity in India may be attributed to degradation of soil quality and excessive/improper use of the scarce water resources. Yet, the increasing demand for food production necessitates increasing crop yields. The strategy is to improve WUE by conserving water in the root zone and minimizing losses.

Table 3.8 Net change in soil organic carbon pool (kg/ha/yr) under different management systems (Adapted from IISS, 2006–07).

Location	Cropping system	Duration (yr)	Control	100% NPK	NPK + FYM
Barrackpur	R-W-J	29	−152	+8	+137
Raipur	R-W	5	−264	−176	0
Akola	S-W	14	−16	+204	+377

R = Rice, W = Wheat, J = Jute, S = Sorghum

Table 3.9 Total annual emission of CO_2 (Mg CE/ha) in the rice-wheat system of the Indo-Gangetic Basin (Adapted from Grace et al., 2003).

		Percent increase with	
Treatment	No fertilizer	Recommended fertilizer	Recommended fertilizer + FYM
CT, Residue retained	3.5	34.3	102.9
CT, Residue burnt	4.0	37.5	100.0
NT, Residue retained	3.0	46.7	123.3

In dry-farming systems of peninsular India, conserving water implies water harvesting and recycling for supplemental irrigation, groundwater recharge, and minimizing losses by runoff and evaporation. In irrigated regions, using appropriate time and method of irrigation, mixing saline and fresh water, using grey or urban wastewater, and decreasing losses by evaporation are important. Identifying viable alternatives to flooded rice in the northwestern region is also important. Direct sowing of wheat after rice and use of crop-residue mulch can save time, labor, and water.

REFERENCES

Abrol, I.P. 1999. Sustaining rice-wheat system productivity in the Indo-Gangetic Plain: Water management related issues. *Agric. Water Manage.* 40:31–35.

Ambast, S.K., N.K. Tyagi, and S.K. Raul. 2006. Management of declining groundwater in the Trans Indo-Gangetic Plain (India): Some options. *Agric. Water Manage.* 82:279–296.

Amthor, J.S. 2001. Effects of atmospheric CO_2 concentration on wheat yield: review of results from experiments using various approaches to control CO_2 concentration. *Field Crops Res.* 73:1–34.

Bhalla, G.S., P. Hazel, and J. Kerr. 1999. *Prospects for India's cereal supply and demand for 2020.* Food, Agric. and Env. Discussion Paper 29, IEPRI, Washington, DC, 24 p.

Bhandari, A.L., J.K. Ladha, H. Pathak, A.T. Padre, D. Dawe, and R.K. Gupta. 2002. Yield and soil nutrient changes in a long-term rice-wheat rotation in India. *Soil Sci. Soc. Am. J.* 66:848–856.

Bhattacharyya, T., D.K. Pal, M. Easter, S. Williams, K. Paustian, E. Milne, P. Chandran, et al. 2007. Evaluating the century carbon model using long-term fertilizer trials in the Indo-Gangetic Plains, India. *Agric. Ecosyst. Env.* 122:73–83.

Bouwan, B.A.M., and T.P. Tuong. 2001. Field water management to save water and increase its productivity in irrigated lowland rice. *Agric. Water Manage.* 49:11–30.

Channappa, T.C. 1986. Role of dryland farming on food production. *Trans. Indian Soc. Desert Tech.* 11:67–41.

Dastane, N.G., M. Singh, S.B. Hukeri, and V.K. Vamadevan. 1970. *Review of work done on water requirements of crops in India.* Poona, India: Nov Bharat Prakashan.

FAO. 1960–2006. *Production Yearbook in Agriculture.* Rome, Italy: FAO.

Grace, P.R., M.C. Jain, L. Harrington, and G.P. Robertson. 2003. The long-term sustainability of tropical and sub-tropical rice and wheat systems: an environmental perspective. In *Improving the Productivity and Sustainability of Rice-Wheat Systems: Issues and Impacts, eds.* J.K. Ladha, 27–43. Madison, WI: Special Publication, ASA.

Gregory, P.J., J.S.I. Ingram, and M. Brklacich. 2005. Climate change and food security. *Phil. Trans. R. Soc.* (B) 36:2139–2148.

Gunnell, Y., and A. Krishnamurthy. 2003. Past and present status of runoff harvesting systems in dryland peninsular India: A critical review. *Ambio.* 30:320–324.

Hobbs, P.R., and R.K. Gupta. 2004. Problems and challenges of no-till farming for the rice-wheat systems of the Indo-Gangetic Plains in South Asia. In *Sustainable Agriculture and Rice-wheat System,* eds. R. Lal, P. Hobbs, N. Uphoff, and D.O. Hansen, 101–119. New York: Marcel Dekker.

IISS. 2006–07. Annual Report. All India Coordinated Research Project "Long-Term Fertilizer Experiments to Study Changes in Soil Quality, Crop Productivity and Sustainability. Bhopal, India: Indian Inst. of Soil Sciences.

Ines, A.V.M., K. Honda, A.D. Gupta, P. Droogers, and R.S. Climente. 2006. Combining remote sensing-simulation modeling and genetic algorithm optimization to explore water management options in irrigated agriculture. *Agric. Water Manage.* 83:221–235.

Jagtap, V., S. Bhargava, P. Streb, and J. Feierabend. 1998. Comparative effect of water, heat and light stresses on photosynthetic reactions in *Sorghum bicolor* (L.) Moench. *J. Exp. Bot.* 49:1715–1725.

Kaur, R., O. Singh, R. Srinivsan, S.N. Das, and K. Mishna. 2004. Comparison of a subjective and physical approach for identification of priority areas for soil and water management in a watershed – A case study of Nagwan watershed in Hazaribagh District of Jharkhand, *India. Env. Modeling & Assessment.* 9:115–127.

Kumar, P. 1998. *Food demand and supply projections for India.* Agric. Econ. Policy Paper 98–01. IARI, New Delhi, India.

Lal, R. 2006. Dryland farming in South Asia. In *Dryland Agricutlure,* 2nd edition, Agronomy Monograph #33, 527–576. Madison, WI: American Society of Agonomy.

Maheswari, M., D.K. Joshi, R. Saha, S. Nagarajan, and P.N. Gambhir. 1999. Transverse relaxation time of leaf water protons and membrane injury in wheat (*Triticum aestivum* L.) in response to high temperature. *Ann. Bot.* 84:741–745.

Mandare, A.B., S.K. Ambast, N.K. Tyagi, and J. Singh. 2008. On-farm water management in saline groundwater area under scarce canal water supply condition in the northwest India. *Agric. Water Manage.* 95:516–526.

Minhas, P.S. 1996. Saline water management for irrigation in India. *Agric. Water Manage.* 30:1–24.

Paroda, R.S., T. Woodhead, and R.B. Singh. 1994. *Sustainability of rice-wheat system in Asia and Pacific.* Publication #1894-11. Bangkok, Thailand: FAO.

Porter, J.R. 2005. Rising temperatures are likely to reduce crop yields. *Nature.* 436:174.

Porter, J.R., and M. Gawith 1999. Temperatures and the growth and development of wheat: a review. *Eur. J. Agron.* 10:23–26.

Porter, J.R., and M.A. Semenov. 1999. Climatic variability and crop yields in Europe. *Nature.* 400:724.

Porter, J.R., and M.A. Semenov. 2005. Crop responses to climate variation. *Phil. Trans. R. Soc. B.* 360:2021–2035.

Porter, J.R., R.A. Leigh, M.A. Semenov, and F. Migilietta. 1995. Modelling the effects of climatic change and genetic modification on nitrogen use by wheat. *Eur. J. Agron.* 4:419–429.

Rajput, T.B.S., and N. Patel. 2005. Enhancement of field water use efficiency in the Indo-Gangetic Plain of India. *Irri. & Drain.* 54:189–203.

Ray, J.D., and T.R. Sinclair. 1997. Stomatal closure of maize hybrids in response to drying soil. *Crop Sci.* 37:803–807.

Reddy, J., and D.N. Kumar. 2008. Evolving strategies of crop planning and operation of irrigation reservoir system using multi-objective differential evolution. *Irri. Sci.* 26:177–190.

Rosegrant, M.W., and S.A. Cline. 2003. Global food security: challenges and policies. *Science.* 302:1917–1919.

Rosegrant, M.W., X. Cai, and S.A. Cline. 2002. *World water and food to 2025: Dealing with scarcity.* Washington, DC: IFPRI.

Samra, J.S., Bijay-Singh, and K. Kumar. 2003. Managing crop residues in the rice-wheat system of the Indo-Gangetic plain. In *Improving the productivity and sustainability of rice-wheat systems: Issues and impacts,* eds. J.K. Ladha, 173–195. ASA Spec. Publ. 65. Madison, WI: ASA, CSSA, and SSSA.

Sharma, D.P., and K.V.G.K. Rao. 1998. Strategy for long-term use of saline drainage water for irrigation in semi-arid regions. *Soil Tillage Res.* 48:287–295.

Sharma, P.B.S., and V.V. Rao. 1997. Evaluation of an irrigation water management scheme – A case study. *Agric. Water Manage.* 32:181–185.

Sidhu, H.S., Manpreet-Singh, E. Humphreys, Yadvinder-Singh, Balwinder-Singh, S.S. Dhillon, J. Blackwell, V. Bector, Malkeet-Singh, and S. Singh. 2007. The Happy Seeder enables direct drilling of wheat into rice stubble. *Aust. J. Exp. Agric.* 47:844–854.

Sidhu, R.S. 2005. Patterns and consequences of intensive agricultural growth in Punjab, India. Proc. GECAFS Indo-Gangetic Plain Workshop, 2–3 May 2005, New Delhi, India. www.gecafs.org

Singh, K.B., P.R. Gogri, and V.K. Arora. 2001. Modeling the affect of soil and water management practices on the water balance and performance of rice. *Agric. Water Manag.* 49:77–95.

Singh, S., M.P. Singh, and R. Bakshi. 1990. Unit energy consumption for paddy-wheat rotation. *Ener. Conver. Manage.* 30:121.

Singh, Y., A.K. Bhardwaj, S.P. Singh, R.K. Singh, D.C. Chaudhary, A. Saxena, V. Singh, and A. Kumar. 2002. Effect of rice (*Oryza sativa*) establishment methods, tillage practices in wheat (*Triticum aestivum*) and fertilization on soil physical properties and rice-wheat productivity on a clay loam Mollisol of Uttaranchal. *Indian J. Agric. Sci.* 72:200–205.

Singh, Y., B. Singh, and J. Timsina. 2005. Crop residue management for nutrient cycling and improving soil productivity in rice-based cropping systems in the tropics. *Adv. Agron.* 85:269–407.

Singh, Y., B. Singh, J.K. Ladha, C.S. Khind, R.K. Gupta, O. Meelu, and E. Pasuquin. 2004. Long-term effect of organic inputs on yield and soil fertility in the rice-wheat rotation. *Soil Sci. Soc. Am. J.* 68:845–853.

Sivanappan, R.K. 1995. Soil and water management in the drylands of India. *Land Use Policy.* 12:165–175.

Srivastava, A.P., J.S. Panwar, and R.N. Garg. 2000. Influence of tillage on soil properties and wheat production in rice (*Oryza sativa*)-wheat (*Triticum aestivum*) cropping system. *Indian J. Agric. Sci.* 70:207–210.

Thuy, N.H., S. Yuhua, Bijay-Singh, K. Wang, Z. Cai, Yadvinder-Singh, and R.J. Buresh. 2008. Nitrogen supply in rice-based cropping systems as affected by crop residue management. *Soil Sci. Soc. Am. J.* 72:514–523.

Tilman, D., J. Fargione, B. Wolff, C.D'Antonio, A. Dobson, R. Howarth, D. Schindler, W.H. Schlesinger, D. Simberloff, D., and Swackhamer. 2001. Forecasting agriculturally driven global environmental change. *Science.* 292:281–284.

Timsina, J., and D.J. Connor. 2001. Productivity and management of rice-wheat systems: issues and challenges. *Field Crops Res.* 69:93–132.

Tyagi, N.K., A. Agrawal, R. Sakthivadivel, and S.K. Ambast. 2005. Water management decisions on small farms: a case study from northwest India. *Agric. Water Manage.* 77:180–195.

Vara Prasad, P.V., P.Q. Craufurd, R.J. Summerfield, and T.R. Wheeler. 2000. Effects of short-episodes of heat stress on flower production and fruit-set of groundnut (*Arachis hypogaea* L.). *J. Exp. Bot.* 51:777–784.

Vincent, L.F. 2003. Towards a small holder hydrology for equitable and sustainable water management. *Natural Res. Forum.* 27:108–116.

Wheeler, T.R., P.Q. Craufurd, R.H. Ellis, J.R. Porter, and P.V. Vara Prasad. 2000. Temperature variability and the yield of annual crops. *Agr. Ecosyst. Environ.* 82:159–167.

Zwart, S.J., and W.G.M. Bastiaanssen. 2003. Review of measured crop water productivity values for irrigated wheat, rice, cotton, and maize. *Agric. Water Manage.* 69:115–133.

Chapter 4

Conserving and optimizing limited water for crop production

D.G. Westfall, G.A. Peterson & N.C. Hansen
Department of Soil and Crop Sciences, Colorado State University, Fort Collins, Colorado, USA

ABSTRACT

Proper soil and crop management systems are critical for sustainable production in semi-arid environments, and there are principles that apply to both dryland (non-irrigated) and limited-irrigation systems. In the non-irrigated environment, crop residue from no-till systems permits more diverse crop rotations with less frequent fallow, which leads to increased precipitation-use efficiency. Soil improvements from no-till systems include decreased soil erosion, increased soil organic matter, improved soil structure, and increased infiltration rate. These soil improvements create a positive feedback loop by making more water available to the crop, increasing yields, and returning more crop residues to the soil. In the Great Plains of the United States, we have found that annualized grain production from no-till systems with less frequent fallow can be increased by 75% with an increase in economic return from 13% to 36% compared with the traditional wheat-fallow cropping system. There is also a need to increase water productivity for irrigated crop production because of competition for water by municipal and industrial users, drought, and declining groundwater supplies. The adoption of cropping systems that use less water and insure economic sustainability must be developed. We have found that limited-irrigation practices that time irrigations with critical growth stages can reduce water use of corn by 50% while reducing yields by only 30%. Alfalfa was found to have great potential for limited irrigation because of its natural drought tolerance and perennial growth habit. Many of the principles of water-conservation practices identified in the United States are adaptable to India's conditions. Soil and crop management systems that use less water and that are sustainable and economically viable in India's limited water environment must be developed.

Address correspondence to D.G. Westfall, Department of Soil and Crop Sciences, Colorado State University, Fort Collins, CO 80523 USA. E-mail: dwayne.westfall@colostate.edu

Reprinted from: *Journal of Crop Improvement*, Vol. 24, Issue 1 (January 2010), pp. 70–84,
DOI: 10.1080/15427520903310605

4.1 INTRODUCTION

About 65% of the arable land in India is farmed under natural rainfall; the majority is in a semi-arid environment with average annual precipitation of less than 700 mm. About 80% of the rainfall occurs during the monsoon period, resulting in long periods with little rainfall. Efficient capture of the seasonal rainfall is difficult and crop failure and low yields are common. Another factor causing the historic low production is that only about 10% of the rainfall is stored in the soil. Soil quality and productivity have been degraded because of continuous crop-residue removal, tillage, and inadequate use of fertilizers.

In the dryland regions of India, cropping systems must be developed and adapted to stabilize crop yields, control soil erosion, and capture a higher percentage of the limited precipitation through improved soil and crop-residue management. In the irrigated regions, pumping water from the groundwater for irrigation is causing dramatic reductions in the water table, which limits future water availability. Increased demand for municipal and industrial use, in combination with declining water tables, is causing great concerns regarding the sustainability of irrigated agriculture in many regions of India. Cropping systems and management practices must be developed that will enhance water-use efficiency to insure the long-term sustainability of India's irrigated agriculture.

Strategies that have been developed in semi-arid environments in the United States may provide a blueprint to guide Indian researchers and farmers in their goal of long-term sustainability of both dryland and irrigated cropping systems. This chapter outlines research on cropping systems in the semi-arid Great Plains of Eastern Colorado.

4.2 WATER LIMITED CROPPING SYSTEMS – U.S. RESEARCH RESULTS

This chapter will be divided into two sections covering our research experiences in the semi-arid environment of the U.S. west-central Great Plains: (1) dryland (non-irrigated) crop production systems and (2) irrigated cropping systems with limited irrigation water availability. The results and conclusions are based upon years of research and production experience and hopefully some concepts and ideas can be applied to the dryland and irrigated cropping systems in India.

4.2.1 Dryland crop production systems

The semi-arid environment of the west central Great Plains of United States is characterized by hot summer days with high sunlight intensity, with a summer rainfall pattern and cold, dry winters (Farahani, Peterson & Westfall 1998). However, the winter conditions in the central Great Plains are very favorable for the production of winter wheat (*Triticum aestivum*). Historically, the limited rainfall (300 to 400 mm long-term average) has required the integration of summer fallow into the cropping system. The traditional winter wheat-summer fallow (WF) system results in fields being in crop production (wheat) for 10 months and fallow for 14 months during a two year period. One crop is produced every two years. During the fallow period, weed control has been accomplished by frequent tillage; however, the advent of glyphosate-based herbicides has made chemical weed control economical.

The frequent tillage that was practiced for decades resulted in the loss of soil organic matter, degradation of soil structure, and a general degradation of soil quality. Associated with these negative effects was an increase in soil erosion by wind and water, with wind being the major problem in this semi-arid environment.

The annual precipitation can vary widely, resulting in bumper crops one year and crop failure the next. Precipitation storage during the 14 month fallow period is critical to crop production that occurs during the 10 month winter wheat-growing period (Farahani, Peterson & Westfall 1998). Because the soils of the Plains were first broken out of native grass in the early 1900s, much progress has been made in increasing precipitation-storage efficiency. The first major step in crop production stabilization occurred after the Dust Bowl disaster in the 1930s, which led to the adoption of summer fallow and conservation tillage. Peterson et al. (1996) summarized the changes that resulted in the increased fallow water-storage efficiency (Table 4.1). During the time when soils were subjected to maximum tillage (7–10 operations during the fallow period), the fallow efficiency (precipitation-storage efficiency during the fallow period) was about 19%. Less than 27 kg of wheat was produced per 25 cm of precipitation. The production per unit of rainfall was doubled by the adoption of improved management practices, such as "stubble-mulch" tillage, minimum tillage (2 to 4 annual tillage practices), and weed control with herbicides in the 1961–75 period. Greb (1979) predicted that by adoption of minimum/no-till in the early 1980s (Table 4.1), the fallow efficiency would be increased to 40%. However, even with

Table 4.1 Progress in fallow systems and winter wheat yields, U.S. Central Great Plains research station, Akron, Colorado.

Years	*Changes in fallow systems*	*Number of tillages*	*Fallow water storage*[a] *mm*	*Fallow efficiency %*	*Winter wheat yields kg ha*$^{-1}$	*Precipitation use efficiency*[b] *kg ha*$^{-1}$ *mm*$^{-1}$
1916–30	Maximum tillage; plow harrow (dust mulch)	7–10	102	19	1070	1.22
1931–45	Conventional tillage; shallow disk, rod weeder	5–7	112	24	1160	1.43
1946–60	Improved conventional tillage; begin stubble mulch 1957	5–7	137	27	1730	2.06
1961–75	Stubble mulch; begin minimum tillage with herbicides (1969)	2–4	157	33	2160	2.78
1976–90	Projected estimate. Minimum tillage; begin no-till 1983	0–2	183	40	2690	3.253

[a] Based on 14 months fallow, mid-July to second mid-September.

[b] Assuming 2 years precipitation per crop in a wheat-fallow system.

Source: Adapted from Greb (1979).

no-till fallow management, we have not been able to achieve his predicted high fallow precipitation-storage efficiency. Under optimum conditions, with no-till fallow management, we have only achieved 35% efficiency.

The increased precipitation-storage efficiency that resulted from adoption of no-till raised the question, can a summer crop be integrated into the traditional WF cropping system? This has great potential because 75% of the annual precipitation is received during the summer growing season (April–September). A more intensive cropping system has several additional benefits: (1) maximization of water use because a crop is present during the greatest rainfall period; (2) improvement of soil quality as a result of the addition of more crop residue to the soil; (3) decrease in erosion potential by wind and water as a result of residue protection of the soil surface; (4) increased annualized grain production; and (5) increased potential economic return.

A long-term no-till cropping systems experiment was initiated in 1986 to address the above five points (Peterson, Westfall & Cole 1993). The experiment included the following variables (Figure 4.1): climate [ET (evapotranspiration) gradient increasing from north to south], soil variables (summit, side slope, and toe slope), and cropping systems (different cropping systems intensity). The cropping systems were WF, wheat-corn (C) ((*Zea mays*)-fallow (WCF), and wheat-corn-millet-fallow (WCMF) [millet (*Panicum miliaceum*)]). Note that as the cropping intensity increases from WF to WCMF, the occurrence of summer fallow decreases from 50% to 25%, and the cropping intensity increases from 25% to 75% over the cropping cycle, one crop per two years (50% cropping system intensity) to three crops per four years (75% cropping system intensity).

During the 20 year period (1986–2006) of this long-term study, a range of climatic conditions was experienced. During the first 12 years (1986–1998), average or above average annual precipitation occurred, but during the subsequent eight years

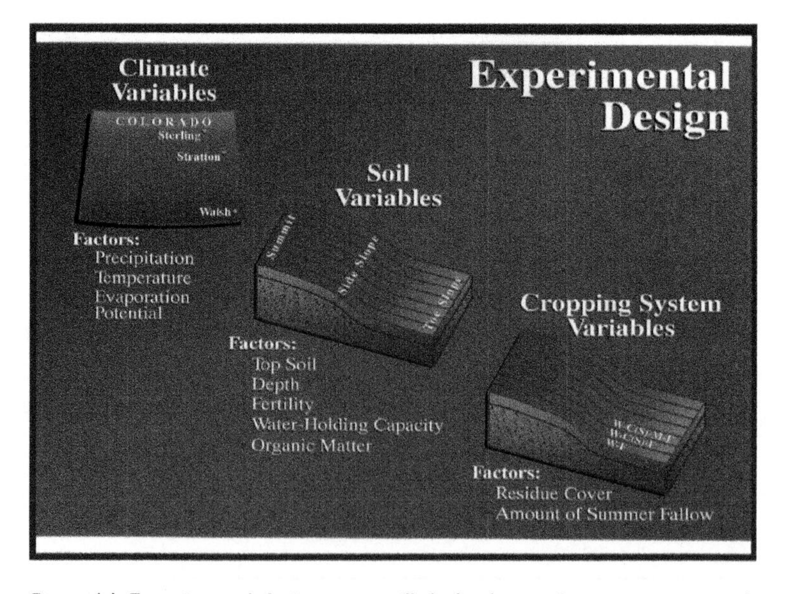

Figure 4.1 Experimental design on no-till dryland cropping systems research project across climatic gradient, soils, and cropping systems in Colorado, United States.

(1999–2006), drought conditions occurred across the Great Plains region. Cropping systems' responses differed for the two time periods. The cropping systems evaluated during these two time-periods are outlined in Table 4.2.

We obtained an average increase of 75% in annualized grain yield (compared to WF) with the intensified WCMF system (Figure 4.2) (Peterson & Westfall 2004). The 75% increase in annualized grain yield resulted in net economic returns ranging from 13% to 36%, depending upon the cropping system (Kaan et al. 2002). Crop residues returned to the soil also increased as annualized grain yield increased (Cantero-Martinez et al. 2006). The increase in residue amount returned to the soil induced a 7% increase in soil carbon (C) content in the 0–20 cm soil depth for the WCF relative to the WF system (Figure 4.2) (Sherrod et al. 2005). The results of our continuous cropping system (CC) (no summer fallow periods), which included an annual forage in the rotation, also are reported in Figure 4.3. Soil C increased 13% in the CC system relative to WF across the 12 year period. The soil C increase in all rotations was directly related to the amount of residue returned to the soil (Peterson & Westfall 2004; Cantero-Martinez et al. 2006). Increased soil C levels improved soil macroaggregate formation (Shaver et al. 2002), which increases water infiltration and precipitation-storage efficiency.

Based upon the success of the intensified cropping systems during the first 12 years, we modified the cropping systems to include more intensive cropping frequency with less summer fallow time (Table 4.2). However, in 1999, the Great Plains entered a drought cycle that lasted for the next eight years. The drought severity, as measured by the Palmer Drought Severity Index, is shown in Figure 4.4, and a negative index indicates drought conditions (Palmer, 1965). The drought index was generally positive during the first 12 years of the study (1986–1998), but after 1998 drought conditions

Table 4.2 Dryland cropping systems evaluated from 1986–1997 and 1998–2006.

Time period		*No-till dryland cropping systems**
1986–1998	Non-drought	wheat-fallow; wheat-corn-fallow; wheat-corn-millet-fallow
1999–2006	Drought	wheat-corn-fallow; wheat-corn-millet; wheat-wheat-corn-millet

*Winter-wheat.

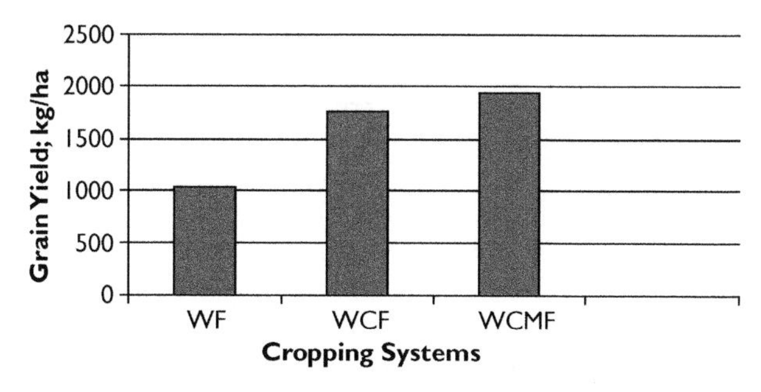

Figure 4.2 Annualized grain yield of three diverse cropping systems after 12 years of production.

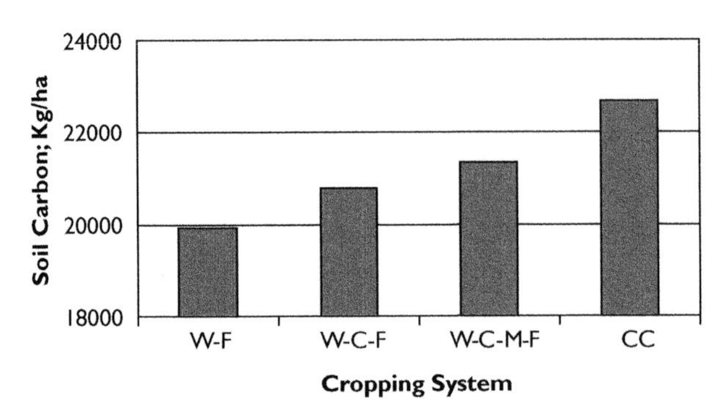

Figure 4.3 Soil organic carbon changes in top 20 cm soil after 12 years in diverse cropping systems.

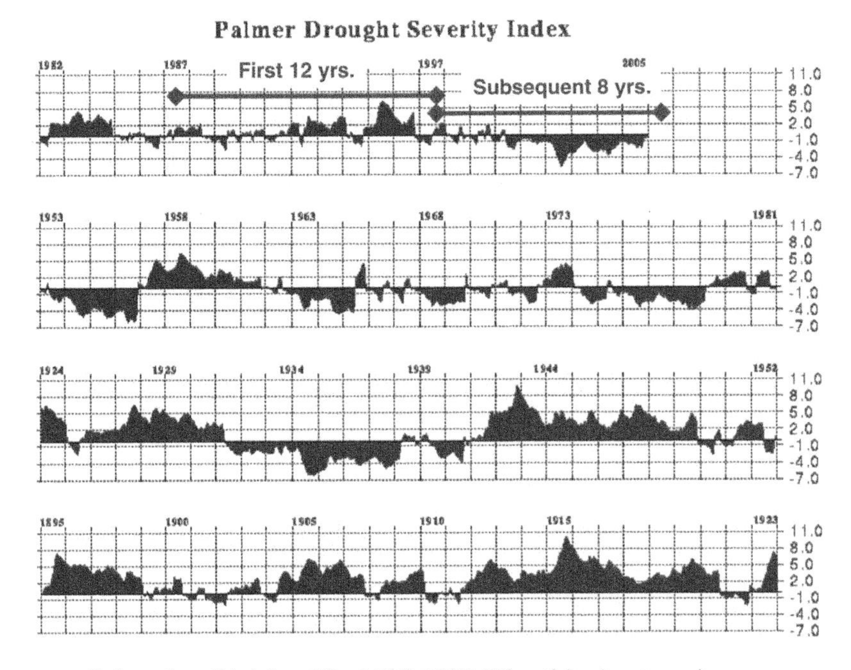

Figure 4.4 Long-Term Palmer Drought Severity Index in the Colorado Great Plain from 1895 to 2005.

existed as shown by a negative index for an eight year time period. The Palmer Drought Severity Index for this eight year period was very similar to the drought severity index during the Dust Bowl in the 1930s. During the Dust Bowl era, crops failed, severe soil erosion by wind occurred and many farmers lost their farms or even abandoned their farms. However, during the 1999–2006 period, which had a similar drought severity index, dust bowl conditions did not exist because farmers had converted their management from intensive tillage (Table 4.1, see 1931–1945 time period) to conservation tillage or no-till cropping systems, where crop residue was retained on the soil surface. Crop-residue retention is the primary protection agent against soil erosion by wind as well as by water.

During the period of drought (1999–2006), our cropping systems experienced decreased yields and increased failure relative to the wetter years (Figure 4.5). For example, during the wetter time period, rotation failure occurred 7% of the years as compared with 14% of the years during the drought period. As cropping intensity increased, the rotation failure increased; WCF failure occurred 7% of the years in the non-drought period, but failed 21% of the years during the drought period. Rotation failure of the continuous cropping systems that did not include fallow (WCM and WWCM) was very high during the drought period, about 35% of the time. Winter wheat yields varied widely during the drought as compared with the non-drought period (Figure 4.6). There were no wheat crop failures for wheat-after-fallow during the 1986–1998 non-drought period, and the average grain yield was 2350 kg ha^{-1}.

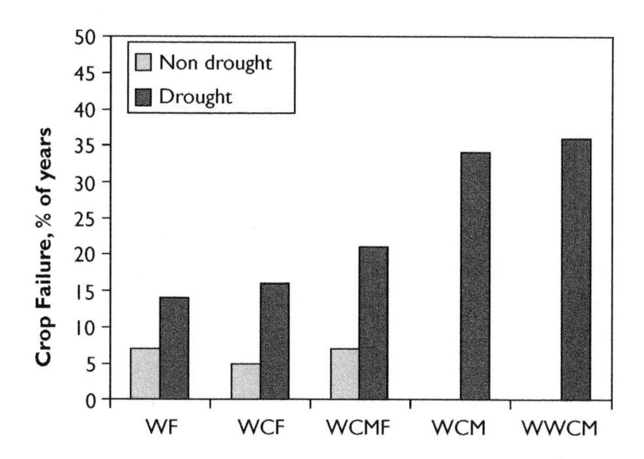

Figure 4.5 Frequency of crop rotation failure during drought and non-drought conditions.

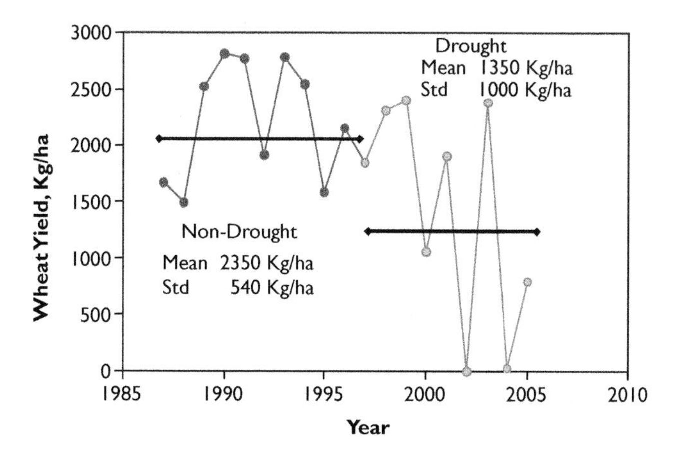

Figure 4.6 Variation in winter wheat yields during the non-drought and drought periods.

During the eight-year drought period, winter wheat crop failure occurred two of the years, and the average yield was 1350 kg ha^{-1}, a 43% reduction in yields. (These analyses were done using all plots with W after F). The standard deviation for wheat grain yield during non-drought was 540 kg ha^{-1}, but was 1000 kg ha^{-1} during the drought period. The greatest occurrence of wheat crop failure was in the WWCM system where the second wheat crop in the rotation had a high incidence of failure and very low yield when a crop was produced. This large deviation in wheat yields makes it financially disastrous for producers.

Based upon our results, it appears that cropping intensity must either include a fallow period before winter wheat or a low water-use crop, such as millet. Some farmers in the region have been successful with a winter wheat-millet cropping system without summer fallow. Others have integrated sunflower (*Heleanthus annuus*) into the system. Sunflower is a very efficient user of stored soil water and exhausts the water profile to a much greater degree than any other adapted crop. Consequently, a one- to two-year fallow period is required following sunflower production to allow the soil to store enough water for the subsequent crop.

4.2.2 Limited irrigation crop production systems

There is increasing competition for a limited water supply throughout much of the United States, as well as in many other parts of the world, particularly India. Increasing competition from urban and municipal water users, declining groundwater levels, and drought are factors that lead to reduced irrigation water quantities for large areas of agricultural land. If water quantity limitations require land to be shifted from irrigated to dryland cropping, it will significantly impact the economic viability of agricultural producers and have far-reaching indirect effects on businesses and communities that support irrigated agriculture. Water-conservation options, other than complete land fallowing (land dry-up), are needed because of the potential adverse economic and environmental concerns associated with conversion to dryland. One approach to reducing consumptive use of irrigation water is adoption of limited irrigation cropping systems. We are conducting a limited irrigation research project in Colorado to develop alternatives to complete dry-up of irrigated land. The project premise is that it is possible to develop alternative water-saving irrigated cropping systems that are economic and significantly reduce the quantity of water needed relative to fully irrigated systems. The objectives of the project are to: (1) develop sustainable agricultural production systems in an environment of increasing competition for a limited water supply; and (2) provide state agencies/regulators, farmers, and urban water users an alternative to the dry-up philosophy that presently exists when agricultural water is transferred to industrial and municipal use.

Several concepts have to be integrated into a limited water-use cropping systems if they are going to be sustainable. They are:

1 Reduced consumptive water use by crops;
2 Management of crop water stress to minimize effect on yield;
3 Growth stage-specific irrigation timing to enhance production per unit of water;
4 Reduced input costs;
5 Rotations with low water-use crops and possibly fallow; and

6 Adoption of water-saving practices that improve water-use efficiency, such as conservation tillage, water-efficient crop rotations, improved irrigation-efficiency practices.

The general concept of the limited irrigation project is shown in Figures 4.7 and 4.8. All six of the above factors were considered during planning and implementation. Our experimental design covers a range of cropping systems that have diverse crops with water-use limitations ranging from fully irrigated to non-irrigated (dryland). The fully irrigated corn-alfalfa (*Metacogo sativa*) cropping system (4 years corn and 4 years alfalfa) is projected to use an average of 60 cm water year^{-1}, whereas the limited irrigation forage-based corn grain-corn silage-alfalfa cropping system would be limited to 40 cm of water year^{-1}. The limited irrigation grain-based wheat-corn-sunflower system would be limited to 30 cm water use year^{-1}. The dryland WF system would only use 13 cm water year^{-1}, all of which is natural rainfall. This range in water use

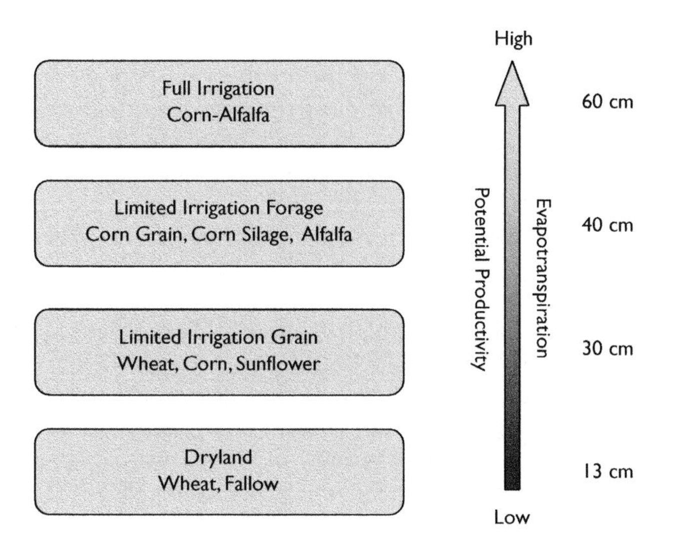

Figure 4.7 General water saving concept of limited-irrigation project.

Water/Cropping Systems Production Scenarios

- Irrigation Regimes
  - Full Irrigation
  - Limited Irrigation
  - Dryland
- Systems management
  - Conventional tillage
  - Conservation tillage
- Cropping systems
  - Grain based
  - Forage based

Alfalfa Irrigation Scenario

- Full irrigation
- Stop after 2nd cutting
- Spring and Fall
- Stop after 1st cutting

Figure 4.8 Irritation and cropping systems combinations included in the limited irrigation project.

Table 4.3 Effects of irrigation level on average (2006–2008) crop yield, evapotranspiration (ET), and water productivity.

		Three-year average		
Irrigation level	*Crop*	*Yield Mg/ha*	*ET cm*	*Water productivity Mg/ha/cm*
Limited	Alfalfa	7.3	45.5	0.16
Full	Alfalfa	9.3	51.8	0.18
Limited	Corn	8.4	38.6	0.22
Full	Corn	10.1	60.7	0.17
Limited	Sunflower	2.5	42.2	0.06
Limited	Wheat	2.3	29.7	0.08
Dryland	Wheat	1.8	27.2	0.07

obviously will cause a range in cropping system production potential; the production potential decreases as water use decreases.

The experiment has been conducted for a three-year period from 2006 to 2008 and results are given in Table 4.3. The most promising results for limited irrigation were found for corn. Limited-irrigation corn had an average yield of 8.4 Mg/ha compared with the full-irrigation yields of 10.1 Mg/ha, representing a yield reduction of 17%. The ET reduction, however, was 36%. Thus, the water productivity was higher for the limited irrigation corn than for full irrigation. When water is more limited or of higher value than land or other inputs, maximizing water productivity is desired. As shown for corn in this study, maximum water productivity does not necessarily correspond to maximum biological yield. For grain crops, water has the largest effect on yield when it is applied during critical growth stages, which begins at tasselling and continues through the mid-grain-fill stage. Water stress during these reproductive growth stages has devastating effects on grain yield. Water stress during vegetative stages has lesser effects on grain yield. Wheat was compared between limited irrigation and dryland. Limited irrigation yields averaged 2.3 Mg/ha and dryland yields averaged 1.8 Mg/ha. Water productivity was not as different between irrigation regimes for wheat compared with corn. Limited-irrigation yields were limited by poor tiller development caused by early water stress. Thus, critical stages for wheat yields include both reproductive stages and tiller development.

4.3 ALFALFA TRIALS

Alfalfa is a resilient crop and has drought- and water-stress tolerance mechanisms that make it biologically suited to limited irrigation. It is a deep-rooted perennial crop with the ability to go into dormancy during drought periods. During dormancy, the aboveground growth is limited, storing energy for rapid growth when water becomes available. Because alfalfa consumes large quantities of water under full irrigation during the growing season, limited irrigation provides a large potential for water savings using limited-irrigation practices. This characteristic allows flexibility in water management.

An experiment was established to determine the adaptability of alfalfa to limited irrigation with the objective to quantify alfalfa yield, consumed water (ET), water-use efficiency, and forage production and quality under full- and limited-irrigation regimes.

When alfalfa was grown under limited irrigation using the alfalfa irrigation scenarios outlined in Figure 4.8, the total yields in 2006 ranged from 18.3 to 8.8 Mg ha^{-1} (8.2 to 3.9 tons/acre) and in 2007 from 19.2 to 15.5 Mg ha^{-1} (8.6 to 6.9 tons/acre) (Figure 4.9). The ability of alfalfa to recover from severe water stress within the growing season is seen in the spring and fall irrigated treatments. In this treatment, alfalfa was allowed to go into drought-induced dormancy in the summer but it recovered after water was applied for the fourth cutting; yields were similar to those obtained in the fully irrigated treatment. Likewise, the similar first-cutting yields in 2007 across all irrigation treatments illustrate its resilience to water stress across multiple crop years. The total season irrigation applied to the fully irrigated treatment averaged 47 cm yr^{-1}. The average total season ET differences across both years ranged from

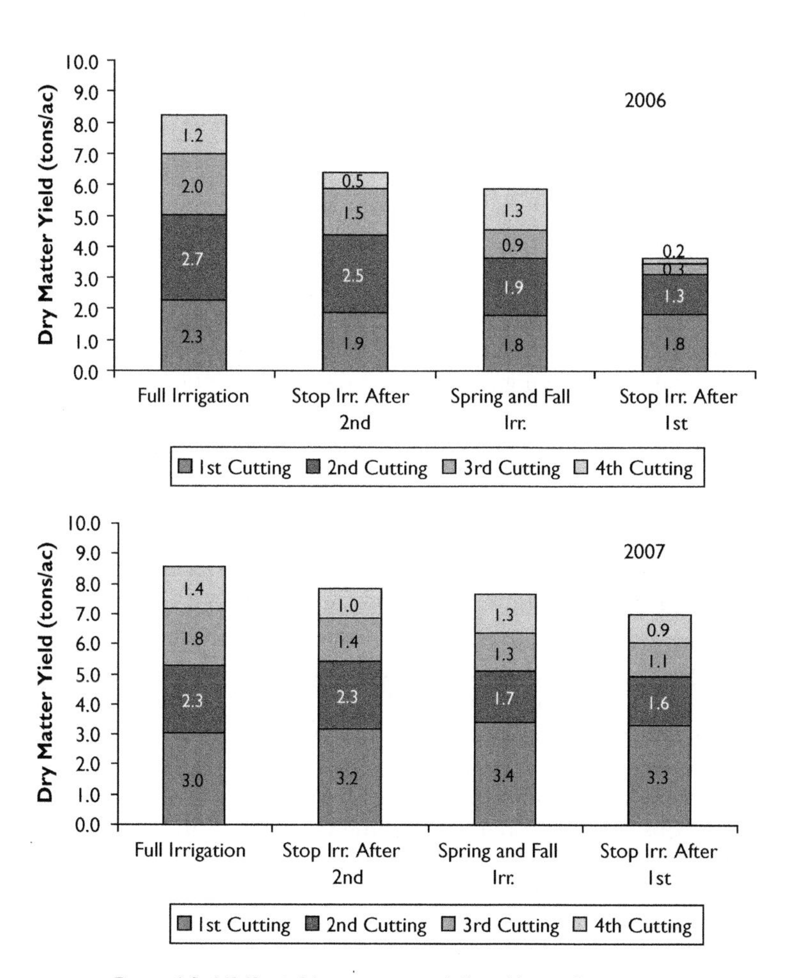

Figure 4.9 Alfalfa yield response to full and limited irrigation.

Table 4.4 Effects of full and limited irrigation treatments on average (2006–2007) Alfalfa seasonal consumptive water savings (ET) and seasonal yield.

Treatment	*Average yield (Mg/ha)*	*Seasonal consumptive water savings (ET cm)*	*Seasonal forage yield reduction (Mg/ha)*
Full Irrigation	18.7	0	0
Stop Irr. After 2nd	15.6	28.2	3.1
Spring and Fall Irr.	15.2	27.2	3.5
Stop Irr. After 1st	12.2	42.2	6.5

28.2 to 42.2 cm yr^{-1}, with yield losses of 3.1 to 6.5 Mg ha^{-1} (Table 4.4). The seasonal consumptive-use savings represent the quantity of water that could be saved by implementing the limited-irrigation strategies and the yield losses can be used to estimate the cost of the water in terms of production losses. A key lesson, learned from this study, is that the highest water productivity is achieved in alfalfa when the crop is provided with all the water needed during periods of the growing season most conducive to growth and then allowed to go dormant. The most efficient water use was achieved in early summer.

4.4 CONCLUSIONS

Water is a driving input of sustainable agricultural systems around the world. In semi-arid regions where precipitation is not adequate to produce crops to their full yield potential, water conserving practices or irrigation must be used to optimize production. Adoption of diverse cropping systems can also have a positive feedback to sustainability.

In the non-irrigated environment, the primary factor that makes diverse cropping systems successful is the retention of crop residue on the soil surface and implementation of no-tillage systems with implementation of chemical weed control and very limited or no-tillage. The residue cover increases precipitation-storage efficiency and decreases evaporation rate, which in turn increases soil carbon levels. The increased soil carbon levels provide positive feedbacks, such as improvements in soil structure and increased precipitation infiltration, which all result in greater precipitation storage and eventual crop yield enhancement. Diversification of the cropping system also improves the chance of getting improved grassy weed control. No-till practices have made it possible to increase cropping intensification beyond the traditional winter wheat-fallow system and, in turn, water-use efficiency has increased by 30% in West Central Great Plains agroecosystems. Cropping intensification has also provided positive feedbacks to soil productivity via the increased amounts of crop residue being returned to the soil and greater economic returns.

When irrigation water is in short supply, limited-irrigation practices can reduce total applied irrigation on corn by as much as 50% while reducing yields by only 30%. Limited-irrigation alfalfa was found to have great potential for limited irrigation.

Because of its natural drought tolerance and perennial growth habit, limited-irrigation alfalfa can generate large savings in water while still maintaining a viable crop-production system.

As agencies around the world make decisions regarding the future of water use and how agricultural water transfers will be used to address growing urban populations, limited irrigation cropping systems should be considered as one means of meeting urban water needs while maintaining viable irrigated-agricultural systems. Agriculture does not have to feel the full brunt of this decision, and government agencies must work with the agricultural community to insure it remains variable. The agriculture water issues in India are not very different than those in the United States. However, parts of India are also confronted with rapidly declining water tables. The principles discussed in water-limited agroecosystems research are also applicable to these situations. Crop system consumptive use must be decreased, which may necessitate a conversion to lower consumptive use, demanding crops and limited-water management practices.

All parties involved must work together to insure that the interests of agriculture are kept on the forefront of all discussions and decisions of water transfer to meet municipal and industry demands. There are viable alternatives to the practice of transfer of all the water and drying up agriculture to meet municipal and industrial demands. Limited irrigation of adapted cropping system must be investigated to allow all parties to make informed decisions that are sustainable to agriculture as well as to municipal and industry interests.

REFERENCES

Cantero-Martinez, C., D.G. Westfall, L.A. Sherrod, and G.A. Peterson, 2006. Long-term crop residue dynamics in no-till cropping systems under semi-arid conditions. *J. Soil Water Cons.* 61: 84–95.

Farahani, H.J., G.A. Peterson, and D.G. Westfall, 1998. Dryland cropping intensification: Fundamental solution to efficient use of precipitation. *Adv. Agron.* 64: 197–223.

Greb, B.W. 1979. *Reducing drought effects on croplands in the west-central Great Plains*, USDA Info. Bull. No. 420. Washington, DC: U.S. Gov. Print. Office.

Kaan, D.A., D.M, O'Brein, P.A. Burgener, G.A. Peterson, and D.G. Westfall, 2002. *An economic evaluation of alternative crop rotations compared to wheat-fallow in Northeastern Colorado*, Technical Bulletin TB 02–1. Fort Collins, CO: Agric. Exp. Stn. Colo. State Univ.

Palmer, W.C. 1965. Meteorological drought. Washington, D.C.: U.S. Weather Bureau, Res. Pap. No. 45.

Peterson, G.A., A.J. Schlegel, D.L. Tanaka, and O.R. Jones, 1996. Precipitation use efficiency as affected by cropping and tillage systems. *J. Prod. Agric.* 9: 180–186.

Peterson, G.A. and D.G. Westfall, 2004. Managing precipitation using sustainable dryland agroecosystems. *Ann. Appl. Biol.* 144: 127–138.

Peterson, G.A., D.G. Westfall, and C.V. Cole, 1993. Agroecosystems approach to soil and crop management research. *Soil Sci. Soc. Am. J.* 57: 1354–1360.

Shaver, T.M., G.A. Peterson, L.R. Ahuja, D.G. Westfall, L.A. Sherrod, and G. Dunn, 2002. Surface soil properties after twelve years on dryland no-till management. *Soil Sci. Soc. Am. J.* 66: 1296–1303.

Sherrod, L.A., G.A. Peterson, D.G. Westfall, and L.R. Ahuja, 2005. Soil organic carbon pools after 12 years in no-till agroecosystems. *Soil Sci. Soc. Am. J.* 1600–1608.

Chapter 5

On watershed management

Hemant Chowdhary
Department of Civil and Environmental Engineering, Louisiana State University, Louisiana, USA

Vijay P. Singh
Department of Biological and Agricultural Engineering, Texas A&M University, Texas, USA

ABSTRACT

Significant developments have occurred in India in the past three to four decades with respect to the state of natural resources and human population, as part of various watershed programs. Much experience has been gained by governmental and non-governmental institutions and by the stakeholders engaged in these endeavors. Although there are many success stories that credit these interventions, much more effort is required to bring the benefits to all regions and remote areas of the country. There have been many management practices and intervention strategies that have evolved from this long learning process, but more pointed research in this regard is required for developing newer technological methods suited for specific situations. Greater research is needed for identifying better institutional strategies applicable to watershed management in different socio-economic conditions and in different physiographic regions of the country. One of the major impediments to the success of most watershed programs is the lack of coordination among the many governmental institutions mandated to run these developmental plans. A sizable amount of about Rs. 1,600 crores is invested annually in improving the conditions of watersheds around the country. Watershed programs can become much more efficient and cost-effective if the roles and responsibilities of all the partners and stakeholders are clearly defined and followed up for ensuring stricter accountability. A countrywide comprehensive and integrated watershed management information system that utilizes the state-of-the-art GIS, database, and web technologies will further consolidate achievements and improve accountability and efficiency of watershed management

Address correspondence to Vijay P. Singh, Department of Biological & Agricultural Engineering, Texas A&M University, 321 Scoates Hall, 2117 TAMU, College Station, Texas 77843-2117, USA. E-mail: vsingh@tamu.edu

Reprinted from: *Journal of Crop Improvement*, Vol. 23, Issue 2 (April 2009), pp. 158–173,
DOI: 10.1080/15427520802646083

programs. This paper draws a few comparisons with watershed programs in the United States and offers suggestions for improving the effectiveness and sustainability of these programs.

Keywords: sustainable agriculture, best management practices, institutional strategies

ABBREVIATIONS

BMPs	best management practices
DDP	Desert Development Program
DPAP	Drought Prone Area Program
DRDAs	district rural development agencies
FPR	flood prone rivers program
GIS	geographical information system
GTZ	the German Agency for Technical Cooperation
IAEPS	integrated afforestation & eco-development project scheme
IWDP	Integrated Wasteland Development Program
IWMP	Integrated Watershed Management Program
MoRD	Ministry of Rural Development, Government of India, India
NGOs	non-governmental organizations
NWDPRA	National Watershed Development Project for Rainfed Areas
PRIs	Panchayati Raj Institutions (local governing bodies in rural India)
PIAs	project implementing agencies
RVP	River Valley Program
USEPA	U.S. Environmental Protection Agency
WDPSCA	Watershed Development Project for Shifting Cultivation Areas
ZP	Zilla Parishads (district councils in India)

5.1 INTRODUCTION

In developing countries, watershed management is becoming increasingly important, and attention is shifting to overall socio-economic welfare along with better water and soil conservation. This is specifically true for India, where the population is continuing to grow at a rapid pace. The ever-increasing pressure on the natural resources of the country is further aggravated by the even faster economic growth the country has witnessed in the past couple of decades. Unprecedented economic activity in areas such as agriculture, industry, power, and communication, is affecting land-use patterns in more ways than anticipated a few years back. More lands are being brought under cultivation and/or irrigation as there is an increased requirement to grow more food. Numerous industries are emerging or expanding because of an astronomical rise in the demand for consumer products, both from within and outside of the country. Such industrial expansion often is not strictly regulated and monitored for achieving desirable environmental and water pollution standards. Various forces of development have also led to expansion of resources for power generation, including hydropower. Major irrigation and hydropower projects are setting significant changes to the hydrological regimes of many water bodies, especially relative to the inability to meet the minimum environmental flow requirements in rivers and streams. Watershed management assumes even greater importance in view of these fast-changing scenarios, which have the potential to adversely impact the already deteriorating available natural resources all around rural India.

Watershed management is essentially a multi disciplinary field involving water, land, agriculture, forest and socio-economic aspects, among others. It is, therefore, not surprising that there are a multitude of governmental agencies, often with overlapping roles and responsibilities, attempting to work in this area. On one hand, it provides a desirable multidimensional approach to the problems at hand; on the other, it results in avoidable duplication of efforts and possibly lesser avenues of accountability. The Ministry of Rural Development and Ministry of Agriculture in India have initiated most of the watershed-management projects in the country and act as the main agencies regarding related issues. Other prominent agencies, such as the Ministry of Environment and Forests, the Ministry of Water Resources, the Ministry of Science and Technology, and a few more related ministries and departments in various states, also run numerous watershed developmental plans. Ideally, the active involvement of so many agencies is profitable in view of the varied experience it can bring, but in reality, the lack of coordination, consultation, and complementary roles lead to inefficient use of resources and time. A strict structuring of the roles of the various agencies in this regard can bring about much-needed progress in this important area of watershed management and development in the country. The past three to four decades of experience in implementation of watershed-development schemes has made it apparent that success of these projects largely depends on the ownership by and active involvement of the local people at the grass roots level. Subsequent governments have realized that in order to ensure sustainability of these projects, public funding is to be augmented by the capital and workforce contributions of local institutions and people. This is reflected in the "Guidelines for Hariyali," which state major objectives as: (a) harvest every drop of rainwater for purposes of irrigation, plantation, and fisheries, etc.; (b) create sustainable sources of income for the village community as well as for drinking water supplies; (c) ensure overall development of rural areas through the Gram Panchayats; (d) generate employment, alleviate poverty, empower communities, and develop human and other economic resources for rural areas; (e) mitigate the adverse effects of extreme climatic conditions; (f) restore ecological balance by harnessing, conserving, and developing natural resources; and (g) encourage village communities toward sustained community action for the operation and maintenance of assets created and further develop the potential of the natural resources in the watershed. It may be noted here that while it is important and useful to ensure participation of people through the local institutions, it is equally important to first ensure that the local institutions are equipped and sufficiently mandated to carry out such responsibility in a systematic, scientific, and sustainable manner. On one hand, the expectations are to have interventions in the watersheds in the most modern and scientific way; on the other hand, the local governing institutions – called the Panchayati Raj Institutions (PRIs) – that are proposed to be planning, implementing, and managing these interventions, seem to be ill-equipped, insufficiently supported, and unprepared for this enormous challenge. However, it is expected that such gaps will gradually be filled in such a manner that PRIs become an effective contributory and operative vehicle for watershed-management projects.

Currently, India invests about Rs. 1,600 crores every year in various watershed-management and developmental schemes, aiming to alleviate poverty, build up sustainable livelihoods, and rehabilitate degraded lands. Of this, about a quarter is being donated by foreign countries (Kotru, 2003). In recent years, the German assistance in this field of watershed management has been visible and sizable at about 20 percent of the total foreign aid. A substantial part of this has been for environmental protection and sustainable management of natural resources, mainly aimed at poverty alleviation. Various projects involving capacity building of non-governmental institutions (NGOs), village developmental committees, PRIs, and future institutions of knowledge centers, such as Water Management Training Institutes, were taken up under Indo-German collaborative projects. The German Agency for Technical Cooperation (GTZ) is one of the main international technical agencies supporting watershed development in India. Several important factors have emerged from GTZ's two decades of in-depth and hands-on experience with watershed-development projects' implementation in a variety of socio-economic setups in eight states of the country. These important constraining factors are: (a) difficulty in setting up a system for continual capacity building of governmental organizations, NGOs, and community-based organizations, especially once the project ends; (b) governmental funding is invariably inadequate and mobilization of other financial resources becomes difficult; (c) lack of baseline and impact data and monitoring guidelines for impact monitoring, more so in the post-project period; (d) lack of clarity with respect to technical and financial responsibilities among state government departments, PRIs, and project-implementing agencies (PIAs); (e) difficulties in garnering support of line departments in view of unclear directives; (f) frequent failures of project benefits reaching the poorest of the poor and women; and (g) concept of equitable distribution of cost and benefit among landless and land owners needs more (research) work and intensive facilitation (Kotru, 2003).

5.2 WATERSHED MANAGEMENT

A watershed is essentially a geophysical unit draining at a point that supplies water to a stream or a river and is made up of physical and hydrologic natural resources as well as human resources. Watersheds vary in size; smaller ones near the ridgeline constitute a small area forming a stream or rivulet, moderate ones form sub-basins, and bigger ones cover vast areas of land, forming mighty river basins. Within any watershed, there are again different portions or stages that exhibit significant variations in characteristics. The headwater region is where water starts converging into small rivulets and has greater slopes affecting more erosion; the middle reaches have deeper stream banks, lakes, and ponds; and the lower portions become larger and sometimes big rivers comprise wetlands and deltas. Each segment of a watershed typically has different physical and ecological features that provide a unique habitat for various species and plants that become most relevant for any sort of intervention strategies.

In India, the watershed approach has conventionally meant the arresting of rainwater runoff, and soil and moisture conservation. The central and various state governments have attempted to achieve these objectives through developmental programs, such as rehabilitating wastelands and degraded lands (MoRD, 2003). These developmental plans have varied on the basis of target areas, but the common feature for all has been the land and water resources management for the sustainable development of natural resources. Various types of best management practices (BMPs) can be evolved for different types of watersheds featuring different hydrometeorological conditions, and cropping and livelihood patterns. For example, BMPs, such as strip cropping, riparian vegetative buffer strips for river banks, silt fencing, and various soil-covering techniques, help control watershed soil erosion.

In the United States, state and tribal water-resource professionals are increasingly turning to watershed management to achieve greater results from their programs, as managing water-resource programs on a watershed basis makes good sense – environmentally, financially, socially, and administratively (USEPA, 1996). For example, by jointly reviewing the results of assessment efforts for drinking-water protection, pollution control, fish and wildlife habitat protection, and other aquatic resource-protection programs, managers from all levels of government better understand the cumulative impacts of various human activities and determine the most critical problems within each watershed. Using this information, priorities for action are set to allow public and private managers from all levels to allocate limited financial and human resources to address the most critical needs. Establishing environmental indicators helps guide activities toward solving those high-priority problems and measuring success in making real-world improvements rather than simply fulfilling programmatic requirements. Besides driving results toward environmental benefits, the approach also results in cost savings by leveraging and building upon the financial resources and willingness of the people with interest in the watershed. Through improved communication and coordination, the watershed approach reduces costly duplication of efforts and conflicting actions. Regarding actions that require permits, specific actions taken within a watershed context (for example, the establishment of pollutant trading schemes or wetlands mitigation banks and related streamlined permit review) enhance predictability that future actions will be permitted and reduce costs for the private sector. As a result, the watershed approach can help enhance local and regional economic viability in ways that are environmentally sound and consistent with watershed objectives. Thirdly, the watershed approach strengthens teamwork between the public and private sectors at the federal, state, tribal, and local levels to achieve the greatest environmental improvements with the resources available. This emphasis gives those people who depend on the aquatic resources for their health, livelihood, or quality of life, a meaningful role in the management of resources. Through such active and broad involvement, the watershed approach builds a sense of community, reduces conflicts, increases commitment to the actions necessary to meet societal goals, and, ultimately, improve the likelihood of sustaining long-term environmental improvements.

5.3 PRINCIPLES OF WATERSHED MANAGEMENT

The United States Environmental Protection Agency (USEPA) promotes the following four core principles for better watershed management practices (reproduced from USEPA, 1994).

5.3.1 Watersheds are natural systems one can work with

It is important to realize that no matter where one lives or works, one is in a watershed, teeming with unique, interrelated natural processes and forces. These natural forces help shape the watershed landscape, its water quality, and – in turn – all lives. In mountainous upland areas, there are unique blends of climate, geology, hydrology, soils, and vegetation shaping the landscape, with waterways often cutting down steep slopes. A variety of activities that influence water quality in the watersheds are the chemicals from the mineral weathering of rocks, from the decay of vegetation, and from groundwater. The water body channel, riparian zone, and upland zone are three zones referred to when discussing watershed management. The vegetation shades the water, influencing temperature and what can live in the water. In an upland plain area, grassy plains, hardy vegetation, and slower moving, meandering streams and rivers can be found. In the coastal area, where ocean meets land, there are again different blends of features and processes shaping the environment. In lowland areas between upland and coastal waters, where tidal wetlands are prevalent, processes serve entirely different functions. In other words, each watershed – indeed each watershed zone – has unique living and nonliving components that interact, with one element responding to the action or change of another. Knowing the watershed means learning the natural processes working within the watershed boundaries. Once a better understanding of these processes is acquired, one can better appreciate how the watershed's ecological processes help sustain life. Working with the watershed also means understanding how most human activities in the watershed can occur in harmony with natural processes. Communities located along streams and rivers, for example, are faced with very basic choices: they can learn how the river functions and learn to draw benefits from it while staying out of harm's way, or they can try to significantly change the river's behavior to accomplish their plans. It may be feasible to change the way a river acts, but this usually means taking on costly and never-ending maintenance of those man-made changes; and, despite all the maintenance, communities may still remain vulnerable to floods and other disasters. In contrast, a community that has made sensible decisions on activities near the river can avoid a costly maintenance burden while sustaining the public use and enjoyment of a healthy river system. Obviously, one would like to live in the latter type of community and pay fewer taxes.

5.3.2 Watershed management is continuous and needs a multidisciplinary approach

Many agencies in the United States have found the need for a more integrated approach to assessing conditions and developing management strategies. Although they have made progress through existing regulations and programs, they are now faced with solving more thorny environmental problems that cut across programs and jurisdictions. Particularly vexing are nonpoint source pollution and habitat degradation. Essentially, a coordinating interdisciplinary management framework leads to coordinated management plans. The emerging framework would not be one that will fit all sizes. It takes working at different geographic scales, weighing multiple management objectives, and addressing unique local concerns. For example, a state agency might be interested in major river basins because it is charged with assessing and managing water quality statewide. A local government wanting to protect its drinking water supply may need to work with neighboring jurisdictions throughout a medium-sized watershed. A federal agency may need to implement a multiple-use management plan on a watershed in public ownership. A local watershed association may be trying to solve a sedimentation problem in a small watershed. If designed well, the watershed approach links all these initiatives with state, local, and regional frameworks that complement and strengthen each other and individual projects rather than competing, conflicting and overlapping with each other.

5.3.3 A watershed management framework supports partnering, using sound science, taking well-planned actions and achieving results

There are three common elements of successful watershed-management frameworks: the geographic management units of the watersheds themselves; that the partners agree upon a common set of units (i.e., watersheds) to provide a functional, practical basis for integrating efforts; and that the stakeholders are involved throughout the process. Stakeholders are those who can impact or are impacted by a decision related to the watershed. Partners agree on a management cycle, including activities they will work on together and a fixed time schedule for sequencing these activities. Importantly, the cycle signals that watershed management is a never-ending job.

5.3.4 A flexible approach is always needed

The true meaning of this final core principle is that one should never look for a rigid, step-by-step "cookbook recipe" for watershed management. One size does not fit all – different regions of the country have watersheds that function in very different ways, and even neighboring watersheds can have major differences in geology, land use, or vegetation, which imply the need for very different management strategies. Different communities vary in the benefits they want from their watersheds. Moreover, watersheds change over time. For example, in the United States, the eastern watersheds cleared of their forests in the first half of the twentieth century had specific management needs during regrowth in the second half of the century, but management needs will likely change again in the 21st century. Changes can even occur on more immediate time scales due, for example, to the appearance of a serious forest pest or disease, a change in water-use patterns, or the arrival of a new community, industry, or enterprise. Watershed management is a dynamic and continually readjusting process that is built to accommodate these kinds of changes.

5.4 EXAMPLES OF SUCCESS STORIES IN THE UNITED STATES

Watershed-management approaches are evolving throughout the United States and are being used to solve tough problems. The USEPA functions as the overarching lead agency with regard to planning, formulation, implementation, and monitoring of the watershed programs in the country. It lays down the guidelines for the distribution of roles and responsibilities for all the other federal, state, and local institutions and the stakeholders expected to engage in such programs. A very strict and objective monitoring and accountability process actively tracks and maintains the progress of various programs and leads them to desired outcomes. The following section highlights five examples of successful watershed-management cases in the United States (reproduced from USEPA, 1994).

5.4.1 Merrimack river initiative, New England: Public and private partners collaborate to build watershed toolbox to aid management decision making

In New England, the Merrimac River Initiative has brought together the states of New Hampshire and Massachusetts with the U.S. EPA and the New England Interstate Water Pollution Control Commission to collaborate on water-quality issues. This has resulted in many joint projects and successes, some of which are collectively referred to as the Watershed Toolbox. These tools not only aid management decision making and implementation, but also make it easier for partners to communicate.

5.4.2 Cooper river corridor project: Corporate community takes the lead in ecological restoration

In 1992, three major chemical companies (Amoco, DuPont, and Bayer) took the lead in forming the Cooper River Corridor Project – a coalition of the U.S. Fish and Wildlife Service, the Wildlife Council, South Carolina's Department of Environmental Protection, citizens, and local corporations – to identify and solve ecological problems in the region. The group first decided to identify weaknesses in a five-square-mile area of the watershed, looking particularly at the habitat of two endangered animal species, two bird species, the longleaf pine, and sweetgrass (a native grass important to a historical basket-weaving cottage industry in the area). The project began a longleaf pine reforestation program, and Amoco planted sweetgrass on many acres of its local land – regenerating sweetgrass and also the local basket-making industry. With these successes from working together, the project, led by Amoco, began a grassroots community strategic-planning process for the entire Cooper River watershed to protect and restore ecosystems and to strengthen local economic opportunities.

5.4.3 Boulder creek, Colorado watershed project: Restoring multiple river corridor values and uses by choosing most cost-effective strategies

A local wastewater treatment plant was targeted for an expensive upgrade to reduce nitrate level thought to be responsible for an ammonia toxicity problem in Boulder Creek. Intensive survey monitoring of Colorado's Boulder Creek indicated that a number of other factors could be contributing to the decline of the diverse fish populations from the Creek's mountain canyon to its high plains. For example, stream monitors found stream habitat so degraded that it was unsuitable for most forms of aquatic life and could be contributing to the buildup of toxic concentrations of ammonia in the water. A physical habitat restoration program was undertaken to restore the complexity of the stream channel, stabilize the stream banks, revegetate the riparian corridor, create buffer strips to reduce agricultural and grazing runoff, and rebuild diversion and return-flow structures to minimize impacts on aquatic habitat. Because of limited funding, key portions of the channel were prioritized and targeted for restoration. The BMPs, habitat restoration, and scaled-back point-source nutrient-control program were successful in reducing ammonia toxicity problems and revitalizing fish populations in the Creek. Boulder Creek now provides the primary corridor for an urban natural area park system.

5.4.4 Occoquan water supply protection: Looking at best use of land throughout watershed, local governments meet multiple objectives

In the mid-1980s, several counties in the rapidly urbanizing area of Virginia developed a comprehensive land-use plan for the Occoquan Reservoir watershed and adopted zoning ordinances regulating the location, type, and intensity of future land uses. This was done after maximizing the limits of treatment technology for the wastewater treatment plants discharging into the tributaries upstream of the reservoir and after intensive data collection and model development. Fairfax County took the lead in working with basin partners to study different land-use development scenarios and how well they met multiple objectives, such as: (a) improved transportation system, (b) economic development, (c) efficient provision of community services, and (d) no degradation of the Occoquan water supply.

5.4.5 North Carolina statewide framework: Innovative, cost-effective solutions through partnerships and leveraging

North Carolina's statewide basin-management approach resulted in more innovative, cost-effective management. In the Tar-Pamlico River Basin, the state water-quality management agency, a consortium of municipal and industrial discharge permittees, a local environmental organization, a national environmental advocacy group, and the state's soil and water agency forged a partnership to implement a pollutant-trading program. The discharge consortium agreed to fund the development of tools to evaluate management alternatives and to provide cost-share funds to implement agricultural BMPs in lieu of more costly nutrient-removal processes at the wastewater treatment facilities. The process was driven by the realization among these parties that point-source controls alone could not solve the problems of most concern, and that forcing more restrictive point-source controls would only yield marginal returns on investment.

5.5 WATERSHED PROGRAMS IN INDIA

Although India has a rich heritage of better water-management practices from ancient and medieval times, it fell behind on this aspect in modern times, especially in these recent times of increased burden because of its large population and unregulated industrial growth. The Ministry of Agriculture, the Ministry of Rural Development, and the Ministry of Environment and Forests have led various programs of land reclamation and treatment in the country (Bhandari et al., 2007). The Ministry of Agriculture initiated the River Valley Program (RVP) in 1962, followed by other programs, such as Watershed Development Project for Shifting Cultivation Areas (WDPSCA) in 1974–75, Flood Prone Rivers Program (FPR) in 1981, Alkali Soil in 1985–86, and National Watershed Development Project for Rainfed Areas (NWDPRA) in 1990–91. Similarly, the Ministry of Agriculture initiated several programs, such as Drought Prone Area Program (DPAP) in 1973–74, Desert Development Program (DDP) in 1977–78, and Integrated Wasteland Development Program (IWDP) in 1989–90. The Ministry of Environment and Forests also started its program, Integrated Afforestation & Eco-development Project Scheme (IAEPS), in 1989–90. Conscious attempts at restoring natural resources in the form of DPAP and DDP were largely carried out in a top-down technical approach, with negligible community participation (Kotru, 2003). People's participation increased marginally in the 1980s with the introduction of the target-area approach in projects such as the IWDP. In the early 1990s, the Government of India initiated the Integrated Watershed Management Program (IWMP) with the prime objective to restore degraded regions by improving and maintaining the natural-resource base to improve conditions in rainfed areas. The IWMP suggested employment of an integrated and coordinated approach across various ministries to promote soil and water conservation by optimizing land-use production systems and use of sustainable low-cost, location-specific technologies (Bhandari et al., 2007). Foreign assistance started in the 1990s with emphasis on the participatory approach to joint forest management, watershed, and eco-development projects. Institutionalization of people's participation in the three developmental programs (DPAP, DDP, and IWDP) was formalized in 2003 in the form of a greater and more formal role given to the Panchayati Raj Institutions (PRIs) by adoption of "Guidelines for Hariyali" – the guidelines proposed and implemented by the Department of Land Resources, Ministry of Rural Development (MoRD, 2003). Such institutionalization of PRIs' role draws strength from the 73rd and 74th constitutional amendments mandating greater role for PRIs in the implementation of the local developmental plans.

"Guidelines for Hariyali" proposes to employ the following criteria for taking up a watershed for possible intervention (MoRD, 2003):

- Watersheds where people's participation is assured through contribution of labor, cash, material, etc., for its development as well as for the operation and maintenance of the assets created.
- Watershed areas having acute shortages of drinking water.
- Watersheds having a large population of scheduled castes/scheduled tribes dependent on it.
- Watershed having a preponderance of non-forest wastelands/degraded lands.
- Watersheds having a preponderance of common lands.
- Watersheds where actual wages are significantly lower than the minimum wages.
- Watershed that is contiguous to another watershed that has already been developed/treated.

- Watershed area may be of a mean size of 500 hectares, preferably covering an entire village. However, if on actual survey, a watershed is found to have less or more area, the total area may be taken up for development as a project.

It also envisages the involvement and lead by the forest department in case there is a significant portion of state forestland becoming part of a watershed under consideration. The policy stipulates District Councils or Zila Parishads (ZP) and District Rural Development Agencies (DRDAs) to play the role of nodal agencies for monitoring, assessing, and certifying the progress of the projects. At the grassroots level, it promotes building up the capacity of PRIs to implement the projects as Project Implementing Agencies (PIAs). Provisions are also made for non-governmental organizations (NGOs) and any other state or central agency to function as a PIA, especially when PRIs are not adequately equipped to take the responsibility.

"Guidelines for Hariyali" (MoRD, 2003) enumerate the following items that may possibly become part of a watershed-developmental action plan.

- Development of small water-harvesting structures, such as low-cost farm ponds, nalla bunds, check dams, percolation tanks, and other groundwater-recharge measures.
- Renovation and augmentation of water sources and desiltation of village tanks.
- Fisheries development in village ponds/tanks, farm ponds, etc.
- Afforestation, agro-forestry, and horticultural development, shelterbelt plantations, sand-dune stabilization, pasture development, etc.
- Land development, including *in situ* soil- and moisture-conservation measures, such as contour and graded bunds, bench terracing, and nursery raising.
- Drainage-line treatment with a combination of vegetative and engineering structures.
- Repair, restoration, and up grading of existing common property assets and structures in the watershed.
- Crop demonstrations for popularizing new crops/varieties or innovative management practices.
- Promotion and propagation of non-conventional energy-saving devices, energy-conservation measures, bio-fuel plantations, etc.

Projects to be funded under "Guidelines for Hariyali" are expected to be of about 500 hectares in size and would cost about Rs. 6000 per hectare. One of the primary criteria for selection of a project is to have local institutions demonstrate 10% of the cost as contribution from their side, as cash, labor, and/or material. This is taken as an indicator of the interest of the local community in owning, developing, and maintaining the project in a sustainable manner. Although the intentions of ensuring this commitment from the local community are good, sometimes it may become an avoidable restriction for the very poor villages and at the same time would become an easy step for the richer village to get their projects sanctioned ahead of the more needy and deserving.

Kerr and Chung (2001) present important findings from a study evaluating watershed-management projects. Arguing in favor of using both the quantitative and qualitative comparative analyses, it is stated that better performance is posted by those projects that had an NGO component. This was true for a range of performance categories, such as soil conservation on drainage lines and common pasturelands, adoption of new crop varieties, and net returns to cultivation. Performance in government project villages, on the other hand, often was not significantly different from that in control villages. This is a vital observation and finding and may have significant implications for the policy of identifying PIAs under the auspices of the "Guidelines for Hariyali," which clearly place the NGOs behind not only the PRIs but even behind the usual line departments, such as agriculture, forestry/social forestry, soil conservation, etc., or other agencies of the state government/university/institute. It is stated in this evaluation report that NGOs usually devote much more time a year, and often across several years, with the target population as compared with not more than a month or so by the governmental institutions. Another, important point that the study highlights is the fact that the level of participation, involvement, and authority offered to various sections of the community, especially the poorer rural community, is significantly different. It is mentioned, for example, that in Maharashtra villages, landless farmers felt harmed by the projects rather than benefited, as they had little say in deciding the closure of common lands to grazing, a livelihood on which they solely depended.

Although significant progress has been made through implementation of scores of watershed-management projects around the country for the past 20–30 years, there are several hurdles still to be overcome and several improvements that could be made. The underpinning theme for these improvements to be possible is to integrate the roles and responsibilities of the various governmental agencies engaged and mandated to plan and run these programs. Improved and objective monitoring and evaluation of the projects and a reduction of bureaucratic framework around the functioning of various partners and stakeholders will ensure faster progress and long-lasting impacts of these restoration programs. A countrywide, comprehensive watershed-management information system integrating the efforts and resources of various institutions and stakeholders, partnering these programs, and serving as a component of a national resource-management system is urgently required. Awareness of these programs and efficiency can be greatly improved by utilizing state-of-the-art capabilities in the field of geographical information system (GIS) databases and Web-based technologies.

Another disturbing aspect of the current watershed-management policies is the fact that in the name of increasing people's participation, governmental institutions are gradually shifting the burden of planning, execution, and monitoring of these programs to the local institutions, such as PRIs and ZPs. Although entrusting decision-making to the lowest governing institution increases interest and ensures participation of local public, this policy must guard against losing the scientific and professional touch in the process while doing so. A countrywide holistic plan needs to be kept in view while drawing regional, divisional, and district-level plans. For optimal results, one central-level institution, with coordination and support of other relevant central, state, and district-level institutions, must lead the whole process of watershed management in the country. It is important for these leading and other main institutions to ensure that all local and district-level plans and designs confirm and fit into the overall plan and standards, which are laid out and prescribed from time to time. In other words, sufficient central and state-level assistance is to be provided in planning, designing, and standardizing watershed-management programs, while at the same time making them adequately responsive to local needs, governance, and aspiration of the people. Tremendous progress has been made in the fields of science and engineering, agricultural research, soil and water conservation and management practices, information and computing technologies, human development and social welfare, and in governance and accounting mechanisms. It is time to synergistically bring all these forces together to provide and deliver an efficient and credible country-wide watershed-management program that aims at achievable targets in the short run and is a thoughtful part of a long-term vision.

5.6 CONCLUSIONS

There has been a huge effort exerted with respect to better watershed-management developmental programs in the country during the past three to four decades. The experience gained by the governmental, non-governmental, and public agencies in general from the various projects is invaluable for taking the country forward in the field of watershed management. The country still faces a daunting task of spreading the few success stories of better water management to remote and difficult-to-reach places in the country. There is an urgent need for better intervention and management strategies on the one hand, and better coordinating and monitoring mechanisms on the other. There is a sincere realization among various partners and stakeholders that watershed management involves a multi disciplinary approach, input, and coordination among multiple agencies, and continuous efforts to maintain and manage natural and human resources in the watersheds on a continuous and sustainable basis. Evaluation studies indicate that the participatory levels achieved by NGOs have been much higher than those by their counterpart governmental agencies. This provides opportunities for governmental agencies to learn from the experiences of the NGOs and incorporate them in their own functioning and structure. It is also suggested that central governmental agencies structure the projects in such a manner that ensures a greater role of non-governmental and private institutions. A few steps aimed at addressing the major constraints experienced during the three to four decades of implementation of scores of watershed management programs are:

- Watershed-management programs across the country should be led at the top by a single overarching central agency in contrast to the present scenario of having more than three to four ministries leading such programs. This will bring the much-needed coherency in the watershed approach and accountability of the results.
- The central lead agency should very closely partner and coordinate with the relevant central and state agencies to conceive, plan, design, fund, implement, and monitor all present and future watershed-management programs, demonstrating the real convergence of all the experiences gained so far.
- The project preparation must also ensure active participation of the end-beneficiaries, local, district, and state institutions. The projects need to be demand-driven rather than supply-driven and, at the same time, adequate expertise must be made available from the top levels to ensure employment of standard practices, utilizing emerging technologies and experiences gained so far.

- A sustainable institutional mechanism is to be developed at the state, district, and local levels by integrating the present governmental institutions in a realistic and objective manner. The whole exercise of watershed-project implementation has to be taken beyond the utilization of funds, and toward achieving the end results that are quantifiable and/or qualitatively convincing. Achievements of most projects are squandered once the project ends and funds dry out. The sustainability of projects beyond their formal end is what has to be objectively planned and ensured.
- After all, the care that has gone into ensuring that the benefits of these projects reach the poorest of the poor, in reality, has not happened in a majority of the cases. The project designs must ensure that benefits are not limited only to the powerful and rich in the communities. The weaker sections, women, and the minorities must be specially safeguarded against biases and discriminatory project plans.
- Accountability and performance evaluation of watershed projects are important concerns and must be given more attention. There are possibly several loopholes due to which various watershed-developmental programs do not achieve the desired results. Efforts must be made to plan, design, and structure the project in such a way that there are comprehensive checks and balances without making the whole system unduly cumbersome to implement and manage.
- Indigenous research is required to develop more promising intervention strategies in specific regions, like hilly areas, deserts, wetlands, and hard-rock regions, among others.

On the whole, it is time for the premier research and scientific institutions and governmental and non-governmental agencies to coordinate and come together to develop a comprehensive watershed-management program for the country that utilizes and embodies the scientific and social developments achieved by modern India so far. Rural India looks forward to receiving visionary ideas, guidance, and technical and monetary support from the leading establishments in the country to plan, design, implement, manage, and sustain coherent watershed-management programs in the country.

REFERENCES

Bhandari, P.M., S. Bhadwal and U. Kelkar, 2007. Examining adaptation and mitigation opportunities in the context of the integrated watershed management programme of the Government of India. Mitig. Adapt. Strat. Glob. Change, 12:919–953.

Kerr, J. and K. Chung, 2001. Evaluating watershed management projects. CAPRi working paper #17, International Food Policy Research Institute, Washington.

Kotru, R. 2003. Watershed management experiences in GTZ supported projects in India. In *Preparing for the next generation of watershed management programmes and projects. Asia*, eds. M. Achouri et al. Proceedings of the Asian Regional Workshop on Watershed Management, Kathmandu, Nepal, Sept. 11–13, 2003.

MoRD, 2003. *Guidelines for Hariyali,* New Delhi: Department of Land Resources, Ministry of Rural Development, Govt. of India.

U.S. EPA, 1994. Distance learning modules on watershed management, Watershed Academy Web, Environmental Protection Agency. http://www.epa.gov/watertrain/

U.S. EPA, 1996. Watershed approach framework. Washington, DC: Environmental Protection Agency.

Chapter 6

Water management in northern states and the food security of India

G.S. Hira

Additional Director of Research, Agriculture, Retd., Punjab Agricultural University, Ludhiana, India

ABSTRACT

India has had a history of famines because of rainfed agriculture and rapid population growth. After the era of Green Revolution that began in the 1960s, India never experienced a famine-like situation; it did, however, experience a drought of the century in 1987. Northern states (Punjab, Haryana and western UP) with a high level of irrigation are contributing almost all of wheat and two-thirds of rice to the central pool of India. Punjab has the largest area (97%) under irrigation, with a corresponding contribution (60% of wheat and 40% of rice) to the central pool.

Large-scale cultivation of rice and its early transplantation facilitated by early (in the month of June) regular supply of electric power to tubewells are currently the major factors responsible for the fall of water table and indebtedness of farmers in Punjab. Shifting rice transplantation from June 10 to June 30 can arrest about two-thirds of the water-table fall. Technologies for making efficient use of irrigation water and enhancing groundwater recharge are available. Shifting of subsidy from input (subsidized power to tubewells) to output (increase in procurement price or bonus on agricultural marketable produce) would encourage farmers to use irrigation water more efficiently.

To feed India's projected population of 1.35 billion in 2025, agricultural production would have to be increased by ~25%. Agricultural production in Punjab, Haryana, and western UP might not be sustainable unless major steps are taken for improved management of groundwater.

Keywords: groundwater recharge, high irrigated area, water table fall, green revolution, famine in India

---

Address correspondence to G.S. Hira at Additional Director of Research, Agriculture, Retd., Punjab Agricultural University, Ludhiana, 141 004 India. E-mail: gurdevshira@yahoo.co.in

Reprinted from: *Journal of Crop Improvement*, Vol. 23, Issue 2 (April 2009), pp. 136–157
DOI: 10.1080/15427520802645432

6.1 INTRODUCTION

Crop failures caused by drought often led to famines in India; the last great famine occurred in 1943 (Bhatia, 1985). At present, India's population is 1.12 billion and is expected to reach 1.35 billion by 2025. By then, agricultural production would have to be increased by ~25%. Predictions are that agricultural production in Punjab—"the granary of India"—might not be sustainable at the current level unless major steps are taken toward improving management of groundwater. An often-asked question is, "Will India be able to feed its growing population?" Wheat import during 2007 (6.5 million tons) has already rung the alarm bell. During the last decade, India's food grain production has oscillated around 205 million tons and that of Punjab (with 1.5% geographical area of India) around 25 million tons. The time has come to plan for the future availability of food grains in India.

In India, area under rain-fed agriculture is 60.5%; the remainder 39.5% area is under irrigation. The state of Punjab leads with its 97% area under irrigation. The next two leading states are Haryana (83%) and UP (68%); all other states of India have less than 51% area under irrigation. Because of the highest levels of irrigation in Punjab, Haryana, and western UP, these states are contributing 98% of wheat and 66% of rice to the central pool of India. The Punjab irrigation system is the largest of its kind in the world. Punjab has played a pivotal role in sustaining India's food security (Hira & Khera, 2000). Punjab has been contributing ~60% of wheat and ~40% of rice to the central pool for the past four decades. This was why India did not have to import food grains during the past four decades.

During the past decade, agricultural growth rate (AGR), under rainfed conditions in India, was negative 10% (Mahapatra, 2007). During the same period, Punjab's AGR decreased from 3.28% to 1.86%. While sustaining agricultural production, Punjab has already depleted its good-quality groundwater resources (GWR), which had been conserved over 105 years, i.e., from 1859 to 1964 with the introduction of a mighty canal network. Punjab agriculture is currently facing the problems of a declining groundwater table (Hira, Jalota & Arora, 2004), increased energy cost for pumping, and deterioration of groundwater quality. These problems are expected to be further aggravated unless interventions are made at this stage. Will it be possible for Punjab to produce wheat and rice at the same, or higher, levels in the future? At present, the answer appears to be negative; but the answer would likely be positive if the solutions suggested here were adopted without loss of time.

A critical evaluation of availability of good-quality groundwater in Punjab has been made. Water availability will be a limiting factor even to sustain the present production levels unless some remedial measures are undertaken immediately. Water management in the northern states (Punjab, Haryana, and UP) will be a key factor for the food security of India. In the current paper, past, present, and future scenarios of GWR-related problems of Punjab have been analyzed. Solutions have been presented to save GWR for sustaining agricultural production in Punjab vis-à-vis food security of India. The groundwater situation in Haryana and western UP is more or less similar to Punjab's. The remedial measures suggested in this paper will also be applicable to Haryana and western UP.

6.2 FOOD SECURITY OF INDIA

As a result of rainfed agriculture, India has been vulnerable to crop failures and famines. Historically, devastating famines every few years were inherent in India. There were 14 famines in India between the 11th and 17th centuries. Five famines occurred during the 18th to 20th century span (Bhatia, 1985). The Bengal famine of 1943 was the last one, however. Two decades later, record production of wheat occurred every year, starting in the early 1960s. The Green Revolution era had begun. Three drought years, 1972, 1987, and 2002, occurred after the start of the Green Revolution. The 1987 drought was one of the worst of the century (Naren, 2008). Of 34 million cattle in Gujrat, 18 million died (Swaminaryan, 1987). India never experienced a famine after the onset of the Green Revolution because of abundant food reserves in Punjab.

6.2.1 Historical perspective

Historically, Punjab's groundwater has been influenced by topography and multipurpose river-valley projects, including the extensive canal network. Each aspect is discussed below.

6.2.1.1 Topography

Punjab is situated between 29' 30" N to 32' 32" N latitude and 73' 55" E to 76' 50" E longitude. Most of Punjab is a flood plain, i.e., adjacent to a river or other body of water. A belt of undulating hills extends along the northeast at the foothills of the Himalayas. Three rivers, the Ravi, the Beas, and the Sutlej, currently flow across the state in the southwesterly direction. It is a part of the Indus-basin and Indian Plate. It has developed from the sea and alluvium brought along by the rivers from the Himalayas. The topography is gentle, with an average slope/water-flow gradient of 0.3 m per kilometer. Annual rainfall is 1,100 mm in the northeast and is 300 mm in the southwestern end; the distance between the two is 350 kilometers. Salt concentration in the groundwater is low in the northeast, but it increases towards the southwest direction. Because of a relatively high rainfall, fresh groundwater reserves exist in flood-plain areas all along the foothills of the Himalayas. The depth of the freshwater layer decreases in the southwest direction. Deep groundwater in the southwest is saline, except along the perennial flow channels. About 40% area of the state, mainly in the southwest, has saline groundwater.

Punjab can broadly be divided into three regions (Figure 6.1): (1) Foot-hill region; mean annual rainfall is 950 mm. Land is sloping and often experiences water erosion. Groundwater is deep but of good quality. (2) Central region: The flood plains have a mean annual rainfall of 650 mm. Groundwater quality is good and fresh-water layer is deep. It covers 50% area of the state and is referred to as "the granary of India." Two-thirds of wheat and rice production of the state occurs in this zone. Two-thirds of tubewells of the state are located in this zone. (3) Southwest region: It accounts for 37% area of the state. Mean annual rainfall is 400 mm. Groundwater is saline. Seventy percent of the area is canal irrigated. Waterlogging and soil salinity problems exist in some areas.

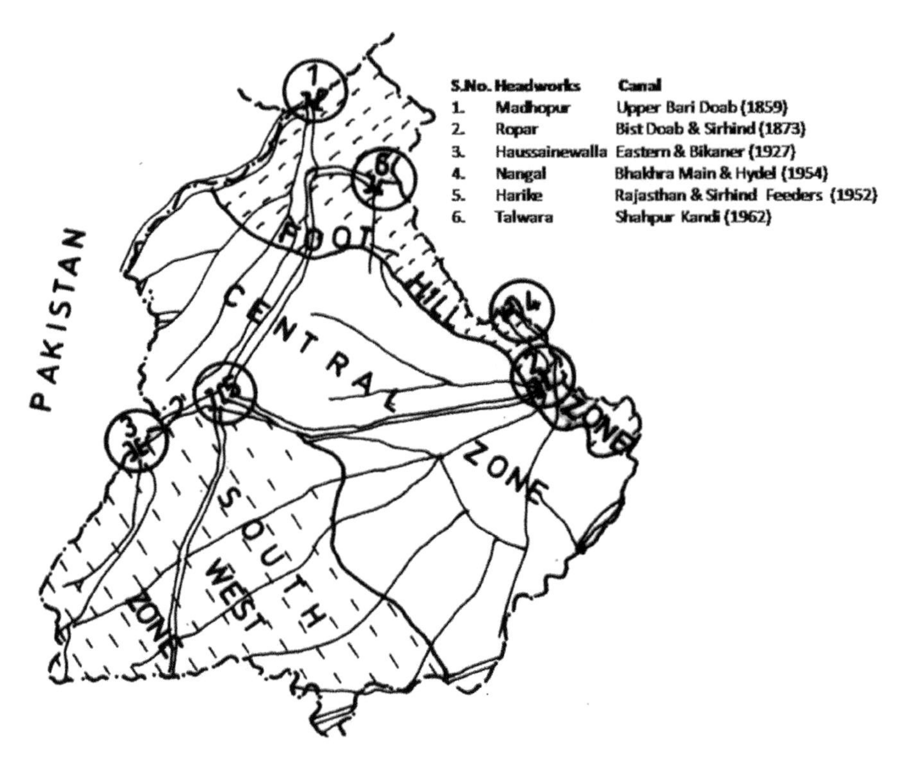

Figure 6.1 Map showing three regions of Punjab, headworks, and canal networks.

6.2.1.2 *Multipurpose river-valley projects*

Punjab was the last state of India to be conquered by the British in 1849. Immediately after that, an Irrigation Directorate was created in 1849 (Dhillon, 1992). The first headworks was constructed at Madhopur in 1859, along with the Upper Bari Doab Canal network. Ropar headworks on the river Sutlej was completed in 1873, along with the Sirhind canal and the Bist Doab canal network. Haussainewalla headworks (near Ferozepur) on the Sutlej was completed in 1927, along with the Eastern canal and the Bikaner canals. Harike headworks (where river Beas joins Sutlej) was constructed in 1952, along with Rajasthan and Sirhind feeders. Nangal headworks on river Sutlej became operational in 1954, along with Bhakha Main canal and Hydel channel. Talwara headworks on river Beas was completed in 1962, along with Shahpur Kandi canal. Headworks and canals are shown in Figure 6.1.

6.2.1.3 *Impact of canal network*

Before the inception of the mighty canal network, groundwater table (GWT) in the Punjab was very deep, except along the riverbeds. The GWT depth ranged from 30.3 m along the foothill zone to 51.5 m toward the southwestern end (Hira, Gupta & Josan, 1998). After the construction of the canal network, GWT started rising. Initially, the

GWT rise was slow but became rapid after the completion of additional canals. With the continuous rise of GWT, the problem of waterlogging appeared in many parts of Punjab. Waterlogging first occurred in the foothill plains along the Upper Bari Doab canal in the early 20th century. Subsequently, it spread to central Punjab. By 1964, about 23% of Punjab was waterlogged. Most of the affected area fell into the central zone (Uppal, 1966). Just before the arrival of the Green Revolution, GWT in most parts of Punjab was within 4.54 m from the surface, except in the southwestern zone (Figure 6.2).

The seepage of water from the canal network resulted in recharging of groundwater. This increased the good-quality (fresh) ground-water reserves near the soil surface. Most of the Punjab farmers had small land holdings. Farmers exploited shallow groundwaters for irrigation using centrifugal pump-type tubewells. For this situation, centrifugal-pump technology was very efficient and cheap. It was, thus, within the reach of the common farmer to install a tubewell for irrigation. Today, Punjab (5 million hectare geographical area) has 1.2 million tubewells.

Continuous waterlogging prevailing in large parts of the Punjab in the mid-20th century caused a buildup of a high concentration of soluble salts in the surface soil (Uppal, 1966). The salt-affected areas in Punjab were 30,000 hectares (ha) in 1950, 180,000 ha in 1955, 350,000 ha in 1960, and 680,000 ha in 1965. Cultivation of rice was taken up initially in high-rainfall and waterlogged areas. Furthermore,

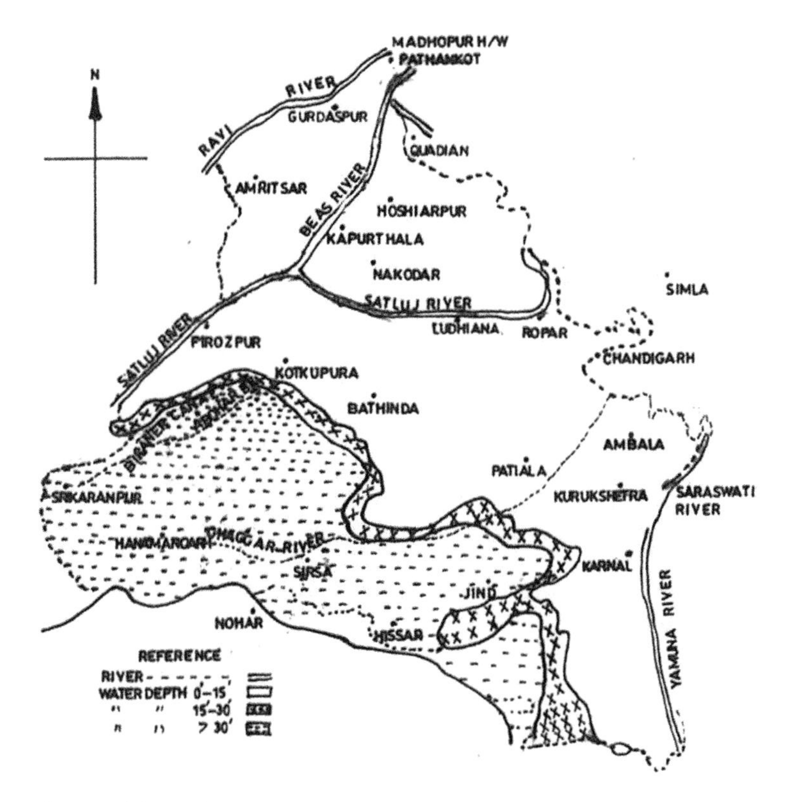

Figure 6.2 Map of undivided Punjab and Haryana showing the groundwater table depth in October 1964 (Uppal, 1966).

rice crop is the best agent for the reclamation of salt-affected soils (Singh, Hira & Bajwa, 1981). As a result, rice cultivation rapidly spread across central Punjab during the Green Revolution era (Hira & Sidhu, 1998). Presently, rice, which earlier was not a traditional crop in Punjab, is cultivated on 85% area of the central zone.

Canal irrigation laid a strong foundation for the Green Revolution to occur in this region of the world (Hira, 1998). It is possible that the Green Revolution might not have occurred had the great canal network not been constructed.

6.3 GREEN REVOLUTION

The era of the Green Revolution started with wheat and rice. In the early 1960s, high-yielding Mexican wheat varieties and later high-yielding rice varieties from International Rice Research Institute (IRRI) were introduced. Tubewell irrigation, both diesel and electric, spread rapidly during this period. There were 98,000 tubewells in Punjab in 1960–61. This number increased to 1 million in 2000–01 (Hira & Khera, 2000). The irrigated cropped area was 54% in 1960–61, which increased to 94% in 2000–01. Because of an increase in tubewell irrigation, there was a large-scale land leveling and a shift in cropping pattern. From 1960–61 to 2000–01, the area under wheat increased from 37% to 78% and that under rice increased from 6% to 60% at the expense of other crops. For example, the area under gram cultivation decreased from 22% to 0.4% and under pulses from 24% to 1.3%. As a consequence of the aforementioned changes during the 30 years (1970–71 to 2000–01), the production of wheat increased from 5.1 million tons to 15.5 million tons and of rice from 0.7 million tons to 9.2 million tons.

6.3.1 Impact of the Green Revolution

A negative aspect of the Green Revolution vis-à-vis food security of India is that it has disturbed the groundwater balance in Punjab. The groundwater table has gone down to dangerously low levels in many areas of Punjab, which, if not reversed, could turn Punjab into a barren land sooner rather than later.

6.3.1.1 Groundwater balance

As a result of the Green Revolution, the water resources of Punjab have been and continue to be overexploited. Availability of water resources is given in Table 6.1 (Hira, Jalota & Arora, 2004).

The deficit of 1.2 million ha m/yr in water availability is being met by overexploiting the groundwater resource in Punjab. The over-withdrawal of groundwater is resulting in the fall of the groundwater table in most parts of the state, except Mukatsar district and some other areas in the southwestern part of the state.

6.3.1.2 Groundwater table fall

The groundwater table-depth data recorded during 1993–2003 showed that out of 140 blocks of the state, water table had fallen in all except a few blocks in the

Table 6.1 Availability of water resources in Punjab.

Source	Amount of water (million hectare meter/year)
Canal water	1.45
Rainfall and seepage	1.68
Total availability	3.13
Water demand	4.33
Deficit	1.20

southwestern part. A largest water-table fall occurred in the central Punjab (Hira, 2005). In this zone, mean annual water-table fall ranged from 0.3 to 1.0 m. In some of the areas, the water-table fall was more than 1.0 m per year. By 2006, the water table had sunk to depths ranging between 15.45 and 27.58 m in the districts of Sangrur (26.97 m), Moga (24.85 m), Patiala (24.24 m), Fatehgarh Sahib (19.70 m), Jalandhar (27.58 m), Ludhiana (25.15 m), Kapurthala (18.48 m), and Amritsar (15.45 m). The mean water-table depth in the central Punjab in 2006 was 22.8 m.

Taking the recent groundwater-table fall and the water-table depth (WTD) data from 2003–07, WTD was predicted for the central part of Punjab for the years 2016 and 2023. Estimates indicate that mean water-table depth, which was 22.8 m in 2006, would likely increase to 34.2 m in 2016 and 42.5 m in 2023.

6.4 THE FUTURE OF PUNJAB AGRICULTURE

The rapid fall of water table in Punjab will have the following negative effects:

1. Increase in energy requirement for pumping groundwater;
2. Increase in the tubewell-infrastructure cost; and
3. Deterioration of groundwater quality.

6.4.1 Increase in energy requirements for tubewells

If the present rate of fall in water-table depth in central Punjab continues, by 2023 the energy cost of pumping groundwater from a depth of 42.5 m would increase by 93% compared with that for 2006. During 2007, Punjab State Electricity Board (PSEB) supplied 7,500 million units of electricity (28% of total consumption of Punjab) to tubewells. Accordingly, by 2023 the PSEB would need to supply 14.5 billion units annually for pumping groundwater required for irrigation. The PSEB is already under debt and is incurring an annual loss of Rs. 20 billion (~US$400 million). In 2007–08, the cumulative loss of PSEB had reached Rs. 115 billion (~US$2.3 billion). If the prevailing situation were allowed to continue, by 2023 the cumulative losses of PSEB would be Rs. 415 billion (~US$8.3 billion).

The main reason for indebtedness of PSEB is that it has been supplying subsidized electricity to tubewells, the annual cost of which is Rs. 24 billion (~US$480 million).

For rice cultivation alone, electricity worth Rs. 14.4 billion (~US$288 million) is supplied free of cost.

6.4.2 Increased infrastructure cost of tubewells

As a result of the fall in water table, Punjab farmers incur a huge cost every year for tubewell irrigation. For centrifugal pumps, the cost involved relates to the deepening/widening of the tubewell pit; and the purchase/repurchase of electric motor, pump, starter, cable, capacitor, voltage stabilizer, etc. Other costs involved are the purchase of diesel generator sets, diesel, and installation of new submersible pumps to replace old centrifugal pumps for irrigation and drinking purposes.

From 1997–2007, there has been a four-fold increase in cumulative fall in water table in the central districts of Punjab. It has been estimated that the total annual expenditure on tubewell infrastructure, both for irrigation and domestic purposes by Punjab farmers, is about Rs. 15 billion (or US$ 300 million). During the same period, Punjab farmers' indebtedness increased four-fold from Rs. 57 billion (~US$1.1 billion) to Rs. 240 billion (~US$4.8 billion; Figure 6.3).

These observations suggest that Punjab farmers' indebtedness is connected with the rapid fall of groundwater table. Will the common farmer be able to bear this regular increase in annual cost for pumping water in the future? A poor farmer would fall deeper into debt. A large number of farmers have committed suicide in the Sangrur district (Kumar, 2007), which has experienced the largest fall in water table during the past four decades. Sangrur district was waterlogged in 1960 whereas water-table depth in 2006 was 27 m. The main reason behind the fall of water table is the early, regular supply of subsidized electric power to tubewells in the month of June. In fact, subsidized power to tubewells is doing more harm than good to the poor peasantry. Subsidized power to tubewells is basically 'a sweet poison,' which is a concept vital for the Punjab farmers to understand.

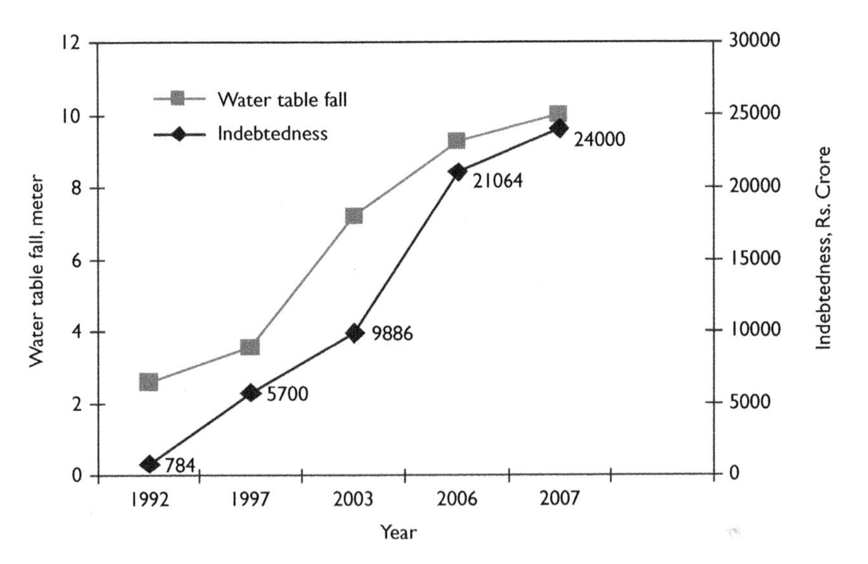

Figure 6.3 Relationship between farmers' indebtedness and fall in water table in the central districts of Punjab (1 crore = 10 million).

6.4.3 Deterioration in groundwater quality

The groundwater quality map (Figure 6.4) shows that salt concentration in the groundwater of Punjab increases from the northeastern to the southwestern end. About 40% area of the state in the southwest part has saline groundwater. By 2023, in the central zone of Punjab where water is of good quality, groundwater table is expected to sink to a depth of 41.2 m in Ludhiana, 44.5 m in Jalandhar, 46.7 m in Sangrur, 51.8 m in Moga, and 57.3 m in Patiala districts. Both Sangrur and Moga districts are very close to the southwest zone where groundwater is highly saline. In these districts, groundwater quality would deteriorate because of "sea water intrusion"—a reverse flow mechanism, which means the saline groundwater from the southwest would flow in the northeast direction (opposite of the natural flow direction). The soil-surface slope is in the direction of northeast to southwest and the direction of groundwater flow is the same. For many centuries, groundwater had been flowing from the northeast to the southwest direction, with a gradient of 30 cm (1 foot) per kilometer. Once the "reverse flow mechanism" starts, it would deteriorate groundwater quality of the central zone of Punjab with disastrous consequences.

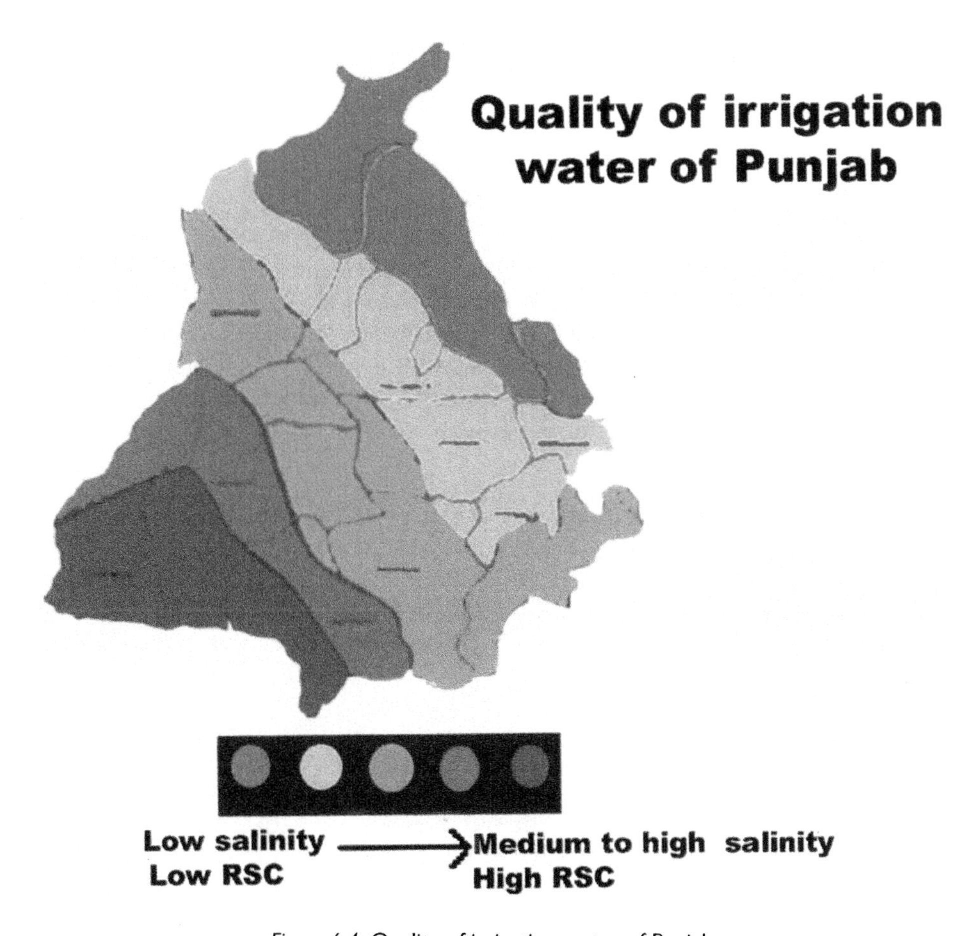

Figure 6.4 Quality of irrigation water of Punjab.

Table 6.2 Changes in groundwater quality with depth and time in Nihalsinghwala block of Moga district of Punjab.

	Depth of tubewell	Percent samples		
Year	Meters	Fit	Marginal	Unfit
1997	41	51	38	11
2004	71	28	43	29

To test this hypothesis of "reverse flow mechanism," a detailed groundwater survey of Nihalsinghwala block of Moga district was conducted in 1997 and 2004 (Table 6.2). During this period, the percentage of water samples falling in the 'fit category' decreased from 51% to 28% and that of the 'unfit category' increased from 11% to 29% (Singh & Bishnoi, 2006.). The mean tubewell depth in 1997 was 41 m, whereas it was 71 m in 2004.

These observations suggest that if the water-table fall in the central districts continues, tubewell irrigation water quality will deteriorate. This is a looming problem that may happen in the near future. Agricultural production would certainly decrease with an increase in soil salinity induced by saline groundwater, as central district areas are primarily tubewell irrigated. Canal irrigation is used on 14% area only.

6.4.4 Doomsday prediction

By 2025, India's production must increase by ~25% to feed its increased population of 1.35 billion from the present level of 1.12 billion. On the contrary, because of the aforementioned reasons (increase in the energy requirement for pumping groundwater and the cost of tubewell infrastructure and deterioration of groundwater quality), predictions are that by 2025 Punjab may not be able to produce even the current levels of wheat and rice, unless some remedial measures are implemented to arrest the fall of the water table. This doomsday could happen in the near future, and India might face famine-like situations during drought years.

6.5 REASONS FOR THE FALL OF THE WATER TABLE

The rise and fall of the groundwater table in any region depends upon the balance between the amount of withdrawal of groundwater and that of water recharged through seepage. The net negative water balance in Punjab is for the following reasons.

6.5.1 Early transplantation of rice

In the 1970s, rice transplanting used to start from June 30, after the onset of monsoon rains. The first fortnight of July was the recommended time for rice transplantation (Anonymous, 1985). Later, the transplantation date had been moved to May. However, the evapotranspiration of rice transplanted on May 20 is 76 cm, on June 10 is 60 cm, and on June 30 is 52 cm (Hira, 1996).

Table 6.3 Paddy yield at different times of transplantation during 2004 and 2005.

Transplantation	Yield, t/ha		
Time	2004	2005	Mean
May 25	9.97	9.21	9.57
June 10	10.86	8.90	9.79
June 25	11.02	8.87	9.95
LSD (0.05)	NS	NS	NS

Mean relative humidity data collected from June 10 to July 10 at Ludhiana from 1970 to 2005 showed that during 1997–2005, relative humidity had increased from 58% to 68%. During the same period, mean groundwater table fell from 3 m to 10 m. Experimental results from Punjab Agricultural University and farmers' experiences showed that a reduction in rice yield had occurred. Are these results/observations real? These observations are not real because the reduction in rice yield occurred due to an increased incidence of insects and pests that infest the late-sown crops. To prove this, a rice crop was raised with five to six insecticide sprays and the results are given in Table 6.3 (S.K. Jalota, personal communication).

These results and earlier observations (Hira, 1996; Hira & Khera, 2000) showed that no reduction in rice yield would occur even if rice were transplanted from the end of June onwards. Thus, we need to ensure that no farmer starts transplanting rice before this date. Early transplanting should be prohibited by law because a farmer who transplants his crop early is essentially committing a crime by reducing the yields of farmers who transplant their crop at the recommended time.

6.5.2 Less canal irrigation and plowing of watercourses

Canal irrigation in the central zone is applied to only 14% area and is much less as compared with the southwestern part (70%). In addition, farmers in the central zone of Punjab have plowed most of the canal watercourses. The main reason for this is subsidized power to tubewells and the availability of fresh groundwater at shallow depths. Because of the above-noted two reasons, there is less recharge of groundwater in the central Punjab.

6.5.3 Shift in cropping pattern

Assured tubewell irrigation has led to the replacement of low-evapotranspiration (ET) crops with high-ET crops (Hira, 2005). Rice (ET 60 cm) has replaced cotton (ET 54 cm), maize (ET 46 cm), and pulses (ET 40 cm). Wheat (ET 40 cm) has replaced gram (ET 30 cm), and oilseed (ET 32 cm). Of the total water demand, rice accounts for 37%, wheat 32%, other crops 14.3%, and fodder, vegetables, and forests 13.3% (Hira, Jalota & Arora, 2004).

6.5.4 Large area under rice

Rice is cultivated on 60% area of Punjab during the summer. In central Punjab, the area under rice is 85% during this season. It is suggested that the area under rice be reduced by 35% in the districts of Patiala, Fategarh Sahib, Sangrur, Ludhiana, and Moga, and by 25% in the Jalandhar, Kapurthala, and Amritsar districts (Singh, 2007).

6.5.5 Early regular supply of subsidized electric power for tubewells

The mother of all the above-mentioned reasons for the fall of the water table is the early regular supply of subsidized power for the tubewells (Hira, 2005). Sixty percent of the annual electricity is supplied to tubewells for rice cultivation during the months of June, July, August, and September. Rice, which was never a traditional crop of this area, is the main culprit for the fall of water table. Large-scale rice cultivation and its early transplanting in the dry and hot summer months (May and June) are practiced only because of early regular supply of subsidized power to tubewells.

It is important to mention that 47% of total annual pan evaporation occurs during the dry/hot summer months of April, May, and June (Hira, 1996). Because of dry-hot conditions during the summer months, most of the diseases, insects, and pests remain under control, but frequent irrigation encourages their spread. Continuously standing water is required for rice cultivation during the early growing season. This increases the relative humidity of the air and the spread of diseases, insects, and pests (Dhaliwal et al., 1985). Only those crops that require limited or no irrigation in the early stages of growth should be sown during the dry-hot summer months. For example, cotton needs to be irrigated six weeks after sowing. Similarly, sorghum and maize also need only limited irrigation during early growth.

The problem of water management in Punjab has serious repercussions for the future of farming, agricultural production, state economy, and food security of India. The key factor responsible for all these problems is the subsidized power to tubewells. It causes an imbalance in the groundwater resources of Punjab. Subsidized power to tubewells encourages withdrawal of groundwater and its loss as evaporation in the hot-dry summer months.

6.6 ARRESTING/REVERSING FALLING WATER TABLE

The falling water table in Punjab can be arrested by (1) decreasing withdrawal and (2) increasing the recharge of groundwater.

6.6.1 The decrease in withdrawal of groundwater can be achieved by:

6.6.1.1 Delayed transplanting of rice

During 2006 and 2007, a mass campaign to educate farmers was begun to delay rice transplanting to June 10 or later and to appeal to PSEB to provide 8 hours of regular supply of electricity to tubewells from this date onwards. This resulted in 18 cm/year arrest of the water table fall (WTF) in the central zone of Punjab during 2005–07, as compared with previous years (2003–05). This WTF is very close to the earlier predictions (Hira 1994, 1996), i.e., 16.4 cm/year WTF can be arrested by delaying rice transplantation to June 10. Following these predictions (Hira, 1994), the government of India decided to shift the start of procurement of rice from September 1st to October 1st during 1995. Accordingly, from 1995, Punjab farmers started rice transplanting from June 10. During 1997, the southwestern region (Mukatsar district) experienced a waterlogging problem. The delay in rice transplantation to 10 June arrested the WTF in the central Punjab but resulted in waterlogging in the southwestern part (Hira, Gupta & Josan, 1998). For such high water table areas, tubewells for irrigation should be installed using multiple well points (2–6) instead of a single bore to pump out good-quality water floating over saline groundwater (Hira & Murthy, 1985).

Based on the actual WTF data of 2003–07, future projections of WTF were made using ET data of rice transplanted on June 15, 20, 25 and 30 (Hira, 1996). It was found that the present WTF of 64 cm/year could be reduced to 51 cm/year, 39 cm/year, 29 cm/year, and 20 cm/year by shifting rice transplanting to June 15, 20, 25, and 30, respectively. Furthermore, future predictions of water table depth (WTD) were also made for the central districts of Punjab (Table 6.4).

It was found that the present (2006) mean WTD of 22.8 m in the central Punjab would reach 42.35 m in 2023. However, with June 30 as the ideal rice transplanting date, WTD will reach only 28.16 m in 2023.

Will it be possible to shift the time of rice transplanting to June 30 without affecting the sowing time of wheat and its yield? The answer is an emphatic *yes*. The reasons are explained below.

First, the recently released rice variety 'PAU 201' matures 15 days earlier and also yields higher than the Pusa 44 variety presently being sown in Punjab. Second, large-scale successful demonstrations on farmers' fields with Happy Seeder machines have been conducted during 2007–08. It has been found that wheat can be sown from October 16 to November 6 with the Happy Seeder immediately after the combine-harvested paddy crop. In addition to the timely sowing of wheat and avoiding burning of paddy straw and cultivation costs, it has been found that the Happy Seeder increases wheat yield by 10% (Sidhu et al., 2007).

Table 6.4 Measured water table depth (WTD) in 2006 and predicted WTD for 2023 in central districts of Punjab with shift of rice transplanting from June 10 to June 30.

District	WTD 2006 measured, m	WTD 2023 predicted, m	
		June 10	June 30
Sangrur	26.97	46.67	33.33
Moga	24.85	51.82	32.73
Patiala	24.24	57.27	34.55
Fatehgarh Sahib	19.70	39.70	25.76
Jalandhar	27.58	44.55	32.12
Ludhiana	25.15	41.21	29.70
Kapurthala	18.48	33.33	23.03
Amritsar	15.45	24.24	18.18
Mean	22.80	42.35	28.16

Sowing of summer green gram (moong) with the Happy Seeder after combine-harvesting wheat has also been successfully demonstrated on the farmers' fields (H.S. Sidhu, personal communication). The delaying rice transplanting to June 30 would not only help in arresting the fall of the water table but would also allow for an additional crop of summer moong after wheat harvest. This, however, should be done with minimal use of irrigation water, which is possible by mulching of the summer moong crop with wheat straw.

6.6.1.2 Crop diversification

Delaying rice transplanting to June 30 would help in crop diversification. Some areas under rice could be shifted to other crops, such as cotton, maize, and sorghum for fodder, as sowing of these crops is done much earlier than June 30. In this way, farmers' idle time after the wheat harvest could be used for the sowing of other crops. Shifting some areas from rice (ET 60 cm) to cotton (ET 54 cm), maize (ET 46 cm), and sorghum for fodder (ET 45 cm) would reduce groundwater withdrawal and help in arresting water-table fall.

6.6.1.3 Efficient use of irrigation water

Using water-saving technologies such as irrigation scheduling, laser land levelers, proper irrigation methods, and mulching can increase the efficiency of irrigation water.

Scheduling irrigation for rice with the tensiometer can save 25% irrigation water (Hira et al., 2002; Kukal, Hira & Sidhu, 2005) without any reduction in yield of rice. Similarly, a laser land leveler can save 25% of irrigation water in rice (Sidhu et al., 2007).

In cotton, broad-bed with spacing of 135 cm and planting cotton in furrows in paired rows increased yield by 44% and saved 40% irrigation water, as compared with row spacing of 76.5 cm in flat-bed system (A.S. Sidhu, personal communication).

Planting of potato on both sides of narrow beds increased tuber yield (24 t/ha) by 25% and saved 20% irrigation water, as compared with the ridge-planting method (20 t/ha) (Thind et al., 2007).

Mulching of chillies with paddy straw increased yield (17.2 t/ha) by 15% and saved 25% irrigation water, as compared with no mulching (14.8%) (Sekhon et al., 2008).

Summer moong yield of 1.6 t/ha with wheat mulch has been observed using the Happy Seeder after combine-harvesting wheat with two to three irrigations (H.S. Sidhu, personal communication).

These water-saving technologies will certainly reduce the energy requirement for pumping groundwater and could help in arresting the falling water table, but it is difficult to ascertain exactly to what level ET can be reduced.

6.6.1.4 Provide economic incentives for the efficient use of irrigation water

No farmer would adopt the previously explained efficient irrigation water-saving technologies unless there is some economic benefit. The economic incentive for efficient irrigation water-use can be given if subsidy is shifted from input to output. The government should provide incentives on marketable agricultural produce instead of giving subsidized electricity to tubewells. Is it really possible to develop an economic incentive program appealing to both farmers and the government? The answer is *yes*. Some economic incentive programs are described below.

One of the economic incentive programs could be to increase the procurement price of wheat and rice. During 2007, the central government announced an increase in the procurement price for wheat from Rs. 7500 to 10,000 per ton for the wheat crop harvested in 2008. This resulted in a net increase in the earnings of Punjab farmers of Rs. 25 billion during 2008.

Another economic incentive program could be to give a bonus on the marketable produce. Presently, the Punjab Government (PSEB) is giving free electric power annually worth Rs. 24 billion to tubewells. Instead of giving free power to tubewells, the government should give a bonus worth Rs. 12 billion on wheat and rice sold in the market. This would equal approximately Rs. 700 per ton for wheat and Rs. 350 per ton for rice. As an additional bonus, other crops/commodities, such as cotton, maize, milk, etc., may be included at a later stage. As soon as subsidized power to the tubewell is stopped, farmers will adopt efficient irrigation technologies. The net annual benefit to both farmers and PSEB will be as follows:

Farmers:

1. Bonus on wheat and rice = Rs. 12 billion.
2. Savings in electricity bill: a) Due to delaying rice transplantation from June 10 to June 30 = Rs. 2.24 billion; b) saving in irrigation water by use of laser land leveler and tensiometer = Rs. 3.2 billion.
3. Savings in annual tubewell infrastructure cost = Rs. 15 billion.
4. Loss due to free electric power withdrawn from tubewells = Rs. 24 billion.
5. Net Gain = Bonus + Savings – Loss = Rs. 8.84 billion.

Government:

1. Savings from stopping free power to tubewells = Rs. 24 billion.
2. Loss by giving bonus on rice and wheat = Rs. 12 billion.
3. Net Gain = Savings – Loss = Rs. 12 billion.

Annual gains to both farmers and Punjab Government = Rs. 20.84 billion.

To encourage efficient use of irrigation water, the central government should give a subsidy for the Happy Seeder and water-saving technologies, such as the laser land leveler and the tensiometer, particularly in the high-intensity irrigated states of Punjab, Haryana, and UP.

6.6.2 Increase recharge of groundwater

There is a great need to increase the recharge of groundwater in the central Punjab. Though the area of central Punjab is only 50%, it produces two-thirds of rice and wheat and two-thirds of tubewells are located in this zone. The canal irrigation is applied to 14% area only and it should be increased to enhance the recharge of groundwater in this water-deficient zone. During monsoon season, all along the Himalayas, the foothills and flood plain area receive high rainfall and this can be used for enhancing recharge of groundwater. There is a great scope for increasing the recharge of groundwater in Punjab by adopting the following measures.

6.6.2.1 *Increase the canal water supply and revive the old canal watercourses in central Punjab*

The major fall in the water table is in the central Punjab. It is these areas where the canal irrigation is insufficient and the farmers have plowed the canal watercourses. An increase in canal water supply and the revival of old canal watercourses in these areas should enhance the recharge of groundwater and help in arresting the falling water table.

6.6.2.2 *Construction of new headworks on river Sutlej*

During July, August, and September, a large amount of water is released downstream of Haussainewala headworks. It has been estimated for the years from 1979 to 1988 that 0.1 million to 1.0 million ha m water per year is going to Pakistan during the monsoon season (Hira, 1998). Headworks on River Sutlej may be constructed between Ludhiana and Ropar to create a large reservoir (a network of a series of lakes) to store the floodwater during the monsoon.

The lake water could be diverted to rice fields through existing Sirhind canal networks or drainage networks. The maximum fall in water table is occurring in Sirhind canal network areas. A larger drainage network exists in the central Punjab. These drains were dug during the 1950s in waterlogged areas to drain the floodwater during the rainy season. Even if we harness 30% of the floodwater, it would raise groundwater table by 22 cm/year in the central districts of Punjab. This is the only way to store the floodwater and divert it to saturate the water-deficient areas of Sangrur, Moga, Ludhiana, Jalandhar, etc.

As such, because of land topography, this project will be very costly and may be an uneconomical proposition from a civil engineering point of view. But a part of the cost could be recovered by establishing a new city on both sides of the embankments of the Sutlej. A new city could be developed on modern ecological lines using a series of lakes to minimize the extreme weather conditions and attract the elite class of society for settlement. In addition, this project would also help in making the best use of frequently flood-prone areas along the Sutlej and, at the same time, it would save the most productive lands presently being brought under urbanization.

The construction of new headworks on the Sutlej and establishment of a new city should be regarded as a multi-utility project because a substantial decrease in river water flow has been observed during the last few decades. Possibly, this is the only way to increase canal water supply in the water-deficient central zone, i.e., Sirhind canal command area.

6.6.2.3 *Foothills*

Areas along the lower Shivalik hills of the Himalayas (Kandi area) receive about 100 cm of rainfall (Hira, 1998). There is a great scope for the recharge of groundwater from these areas. Rainwater, once it percolates into the soil, joins the groundwater and moves laterally towards the central zone. An estimated 40% of rainwater is lost as runoff in the Kandi areas. Runoff losses from this area can be reduced by adoption of *in situ* soil-conservation measures and the construction of check dams.

6.6.2.4 Areas with restricted layer

Large areas of districts of Patiala and Fatehgarh Sahib have a thick restricted layer below the soil surface. It does not allow the rainwater to percolate and join groundwater. Surface flow of floodwater is common in these areas during the monsoon season. Dug wells could be constructed in these areas to divert the floodwater from agricultural fields. However, care must be taken that this practice is adopted only for the monsoon season. At times, farmers resort to such drainage from wheat fields that contain a lot of nitrate nitrogen in the winters. This can be avoided by sowing the wheat crop using the bed-planting method (Kukal et al., 2008).

6.6.2.5 Rainwater harvesting in cities

Rainwater harvesting projects should be implemented in cities, especially in high-rainfall areas. Three-tier strategies could be adopted. First, areas along the roadside and parking lots can be put under grass-saver tiles. Common grass can be planted in holes that would allow the rainwater to percolate rapidly into the soil. This should help in reducing the sewage water load and avoiding the flooded conditions on the city roads during monsoon rains. The author has already successfully tested the usefulness of this technology. Second, adoption of rooftop rainwater-harvesting technology can be made mandatory for new structures in various cities. Third, treated sewage water could be diverted for irrigating parklands and crops (Anonymous, 2006).

6.6.2.6 Renovation of village ponds

In central Punjab, before the Green Revolution, village ponds used to be a source of drinking water for animals. These ponds are now malodorous but contain many nutrients (Hira et al., 2001). It has been estimated that one irrigation from village-pond water (7.5 cm) adds 17.2 kg N, 1.7 kg P, and 75.3 kg K per ha (J.S. Brar, personal communication). These ponds also have a 4–5 ft thick layer of sediments, which is a rich source of nutrients and organic matter. Renovation of ponds, including diversion of the sediment and water to agricultural lands, should be a good source of nutrients and should result in an increase in the recharge of groundwater.

The central government should provide financial support for the above-explained measures to enhance the recharge of groundwater in the high-intensity irrigated areas of Punjab, Haryana, and western UP.

6.7 MESSAGE

Now is the time for farmers, high-irrigated northern states of India, and the central government to act to arrest the groundwater table fall in Punjab, Haryana and western UP, or wait for the doomsday to come, which is not far away. Farmers of these states must understand that the early regular supply (in the month of June) of subsidized electric power to tubewells is a sweet poison because it is the main reason behind the rapid fall of water table and the indebtedness of farmers. Agriculture all over the world is subsidized. The shifting of subsidy from input (subsidized power to tubewells) to output (increase in procurement price or bonus on marketable agricultural produce) should encourage farmers to make efficient use of irrigation water.

REFERENCES

Anonymous. 1985. *Package of practices for Kharif crops of Punjab*. Ludhiana, India: Punjab Agricultural University.

Anonymous. 2006. Pollution of Buddah Nallah, Ludhiana—Report of Expert Committee, before the Punjab State Human Rights Commission at Chandigarh.

Bhatia, B.M. 1985. *Famines in India: A study in some aspects of the economic history of India with special reference to food problem*. Delhi: Konak Publishers.

Dhaliwal, G.S., G.L. Raina, and G.S. Sidhu. 1985. Second generation rice insect and disease problems in Punjab, India. Int. Rice Commn. News. 34(4): 54–65.

Dhillon, G.S. 1992. Canal system of Punjab region. Punjab Agricultural University Ludhiana Bulletin No.: PAU/1992/F/525E. Directorate of Extension Education, Punjab Agricultural University, Ludhiana, India.

Hira, G.S. 1994. The scarcity of water and waterlogging in Punjab—A historical perspective. Proc. National Conference on Soil and Water Conservation for Sustainable Production and Panchayat Raj held at Chandigarh, Jan. 28–30, pp. 25–27.

Hira, G.S. 1996. Evaporation of rice and falling water table in Punjab. In *Proceedings of International Conference on Evapotranspiration and Irrigation Scheduling held at Texas, USA*, Nov. 3–6. eds. C.R. Camp, E.J. Sadler, and R.E. Yoder, 579–584. American Society of Agricultural Engineers, USA.

Hira, G.S. 1998. Water management for crop production and groundwater table in Punjab. In *Proceedings of International Conference on Ecological Agriculture: Towards sustainable development held at Chandigarh*. Nov. 15–17. eds G.S. Dhaliwal, N.S. Randhawa, R. Arora, and A.K. Dhawan, vol. 1, pp. 240–251. Indian Ecological Society, Punjab Agricultural University, Ludhiana, India.

Hira, G.S. 2005. Depleting groundwater, causes and remedial measures in Punjab. Theme paper, seminar on management and conservation of irrigation water, Mohali, 27 April 2005.

Hira, G.S., P.K. Gupta, and A.S. Josan. 1998. *Waterlogging—Causes and remedial measures in southwest Punjab*. Research Bulletin No 1/98. Ludhiana, India: Department of Soils, Punjab Agricultural University.

Hira, G.S., S.K. Jalota, and V.K. Arora. 2004. *Efficient management of water resources for sustainable cropping in Punjab*. Ludhiana, India: Department of Soils, Punjab Agricultural University.

Hira, G.S., and K.L. Khera. 2000. *Water resource management in Punjab under rice-wheat production system. Research Bulletin No. 2/2000*. Ludhiana, India: Department of Soils, Punjab Agricultural University, Ludhiana.

Hira, G.S., and V.V.N. Murthy. 1985. An appraisal of waterlogging and drainage problems in southwest parts of Punjab. Tech. Bull. Dept. of Soil and Water Eng., Punjab Agricultural University, Ludhiana.

Hira, G.S., and A.S. Sidhu. 1998. Development of paddy cultivation in Punjab—Water table, soil, and environmental effects. Proc. International Symposium on Lowland Technology/ Saga University. Japan, pp. 395–404.

Hira, G.S., A.S. Sidhu, and S.S. Thind. 2001. Traditional soil nutrient management techniques practiced in Punjab. In *Indigenous Nutrient Management Practices – Wisdom Alive in India*, eds. C.L. Acharya, P.K. Ghosh, and A. Subba Rao, 13–30. Nabibagh, Bhopal, India: Indian Institute of Soil Science.

Hira, G.S., R. Singh, and S.S. Kukal. 2002. Soil matric potential: A criterion for scheduling irrigation to rice. *Indian Journal of Agricultural Sciences*. 72: 236–237.

Kukal, S.S., G.S. Hira, and A.S. Sidhu. 2005. Soil matric potential-based irrigation scheduling to rice. (*Oryza sativa*). *Irrigation Science* 23: 153–159.

Kukal, S.S., E. Humphreys, Yadvinder-Singh, Balwinder-Singh, Sudhir-Yadav, Amanpreet-Kaur, et al. 2008. Permanent beds for rice-wheat in Punjab, India. Part 1: crop performance. In *Permanent beds and rice-residue management for rice-wheat systems in Indo-Gangetic plains*. Proceedings of International Workshop held at PAU Ludhiana, India from 7–9 September 2006. ACIAR Proceedings No. 127, eds. E. Humphreys and C.H. Roth. Australian Centre of International Agricultural Research, Canberra, Australia.

Kumar, B. 2007. Farming woes-behind water scarce Punjab's falling growth. In *Down to Earth*, ed. S. Narain, 9–10. New Delhi, India: Centre for Science and Environment.

Mahapatra, R. 2007. Horns of a dilemma. In *Down to Earth*, ed. S. Narain, 36–43. New Delhi, India: Centre for Science and Environment.

Naren, K. 2008. Understanding drought in India. In *Programme 2001–2008 Drought in India: Challenges and Initiatives Poorest Areas Civil Society (PACS)*, pp. 1–15. PACS, New Delhi, India.

Sekhon, N.K., C.B. Singh, A.S. Sidhu, S.S. Thind, G.S. Hira, and D.S. Khurana. 2008. Effect of straw mulching, irrigation and fertilizer N levels on soil hydrothermal regime, water use and yield of hybrid chilli. *Archives of Agronomy and Soil Science. 54:* 163–174.

Sidhu, H.S., Manpreet-Singh, E. Humphreys, Yadvinder-Singh, Balwinder-Singh, S.S. Dhillon, J. Blackwell, V. Bector, Malkeet-Singh, and Sarbjeet-Singh. 2007. The Happy Seeder enables direct drilling of wheat into rice stubble. *Australian Journal of Experimental Agriculture* (forthcoming).

Singh, B., and S.R. Bishnoi. 2006. Changes in quality of underground irrigation waters in Nihal Singh Wala block of Moga district of Punjab. J. *Res. (PAU). 43:* 19–20.

Singh, K. 2007. Punjab – The dance of water-table. The Punjab State Farmers Commission, Punjab State Farmers Commission Government of Punjab, Chandigarh, India.

Singh, N.T., G.S. Hira, and M.S. Bajwa. 1981. Use of amendments in reclamation of alkali soils in India. *Agrokemia ES Talajtan Tom. 30(Supplementum)*: 158–177.

Swaminaryan, Sanstha, BAPS. 1987. *Famine of Saurashtra, India*. Available online at http://www.baps.org/activities/relief/famine1987.htm

Thind, S.S., A.S. Sidhu, N.K. Sekhon, C.B. Singh, Tarinder-Kaur, and Amanpreet-Kaur. 2007. Effect of different irrigation geometry on water and nutrient use efficiency of potato-sunflower sequence. Proceedings of Environmental and Livelihood Security through Resource Management in Northern India (ELSTRM, 2007).

Uppal, H.L. 1966. Reclamation of saline and alkali soils. Intern. Commission on Irrigation and Drainage, 6th Congress, Vol. 1: 381–440.

Section III

Sustainable soil management strategies

Chapter 7

Sustainable management of Vertisols in central India

R.P. Rajput, D.L. Kauraw, R.K. Bhatnagar & Manish Bhavsar
The Jawahar Lal Nehru Agricultural University, Jabalpur, India

M. Velayutham
M.S. Swaminathan Research Foundation, Chennai, India

R. Lal
Carbon Management and Sequestration Center, The Ohio State University, Columbus, Ohio, USA

ABSTRACT

A community-based field operational research project was implemented in the Vertisols of Madhya Pradesh (MP), India, to demonstrate the best management practices (BMPs) of land use and soil-fertility management for enhancing productivity. The raised-sunken bed system (RSBS) of land treatment was used to enhance *in situ* rainwater conservation and minimize soil erosion and nutrient losses. Grain yields of wheat (*Triticum aestivum*) and chickpeas (*Cicer arietinum*) were higher in this system than in the flatbed system (FBS) of planting. Soybean (*Glycine max*) yield increased nearly 100% with the ridge-furrow system (RFS) and about 55% in broad-bed and furrow system (BBFS) compared with the FBS. Adoption of integrated nutrient management (INM) based on soil testing increased soybean and wheat yields by 71% over farmers' practice at Narsinghpur, compared with about 100% for soybean and 187% for wheat at Hoshangabad. Inter-cropping of soybean with pigeon pea (*Cajanus cajan*) in 4:2 ratio produced higher net return (Rs. 27,620/ha) and benefit-cost ratio (3.3:1) than either of the monocropping system. An aquaculture system was a better alternative to traditional monocropping (Haveli system) in the monsoon season. Aquaculture in the ponded water in the bunded field during monsoon season and growing of wheat or chickpeas in the winter season proved successful, and is being adopted in the region. The concepts of 'seed farmers' and 'seed village' were promoted to ensure seed replacement and availability of quality seeds of high-yielding varieties of soybean, chickpea, and wheat. Seed replacement increased mean yield in the participating villages by 30% to 50%.

Address correspondence to R. Lal, Carbon Management and Sequestration Center, OARDC/FAES, 2021 Coffey Road, The Ohio State University, Columbus, Ohio, 43210 USA. E-mail: lal.1@osu.edu

Reprinted from: *Journal of Crop Improvement*, Vol. 23, Issue 2 (April 2009), pp. 119–135, DOI: 10.1080/15427520802643718

Keywords: Vertisols, soil conservation, raised sunken bed system, participatory research, integrated nutrient management

ABBREVIATIONS

AWC	available water capacity
CEC	cation exchange capacity
EC	electrical conductivity
GWT	ground water table
ha	hectare
I	irrigation
INMP	integrated nutrient management practice
Mg	megagram (metric ton)
Rs	Indian Rupees

7.1 INTRODUCTION

Vertisols (black soils) occupy 7% of the arable lands in the semi-arid tropical region of India and cover large areas under dry farming in central and southern India (Figure 7.1). Vertisols are derived from base-rich rocks (basalt) or the related colluvium or alluvium parent materials. These soils have a high clay content (30%–70%),

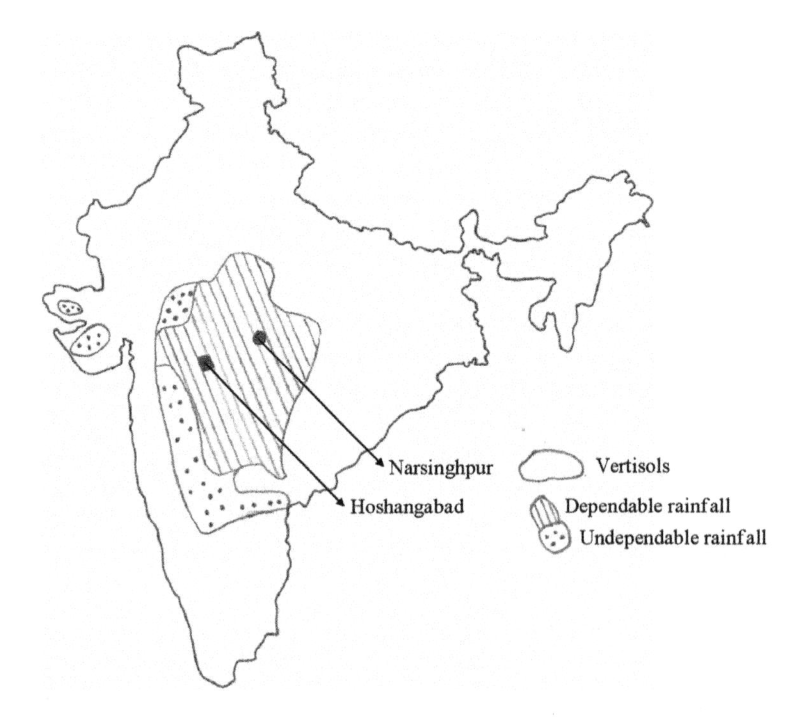

Figure 7.1 Location of the project sites, and distribution of Vertisols in Central India.

vary in depth (15–200 cm), and exhibit distinct cracking patterns. Vertisols are also highly prone to sheet erosion. Surface runoff varies from 10% to 40%, and increases with increase in rainfall. Heavy texture and waterlogging make it difficult for early and rapid seedbed preparation. Important soil groups in the region are Chromusterts and Pellusterts (Murthy et al., 1981; Swindale, 1982; Venkateswarlu, 1987).

In Central India, Vertisols have a large potential of increasing agricultural production, provided appropriate technologies for conservation and management of natural resources, particularly soil and rainwater management, are widely implemented. Upland crops grown on these soils in high rainfall areas (>1000 mm), mainly soybean and maize (*Zea mays*), are prone to temporary waterlogging and anaerobiosis.

In India, about 18 million ha (Mha) of Vertisols (12 Mha in the state of Madhya Pradesh [MP]) are left fallow during the rainy/monsoon season. Therefore, these soils are cropped only during the post-rainy season on profile-stored soil moisture. The rainy season fallowing, locally called the "Haveli system" of cultivation, leaves the land unutilized and prone to severe erosion and runoff. The low water infiltration rate (3–5 mm/h) is attributed to high clay content (>40%), which are predominantly expanding-type clay minerals. Through adoption of best management practices (BMPs), there is a vast scope to improve utilization of soil and water resources in the region for intensification of these inherently fertile soils.

The M.S. Swaminathan Research Foundation (MSSRF), Chennai, India, The Ohio State University, (OSU) United States, and the Jawaharlal Nehru Krishi Vishwa Vidyalaya (JNKVV) Jabalpur, MP, India, implemented a community-based field operational research project. Specific objectives of the project were to: 1) promote adoption of BMPs for rainfed and irrigated agriculture in central India through demonstration of soil- and water-conservation measures and improved agronomic practices; 2) demonstrate the usefulness of integrated nutrient management practices (INMP); and 3) provide training opportunities to researchers, extension workers, and farming communities.

7.2 PROJECT SITES

On-farm demonstrations were established on two representative series of Vertisols in Narsinghpur and Hoshangabad districts of MP (Figure 1; MSSRF, 2006). These two districts come under 10.1- (Malwa Plateau, Vindhyan and Narmada Valley) hot, dry, subhumid eco-sub region (Velayutham et al., 1999). The subregion is characterized by subhumid climate with dry summers, mild winters, and ustic soil moisture regime. Temperature ranges from 31°–40°C in summer and 9°–19°C in winter. Annual rainfall ranges from 1000 mm to 1500 mm, and length of the crop-growing period ranges from 150 to 180 days. Soils of the demonstration sites were shallow to moderately deep, moderately well drained, slowly permeable, clayey vertic Ustochrepts and Ustorthents in the gently to very gently sloping uplands, and very deep, clayey Chromusterts in the nearly level uplands (Tamgadge et al., 1999).

The groundwater table (GWT) was below 10 m at Narsinghpur, and 5–10 m deep at the Hoshangabad sites. Groundwater was the source of irrigation at the Narsinghpur site, whereas canal irrigation from Tawa Command was used at the Hoshangabad site. The GWT has started declining at the Narsinghpur site at the rate of 15–20 cm per annum, whereas the GWT in 'B' zone of Tawa Command at Hoshangabad is rising at the rate of 20 cm per annum. Both areas need judicious use of available water resources for sustaining high agricultural productivity. The characteristics of the soil at the two sites are given in Tables 7.1 and 7.2 and Figure 7.2. Total soil organic carbon (SOC) pool to 120 cm depth was 89.6 Mg ha^{-1} for the soil at Narsinghpur site and 74.2 Mg ha^{-1} for the Hoshangabad site. The clay content ranged from 42.1% to 46.1% for the Narsinghpur site and 43.5% to 50.0% for the Hoshangabad site. Both soils had a high water-holding capacity at 33 K Pa suction or field-moisture capacity (30 to 40% by weight). Total available water capacity to 120 cm depth (computed on volumetric basis for the specific depths) was 27.5 cm for the Narsinghpur site and 25.8 cm for the Hoshangabad site (Tables 7.1 and 7.2).

These soils are characterized by typical morphological features called 'slicken sides,' which are caused by the presence of swell-shrink minerals like smectites. Vertisols are difficult to plow during the rainy season, and need proper land configuration and soil-conservation measures for effective irrigation and crop production.

Table 7.1 Properties of a Vertisol at Narsinghpur site.

	Soil depth (cm)				
Soil properties	*0–15*	*15–30*	*30–60*	*60–90*	*90–120*
Sand (%)	17.6	17.1	16.1	15.8	15.0
Silt (%)	39.9	40.3	40.7	40.3	38.9
Clay (%)	42.5	42.6	43.2	43.9	46.1
Bulk Density (Mg m^{-3})	1.40	1.42	1.48	1.50	1.50
pH (1:2.5)	7.1	7.1	7.4	7.5	7.7
E.C. (1:2.5)(dSm^{-1})	0.40	0.41	0.39	0.38	0.38
Water retention (%)					
33 kPa	32.4	33.0	33.2	33.5	33.5
1500 kPa	17.5	17.6	17.6	17.7	17.7
Organic Carbon (%)	0.69	0.54	0.51	0.49	0.42
Carbon Pool (Mg ha^{-1})	14.5	11.5	22.6	22.1	18.9
C.E.C. (me/100g soil)	37.1	38.6	39.3	39.0	41.1
Avail. N (kg/ha)	242	228	–	–	–
Avail. P_2O_5 (kg/ha)	22.4	20.3	–	–	–
Avail. K_2O (kg/ha)	471	465	–	–	–
Zinc (ppm)	0.36	0.37	–	–	–
Cu (ppm)	1.59	1.66	–	–	–
Fe (ppm)	5.4	5.2	–	–	–
Mn (ppm)	3.8	3.7	–	–	–
S (ppm)	4.6	4.5	–	–	–
†AWC (cm)	3.1	3.3	6.9	7.1	7.1

Land capability sub-class IIIes.
Irrigability class, 3rd.
†AWC = available water capacity.

Table 7.2 Properties of a Vertisol at Hoshangabad site.

	Soil depth (cm)				
Soil properties	*0–15*	*15–30*	*30–60*	*60–90*	*90–120*
Sand (%)	23.2	22.0	22.0	21.0	21.0
Silt (%)	33.3	33.5	32.3	29.8	29.0
Clay (%)	43.5	44.5	45.7	49.2	50.0
Bulk Density (Mg m^{-3})	1.42	1.45	1.48	1.58	1.59
pH (1:2.5)	7.7	7.9	7.8	7.9	8.0
E.C. (1:2.5)(dSm^{-1})	0.2	0.3	0.3	0.3	0.3
Water retention (%)					
33 kPa	32.5	33.1	33.2	34.6	33.7
1500 kPa	18.6	18.7	18.9	20.0	20.1
Organic Carbon (%)	0.57	0.49	0.45	0.34	0.32
Carbon Pool (Mg ha^{-1})	12.1	10.7	20.0	16.1	15.3
C.E.C. (me/100g soil)	36.9	36.0	36.2	35.5	35.4
Avail. N (kg/ha)	230	225	–	–	–
Avail. P_2O_5 (kg/ha)	25.4	25.9	–	–	–
Avail. K_2O (kg/ha)	571	565	–	–	–
Zinc (ppm)	0.5	0.4	–	–	–
S (ppm)	3.5	2.8	–	–	–
†AWC (cm)	3.0	3.1	6.3	6.9	6.5

Land capability sub-class, IIIs.
Irrigability class, 3rd.
†AWC = available water capacity.

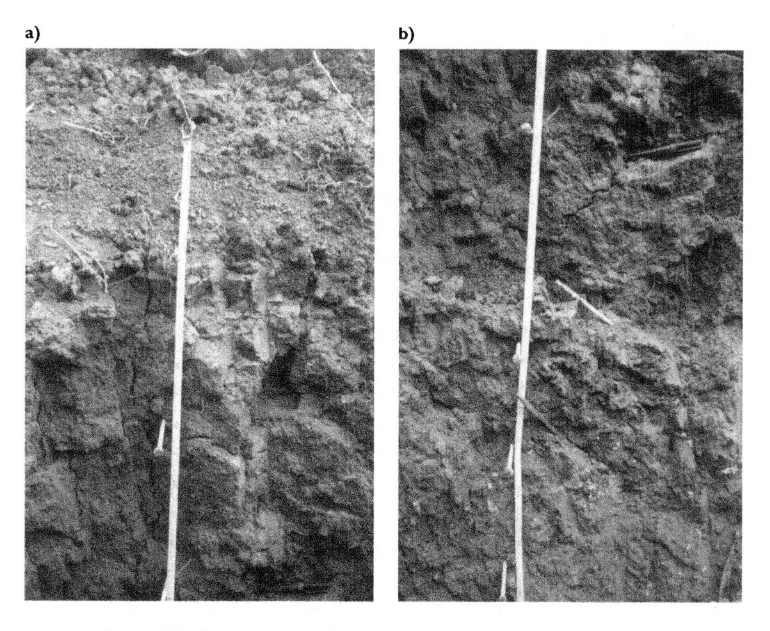

Figure 7.2 Soil profiles: a) Narsinghpur, and b) Hoshangabad.

Field demonstrations were established on ten farms in three villages (viz., Dangidhana, Bagpodi, and Murlipodi) in Narsinghpur district and expanded to twenty farms later. Similarly, ten farmers' fields were selected in three villages (viz., Mongwari, Bhairakhedi, Dolariya) in Hoshangabad district and expanded to forty farms during the second and third years.

7.3 LAND TREATMENT AND SOIL CONSERVATION

The raised/sunken bed system (RSBS) of land treatment was established on farmers' fields. The system consists of an array of raised and sunken beds of 8 m width, with elevation difference of 30 cm. The system is created by mechanically moving soil from demarcated 6 m wide strips, designated as sunken beds. Beds thus created are tied across with small earthen bunds of about 10 cm height at 20 m intervals to ensure uniform spread of runoff in sunken beds. The runoff from raised beds is diverted and captured in the adjacent sunken beds. The RSBS (Figure 7.3) facilitates drainage for growing upland crops, encourages *in situ* rainwater conservation, and minimizes soil erosion and nutrient losses.

During the winter season, grain yields of wheat sown in sunken bed and chickpea in raised bed were compared with those from the flatbed system (FBS) of cultivation. During the monsoon season, soybeans were sown in the raised bed and rice (*Oryza sativa* L.) in the sunken bed. The results (Table 7.3) indicate the beneficial effects of RSBS over the FBS on the yield of wheat and chickpea. The RSBS, once formed, stabilizes over time (cropping cycles) and increases the benefit-cost (B:C) ratio over the FBS.

7.4 INTEGRATED PLANT NUTRIENT MANAGEMENT PRACTICE (INMP)

Eight trials were conducted at the Narsinghpur site to demonstrate the effectiveness of INMP for wheat and chickpea. The results (Table 7.4) indicate that INMP treatment produced the maximum yield in comparison to the general recommended fertilizer alone. With INMP treatment, an additional income of Rs. 6978 ha^{-1} and Rs. 11976 ha^{-1} was obtained for wheat and chickpea, respectively, over the farmers' practice.

Similar results were obtained with soybean trials established in fifteen locations at the Narsinghpur site. The highest soybean yield (2.58 Mg ha^{-1}) was obtained under INMP at +4 Mg ha^{-1} with application of farmyard manure (FYM; Table 7.5), which generated an additional income of Rs. 6847/ha compared to the farmers' practice.

7.5 HOSHANGABAD SITE

Ten demonstration trials were conducted at the Hoshangabad site to assess the BMPs for wheat var. DL-788–2 in comparison with farmers' practice. The results (Table 7.6) indicate the high yield (4.0 Mg ha^{-1}) with adoption of BMPs. Similar results were obtained with soybean (Table 7.7), which indicates the doubling of yield with the adoption of INMP.

Figure 7.3 Raised bed-sunken bed system; soybean-rice system.

Table 7.3 Effect of raised-sunken bed system on yield of wheat and chickpea.

		Yield (Mg ha^{-1})	
Plots	*Treatment*	*Wheat*	*Chickpea*
1	Raised bed	–	1.95
2	Sunken bed	3.57	–
3	Flat bed	3.45	1.90

Table 7.4 Residual effect of Integrated Nutrient Management on yield of wheat and gram at Narsinghpur.

	Wheat		*Chickpea*	
Treatments	*Grain yield (Mg ha^{-1})*	*% increase over the farmers' practice*	*Grain yield (Mg ha^{-1})*	*% increase over the farmers' practice*
Farmers Practice (FP)	1.9	–	1.3	–
INMP	3.1	60.8	2.2	73.0
GRD	2.6	36.6	1.9	45.5

GRD = General recommended dose.

Two land configuration treatments, viz., RFS and BBFS, were compared with the FBS of planting soybean. The ridges were formed 15–20 cm high on 0.5% grade (Figure 7.4). The BBFS involved erection of 150 cm wide and 15 cm high raised beds on 0.4% grade. The beds were separated by 50 cm wide furrows that drained into

Table 7.5 Effect of INMP on yield of soybean at Narsinghpur District (Monsoon Season 2002).

Treatment	*Yield (Mg ha^{-1})*	*% increase over farmers practice*
Farmer's Practice	1.4	–
INMP	2.5	71.5
INMP +4 Mg FYM ha^{-1}	2.6	79.0

Table 7.6 Effect of full package of practices on wheat at Hoshangabad.

Treatments	*Yield (Mg ha^{-1}) (Av. of 10 farmers)*	*% increase over F.P.*
Farmers Practice	1.4	–
Full Package	4.0	187

Table 7.7 Effect of integrated plant nutrient management on yield of soybean at Hoshangabad.

Treatment	*Soybean yield (Mg ha^{-1}) (Av. of 11 farmers)*	*% increase over control*
Control	8.7	–
INMP Full Package	17.2	97.6

Figure 7.4 Soybean under ridges and furrows system.

grassed waterways. The results (Tables 7.8 and 7.9) indicate the advantage of these two systems in improving moisture conservation and soil moisture availability and yield of soybean. Increase in soybean yield was nearly 100% with RFS and 55% with BBFS compared with the farmers' practice. Selvaraju et al. (1999) reported that BBF and compartmental bunding land configuration could be adopted on Vertisols for better water conservation and crop productivity.

One of the improved agronomic practices in soybean cultivation is intercropping of soybean with pigeon pea. The results of demonstrations conducted in a farmer's field (Table 7.10) indicate that this system is compatible in 4:2 ratio (four lines of soybean and two lines of pigeon pea) and produced higher net return (Rs. 27,620 ha^{-1}) and B:C ratio (3.3:1) over either of the two monocultures.

One of the improved agronomic practices in rice cultivation is water management or the frequency of flooding. In general, farmers continuously flood rice fields for the entire growing period. Through irrigation at critical stages of the crop rather than continuous flooding, there is scope for water saving without causing yield reduction. Thus, a demonstration experiment on the effect of irrigation management on rice was established. The data showed that with three irrigations at critical stages, rice yields were equivalent to those for continuous flooding (Table 7.11).

Table 7.8 Effect of ridge planting on soybean yield.

Treatment	*Grain yield (Mg ha^{-1}) (Av. of 5 farmers)*
Ridge	1.7
Flat bed	0.8
% increase over flat bed	109.3

Table 7.9 Yield of soybean as influenced by BBF and RF system.

Treatment	*Yield (Mg ha^{-1})*	*% increase over flat bed system*
Flat bed system	1.02	–
Broad bed & Furrow System (BBF)	1.58	54.9
Ridge & Furrow system (RFS)	1.95	91.3

Table 7.10 Results of soybean-pigeon pea intercropping.

	Yield (Mg ha^{-1})		*Pigeon pea equivalent yield*	*Gross income*	*Cost of cultivation*	*Net return*	*Benefit:*
Treatment	*Soybean*	*Pigeon pea*	*(Mg ha^{-1})*	*(Rs. ha^{-1})*	*(Rs. ha^{-1})*	*(Rs. ha^{-1})*	*cost ratio*
Sole soybean (JS-335)	1.5	–	0.8	16060	7842	8218	1.05: 1
Sole pigeon pea (JKM-7)	–	1.3	1.3	25600	6471	19129	2.97: 1
Soybean + pigeon pea (4:2)	1.2	1.2	1.8	36080	8460	27620	3.27: 1

Table 7.11 Effect of irrigation management on rice yield.

Treatment	*Rice yield (Mg ha^{-1})*	*% increase over check*
Continuous flooding	3.50	116
Irrigation at tillering + flowering + grain filling stage	3.75	131

This demonstration has created awareness among farmers in the region about the benefits of water-conservation technology for growing rice.

7.6 AQUA-AGRICULTURE—AN ALTERNATIVE TO MONOCROPPING

The 'Haveli system' of cultivation of Vertisols in central India entails ponding the rainwater in bunded fields during the rainy season (to minimize losses by runoff and erosion), and growing wheat or chickpea in the post-rainy season on stored profile moisture. The available rainwater is not effectively utilized, and the land is left fallow. Raising fish in the rainy season in these fields and growing crops in the post-rainy season can augment farmers' income and enhance employment throughout the year.

A 2 ha field was selected for demonstration and three species of fish (i.e., Mrigal, Rohu, and Katla) were introduced (Figure 7.5) according to the standard management practices. Wheat and chickpea were sown in the same field after fish harvest in October. Fish, wheat, or chickpea rotation in the field increased the net returns considerably more than fallowing during the rainy season and cropping only during the winter (Table 7.12). The data highlight the feasibility and economic advantage of adopting aqua-agriculture in the vast Vertisol region of MP. The demonstration trials of this type conducted in farmers' fields provided excellent loci for dissemination of this important technology.

In the villages, ponds serve as common property resources. Two old, unutilized ponds (tanks) were cleaned by the farmers and, with the assistance of the project staff, the farmers took up fish culture after treating the pond with lime and superphosphate. Fresh cow dung and oilcakes were used daily as fish feed. After harvest of fish, water in the pond was used for pre-sowing irrigation for cultivation of wheat and chickpea. The returns (Table 7.13) demonstrate the economic advantage of this diversified production system, in the form of aquaculture in tanks in rainy season and crop growing with the harvested water in the subsequent winter season.

7.7 INTEGRATED RESOURCE MANAGEMENT OF CROPPING SYSTEMS

Two field experiments to maximize the interactive effect of water and nutrient resources were conducted on soybean-wheat cropping systems under varying fertility

Figure 7.5 Ponded agricultural field for fish culture.

Table 7.12 Results of fish-wheat/gram rotation.

Treatment	*Production (Mg ha^{-1})*	*Cost of cultivation (Rs.)*	*Gross income (Rs. ha^{-1})*	*Net return (Rs. ha^{-1})*
Fish	1.1	12576	27500	14924
Wheat	2.5	7092	14880	7788
Fish	1.1	12576	27500	14924
Chickpea	1.9	6616	21737	15121
		Total: 19192	49237	30045

Table 7.13 Results of water harvesting and fish culture and soybean wheat/chickpea rotation.

Land Use	*Production (Mg ha^{-1})*	*Cost of Cultivation (Rs.)*	*Gross income (Rs. ha^{-1})*	*Net return (Rs. ha^{-1})*
Fish culture (0.15 ha)	0.450 Mg/ W.H. tank	5576	9000	3424
Soybean	1.8	8542	18543	10000
Wheat	2.8	6949	16830	9880
		Total: 21067	44373	23304
Fish culture (0.15 ha)	0.38 Mg/ W.H. tank	5576	7600	2024
Soybean	1.7	8592	18070	9477
Chickpea	2.0	6531	225470	16008
		Total: 20700	28210	27509

W.H. = water harvest.

and irrigation levels. Three moisture regimes (i.e., irrigation at seeding/transplanting [I_0], moderate moisture regime [I_1] and optimum moisture regime [I_2]) in combination with nine nutrient sources (S_1 to S_9) were evaluated. The irrigation schedule for the three treatments for the two crops was as follows:

Irrigation	I_0	I_1	I_2
Soybean (Monsoon)	No irrigation	Irrigation at 60 DAS	Irrigation as per need
Wheat (Winter)	Come up irrigation + 21 DAS	Irrigation at 21 & 45 DAS	Irrigation at 21, 45, 75 & 90 DAS

DAS = Days after sowing.

The nine nutrient levels were: 100% NPKS optimum dose as per soil test (S_1); 75% of optimal dose +5 Mg ha^{-1} of FYM (S_2); 50% of optimal dose +5 Mg FYM ha^{-1} (S_3); S_2 + rhizobium/Azotobacter @ 1.5 kg ha^{-1} (S_4); S_2 + phosphate solubilizing bacteria (PSB) @ 1.5 kg ha^{-1} (S_5); S_2 + PSB + rhizobium (S_6); S_6 + Zn @ 10 kg zinc sulfate/ha (S_7); S_6 + Mo as ammonium molybdate @ 0.5 kg ha^{-1} (S_8), and S_6 + Zn + Mo (S_9).

The grain yields of soybean and wheat are given in Figures 7.6 and 7.7, respectively. The highest yields of soybean (2.4 Mg ha^{-1}) and wheat (4.4 Mg ha^{-1}) were obtained under I_2 moisture regime. In the case of nutrient sources, grain yield was significantly higher in S_9 than in other treatments. The treatment combination I_2 S_9

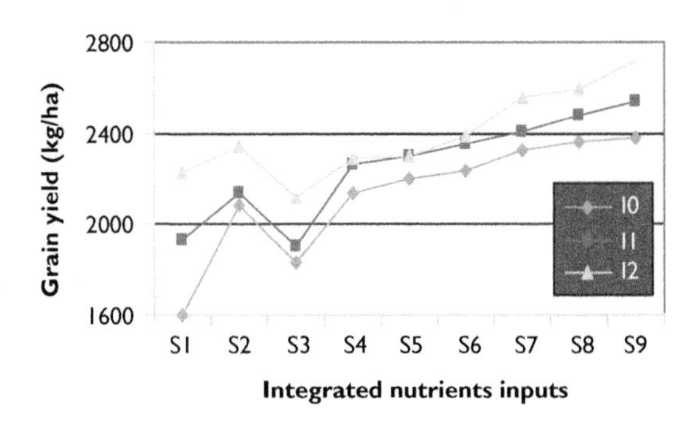

Figure 7.6 Grain yield of soybean (kg ha^{-1}).

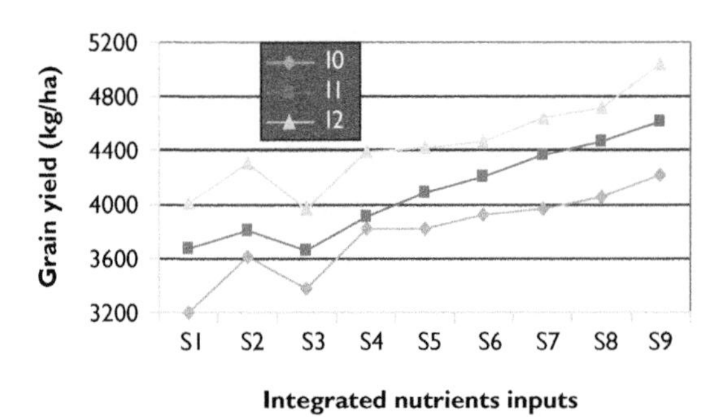

Figure 7.7 Grain yield of wheat (kg ha^{-1}).

(optimum moisture level and integrated nutrient sources) was the best option for increasing grain yield of both crops.

7.8 'SEED VILLAGE' CONCEPT FOR SEED REPLACEMENT

The availability of good, quality seeds of high-yielding varieties (HYV) of crops is a major constraint among farmers to realizing the full benefits of HYV technology. The use of harvested grains as seed, in successive seasons, reduces the seed vigor and crop yield. Mobilizing farmers in a village as 'seed farmers' to multiply the breeder's seed and make seed replacement by all farmers in the village provides the best quality seed delivery system at the local level. The project staff assisted this process in the selected villages and the impact of seed replacement and the increases in mean yield in the villages are given in Tables 7.14 and 7.15, respectively. In three years, the production and distribution of HYVs of crops increased the mean yields of soybean, wheat, and chickpea by 50%, 30%–50%, and 50%, respectively. Such an approach is important to achieving self-sufficiency in quality seed production and distribution at the local level.

7.9 CONCLUSIONS

Soil and rainwater management are the key factors for increasing the production potential of Vertisols in the high-rainfall region of central India. The farmer participatory operational research project in the villages proved successful in demonstrating

Table 7.14 Multiplication of seeds of high yielding varieties in 'Seed Villages'.

	Narsinghpur site		
	Mean of three villages (Dangidhana, Murlipodi, Bagpodi)		
Village	*2000–2001*	*2001–2002*	*2002–2003*
Seed replacement (Breeder seed) in three villages (Mg)			
Soybean	0.40	7.2	21.5
Wheat	0.40	40.9	55.7
Chickpea	0.35	24.6	28.5
	Hoshangabad site		
	Mean of three villages (Mongwari, Baihrakhedi, Dolaria)		
Village	*2000–2001*	*2001–2002*	*2002–2003*
Seed replacement (Breeder seed) in three villages (Mg)			
Soybean	–	1.0	15.0
Wheat	0.4	27.2	30.6
Chickpea	–	0.6	13.2

Table 7.15 Mean yield of crops in 'Seed Villages'.

	Narsinghpur site		
	Mean of three villages (Dangidhana, Murlipodi, Bagpodi)		
Village	*2000–2001*	*2001–2002*	*2002–2003*
Productivity (Mg ha^{-1})			
Soybean	1.0	1.8	2.1
Wheat	1.9	2.6	2.6
Chickpea	1.3	1.8	1.8
	Hoshangabad site		
	Mean of three villages (Mongwari, Baihrakhedi, Dolaria)		
Village	*2000–2001*	*2001–2002*	*2002–2003*
Productivity (Mg ha^{-1})			
Soybean	1.0	1.1	1.5
Wheat	1.4	2.7	3.1
Chickpea	1.2	1.4	1.7

best management practices, such as proper land configurations and soil fertility enhancement. Land treatments (raised-sunken bed system, ridges and furrows, broad bed and furrows) increased *in situ* soil moisture conservation, minimized runoff, and increased the yield of principal crops grown in the region (soybean, wheat, chickpea, and rice). Integrated nutrient management based on soil testing and application of well-decomposed farmyard manure and biofertilizers increased the yield of soybean and wheat. Intercropping with soybean and pigeon pea in 4:2 ratio was a superior cropping system to the traditional monocropping of soybean. The concept of 'seed village' for decentralized production and distribution of quality seeds of high-yielding varieties of crops among the farming community proved successful in increasing the mean yield of crops in the project villages. Aquaculture rather than fallow during the monsoon season is a feasible aqua-agriculture system for the region.

REFERENCES

M.S. Swaminathan Research Foundation (MSSRF). 2006. Sustainable management of natural resources for food security and environmental quality. Project achievements, Chennai, India: M.S. Swaminathan Research Foundation.

Murthy, R.S., L.R. Hirekarur, S.B. Deshpande, B. Venkat Rao, and K.S. Sankaranarayana, eds. 1981. Benchmark soils of India. Nagpur, India: National Bureau of Soil Survey and Land Use Planning.

Selvaraju, R., P. Subbian, A. Balasubramanian, and R. Lal, 1999. Land configuration and soil nutrient management options for sustainable crop production on Alfisols and Vertisols of southern peninsular India. *Soil & Tillage Res.* 52: 203–216.

Swindale, L.D. 1982. Distribution and use of arable soils in the semi-arid tropics. In *Managing Soils Resources—Plenary Session Papers, 67–100.* Proc. 12th Intl. Congress of Soil Science, February 1982, New Delhi, India.
Tamgadge, D.B., K.S. Gajbhiye, M. Velayutham, and G.S. Kaushal, 1999. *Soil series of Madhya Pradesh.* Pubn. No. 78. Bhopal, India: National Bureau of Soil Survey and Land Use Planning, Indian Council of Agricultural Research in co-operation with Dept. of Agriculture, Govt. of Madhya Pradesh.
Velayutham, M., D.K. Mandal, M. Champa, and J. Sehgal, 1999. *Agro-ecological sub-regions of India for planning and development* Pubn. No. 25, Nagpur, India: National Bureau of Soil Survey and Land Use Planning.
Venkateswarlu, J. 1987. Efficient resource management systems for drylands of India. *Advances in Soil Science* 7: 165–221.

Chapter 8

Sustainable management of dryland Alfisols (red soils) in South India

S.P. Palaniappan & R. Balasubramanian
M.S. Swaminathan Research Foundation, Chennai, India

T. Ramesh
Tamil Nadu Agricultural University, Coimbatore, India

A. Chandrasekaran, K.G. Mani & M. Velayutham
M.S. Swaminathan Research Foundation, Chennai, India

R. Lal
Carbon Management and Sequestration Center, The Ohio State University, Columbus, OH, USA

ABSTRACT

A community based cooperative research project was implemented on farmers' fields on some dryland Alfisols in Tamil Nadu, India, to demonstrate and validate improved dry-farming technologies, such as: 1) soil and water conservation and water harvesting; 2) cropping systems, including intercropping and double cropping; 3) recycling of processed agricultural wastes and byproducts; and 4) low-cost drip irrigation. Disc plowing to 30 cm depth during summer and contour bunding enhanced soil moisture storage in the profile, and facilitated harvesting of runoff water into a community pond. Under the bimodal rainfall pattern in the region, among the medium- and long-duration varieties of pigeon pea, short-duration varieties of blackgram and greengram evaluated, long-duration variety of pigeon pea was the most suited to this region. The pigeon pea variety 'VBN2', blackgram variety 'VBN3', and greengram variety 'VBN2' produced the maximum grain yields. Increasing the land equivalent ratio (LER) by intercropping of pigeon pea with pulses and oilseeds enhanced agronomic productivity. Pigeon pea + groundnut was the best intercropping system. Pigeon pea + lablab and pigeon pea + groundnut intercropping system produced the highest yields. In years with normal rainfall, green manuring with sunnhemp and raising a pulse crop, horsegram, increased soil fertility. Application of compost enriched with rock phosphate (produced by using locally available crop residues of cotton, pigeon pea, sugarcane, and raw pressmud) significantly enhanced the yield of pigeon pea, groundnut, onion, and okra in these degraded Alfisols with

Address correspondence to R. Lal, Carbon Management and Sequestration Center, The Ohio State University, 2021 Coffey Road, Rm 210 Kottman Hall, Columbus, Ohio 43210, USA. E-mail: lal.1@osu.edu

Reprinted from: *Journal of Crop Improvement*, Vol. 23, Issue 3 (July 2009), pp. 275–299, DOI: 10.1080/15427520902809888 high phosphate fixation capacity. Tied ridging and mulching with groundnut residues produced the maximum yield of cowpea even in seasons with below normal rainfall. Tied ridging produced the highest net returns from pigeon pea + greengram intercropping system. A low-cost, zero-energy drip-irrigation system produced the highest yield of 23.2 Mg/ha of tomato, with a saving in water of 73.8% compared with the control. Introduction of arid horticulture with amla, sapota, and mango, and water management through pitcher-pot irrigation and mulching with coconut husk as a means of diversification of land use management, provided employment during the off-season and enhanced household income.

Keywords: rainfed agriculture, water conservation, Alfisols, participatory research, intercropping, tied ridges, enriched compost

8.1 INTRODUCTION

About 58% of the 142 Mha of arable land in India depends entirely on the rains received during the southwest and northeast monsoons of South Asia. The soils of the dryland regions of India comprise: 1) Sierozems (Aridisols); 2) sub-montane soils (Mollisols, Inceptisols, and Entisols); 3) Alluviums (Inceptisols, Alfisols, and Entisols); 4) red soils (Alfisols); 5) black soils (Vertisols); and 6) Laterites (Oxisols) (Swindale, 1982; Venkateswarlu, 1987). Genesis and properties of the soils of India are discussed by Murthy et al. (1981).

Crop yields under dryland conditions are low and highly variable. The Green Revolution of the 1960s and 1970s occurred on irrigated soils of the Indo-Gangetic Basin. In contrast, Alfisols of southern India have numerous constraints to adopting intensive cropping and enhancing crop yields. Effectiveness of highly variable and erratic monsoons is extremely low because of high intensity that generates a large volume of surface runoff and high temperatures that cause severe losses by evaporation. Consequently, crops suffer from frequent and severe drought stress at critical stages of growth. Adverse impacts of drought stress are exacerbated by the severe problem of soil degradation. Alfisols have inherently low soil fertility characterized by low nutrient reserves (e.g., N, P, K, Ca, Mg, Zn, Cu), low soil organic matter (SOM) content, low available water capacity (AWC), and consequently low microbial biomass C. With predominantly coarse texture and scanty vegetation cover, crops suffer from supra-optimal soil temperatures exceeding 50°C in the top 0 to 5 cm layer. Adverse impacts of high soil temperatures and recurring drought stress are accentuated by the root-restrictive ground layer at shallow depth of 30 to 40 cm. Thus, alleviating these soil-related constraints by improving soil quality, conserving water in the root zone, moderating soil temperature, improving soil fertility, and enhancing SOM content are essential to increasing agronomic production. Agronomic benefits of improving soil quality can be realized only through adoption of improved cultivation and cropping systems. In this regard, the importance of intercropping and legume-based rotations cannot be overemphasized. The strategy is to enhance agronomic production per unit land area, per unit time, and per unit input of water and other off-farm resources.

Thus, a community-based project was initiated by the M.S. Swaminathan Research Foundation (MSSRF), Chennai, India, and the Ohio State University (OSU), Columbus, Ohio, to promote adoption of proven soil- and crop-management practices to enhance local food security and improve environmental quality on rainfed Alfisols in southern India (MSSRF, 2006). The specific objectives of the project were to:

1. Test, validate, and extend adoption of management options to alleviate soil-related constraints through a participatory research approach;
2. Enhance production diversification by introducing legumes/pulses, vegetables, livestock, agro-forestry, and aquaculture through on-farm demonstration and validation;
3. Develop and introduce management practices for efficient utilization of soil, water, and organic resources that would lead to sustainable high productivity; and
4. Restore degraded soils, enhance water recharge and soil organic carbon (SOC) concentration.

8.2 SITE

Field demonstrations were established at Ariyamuthupatti village in Kudumianmalai Panchayat in Pudukottai district of Tamil Nadu, India (Figure 8.1). The population is entirely dependent on agriculture, and predominant soils of the district comprise degraded, infertile red soils (Alfisols- Vayalogam series) under semi-arid climate. The district is located between 8° 30′ to 10° 40′ N latitude and 78° 24′ to 79° 40′ E longitude in East Central Tamil Nadu, India, and falls in the 8.3 agro-eco sub-region of India (Natarajan et al., 1997; Velayutham et al., 1999). The district has a bi-modal rainfall distribution, with an average annual rainfall of 685 mm. About 40% of rainfall is received during southwest monsoon season (July–September), 44% during northeast monsoon (October–December), 7% during winter (January–February), and 9% in summer (March–June).

Based on the participatory rural appraisal approach (Pretty, 1994; Velayutham, Ramamuthy & Venugopalan, 2002) and interaction with the farmers, an on-farm work plan for five years (2001–05) was developed. It consisted of: (i) crop diversification; (ii) improved crop varieties; (iii) enriched compost made from locally available crop residues and organic wastes; (iv) soil and water conservation, water harvesting, and micro-irrigation techniques; and (v) agriculturally based income-generating activities.

Demonstrations were also conducted at Pudupatti village, near Kannivadi, in Dindigul district. It lies between 10° 3′ and 10° 48′ N latitude and 77° 15′ and 78° 20′ E longitude. Predominant soils of the study site also represent Alfisols (red soils). The soil of the project site belongs to Irugur series. These soils are used for intensive cultivation of vegetables by surface irrigation methods. Accelerated erosion by water runoff and shallow soil depth are the major soil-related problems. The soil is sandy clay in texture with pH of 7.5 and SOC concentration of 0.58% in the 0–15 cm depth. The demonstrations were focused on drip irrigation.

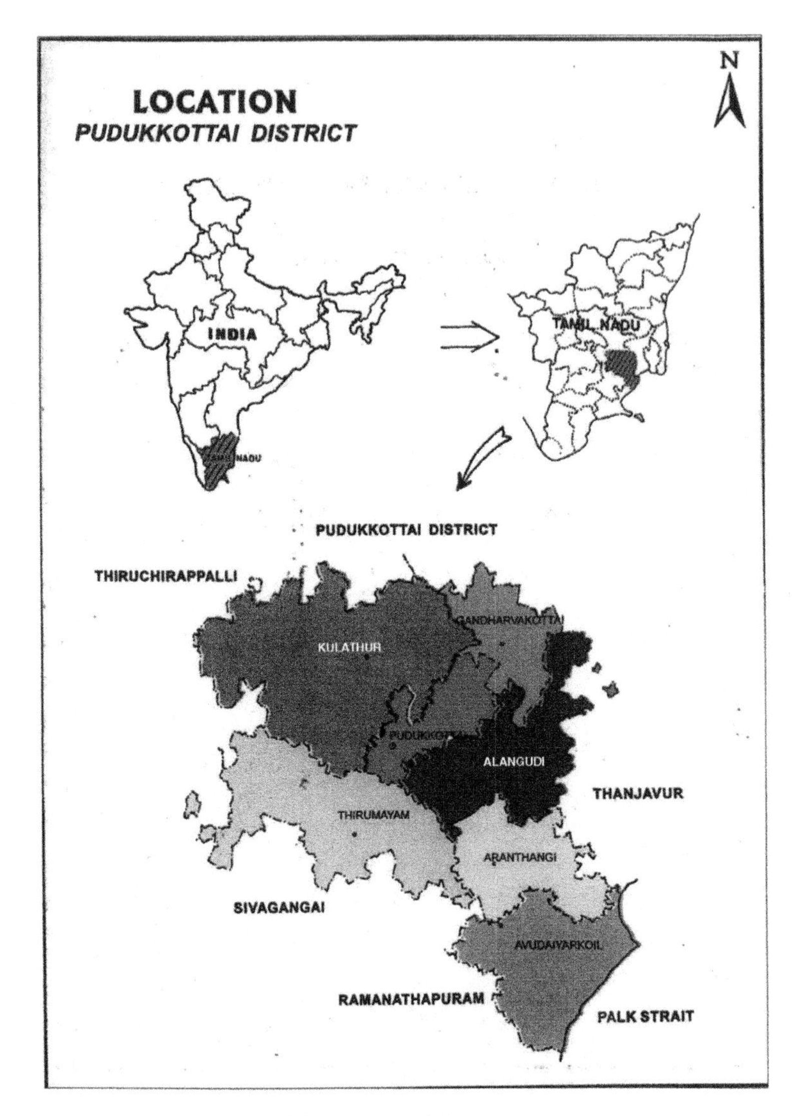

Figure 8.1 Location of the project site.

8.3 WATER MANAGEMENT TECHNOLOGY

At the start of the project, most of the rainfed soils in Ariyamuthupatti village had been uncultivated and left fallow for the past two decades. The farmers were reluctant to undertake high risks of crop failure involved in cultivating these soils under dry farming conditions. Consequently, the village water pond (locally called 'tank') was uncared for several years and had been silted up by sedimentation because of uncontrolled and accelerated soil erosion. The pond was overgrown with weeds and shrubs (e.g., *Prosopis spp.*).

Farmland area of 5.1 ha leased from the community was chosen as the experimental-cum-demonstration farm. Initial soil analysis was done to establish baseline soil properties (Jackson, 1975). The data in Table 8.1 show a low level of plant available nutrients of only 146 kg/ha of NPK, and a low cation exchange capacity (CEC) of 4.6 C mol (t)/kg of soil. The SOC concentration (0.26%) was below the critical level of 1.1% for soils of the tropics (Aune & Lal, 1998). The effective rooting depth is shallow, because the soil is underlain by a thick gravelly horizon (Figure 8.2).

Table 8.1 Nutrient status (0–20 cm) of the soil at Ariyamuthupatti village site just before sowing in 2001.

Parameter	*Value*	*Remarks*
Available nitrogen (kg/ha)	52	Low
Available phosphorus (kg/ha)	14	Medium
Available potassium (kg/ha)	80	Low
Soil organic carbon (%)	0.26	Low
Extractable calcium (cmol/kg soil)	2.26	Low
Extractable magnesium (cmol/kg soil)	1.10	Low
Extractable sodium (cmol/kg soil)	0.34	Low
Cation exchange capacity (cmol/kg soil)	4.65	Low
Base saturation (%)	82.15	High
Exchangeable sodium (%)	7.3	Low
†DTPA extractable zinc (ppm)	0.44	Low
DTPA extractable iron (ppm)	10.03	High
DTPA extractable copper (ppm)	0.62	Low
DTPA extractable manganese (ppm)	9.79	High

†DTPA = diethylene triamine pentaacetic acid.

Figure 8.2 Soil profile at Ariyamuthupatti, site at the start of the project.

The surface soil has low AWC, and is prone to crusting, moderate to severe erosion, and acidification. The soils have a high fixation capacity of P because of the presence of iron and aluminum hydroxides.

Improved dry-farming technologies evaluated were: 1) soil and water conservation techniques and water harvesting; 2) cropping systems, including intercropping and double cropping; 3) recycling of processed agricultural wastes and byproducts; and 4) drip irrigation.

8.4 WATER HARVESTING

Seedbed preparation during summer was done by disc plowing to about 20 cm depth. Bunds were erected across the slope around individual fields for diverting the excess rainwater into the channels. These channels, about 0.5 m deep with strengthened bunds and installed on the contour, were connected to a main waterway. The latter with a gentle gradient and a depth of 0.75 to 1.0 m, drained into the village pond. The pond was excavated to more than 1 m depth by farmers for storing the excess runoff water carried off from the fields through a network of channels and waterways (Figure 8.3). Periodic desilting and deepening of the pond was done by the village community throughout the study period. In addition to serving as a water reservoir, the tank also recharged the groundwater in its vicinity.

Figure 8.3 Water harvesting—rain water stored in Ariyamuthupatti tank after desilting and deepening.

8.5 CROPS AND CROPPING SYSTEMS

Improved varieties of pigeon pea (*Cajanus cajan* L.) tested involved those of varying growth duration. These included medium-duration (135 days) or MD varieties (APK 1, CO 5, and VBN 1), and long-duration (165–180) or LD varieties (VBN 2, LRG 30, and ICPL 87119). The data in Table 8.2 show that LD varieties were better suited for this eco-region with a bimodal rainfall pattern than MD varieties. Furthermore, among LD varieties, the highest grain yield of 263 kg/ha was obtained with the VBN 2 variety. Four varieties of black gram (*Vigna mungo* L. Hepper; VBN 2, VBN 3, CO 6, T 9) and three varieties of greengram (*Vigna radiata* L.R. Wilczek; VBN 2, CO 6, K 851), with crop duration range of 65–70 days, were also tested for their performance. Variety VBN 2 of greengram and VBN 3 of blackgram performed relatively well in these soils and yielded 188 and 180 kg/ha, respectively. The prolonged dry spell in August coincided with the grain-filling stage and resulted in low yields.

8.6 INTERCROPPING WITH PIGEON PEA

The LD pigeon pea variety (VBN 2) was tested for its productivity in an inter-cropping system (Figure 8.4). Both groundnut (*Arachis hypogaea* L.; TMV 3) and cowpea (*Vigna unguiculata* L.; VBN 1) were grown as intercrops. Pigeon pea was sown at a row-to-row spacing of 1.0 m. In between the rows of pigeon pea, three rows of cowpea and groundnut were sown at a spacing of 30 × 10 cm. The data in Table 8.3 indicate that intercropping of pigeon pea is climatically adaptable (Velayutham, 1999) and economically advantageous for this region. Pigeon pea + groundnut was the best intercropping system in net income (Rs. 4145/ha).

8.7 DOUBLE CROPPING

Double cropping, or sequential cropping, was evaluated by disc plowing during the summer to 30 cm depth. In addition, water harvesting using diversion channels and waterways and supplemental irrigation were also used. Enriched pressmud was used

Table 8.2 Performance of pigeon pea varieties of different growth duration.

Parameter	*APK 1*	*Co 5*	*VBN 1*	*VBN 2*	*ICPL 87119*
Duration (days)	126	160	121	183	178
Days to 50% flowering	70	80	70	152	127
No. of pods/plant	14	25	18	66	81
No. grains/pod	5	4	4	4	4
No. of plants/m^2	33	33	33	28	28
100 grain weight (g)	6.5	8	6.7	7.3	11.4
Dry matter production (Mg/ha)	3.0	3.0	2.5	7.0	5.0
Grain yield (kg/ha)	–	140	–	263	198

Figure 8.4 Red gram + green gram intercropping under mulching moisture conservation.

Table 8.3 Agronomic performance of intercropping of pigeon pea (var. VBN2) with cowpea and groundnut.

			Intercrop	
Parameter	*Monoculture*	*Main crop*	*Cowpea*	*Groundnut*
Population (plants/m^2)	28	28	72	78
Duration (days)	183	187	65	98
Days to 50% flowering	152	151	41	41
No. of pods/plant	66	40	9	11
No. of grains/pod	4	4	11	2
100 grain weight (g)	7.2	7.2	13.5	36
Dry matter production (Mg/ha)	6.0	5.5	3.1	3.2
Grain yield (kg/ha)	263	109	23.5	313
Net income (Rs./ha)	1315	545	235	3600

Net income (Rs./ha): pigeon pea (pure) – 1315; pigeon pea + cowpea – 780; pigeon pea + groundnut – 4915.

to improve soil fertility. Compost was used in conjunction with foliar application of di-ammonium phosphate. Integrated pest management (IPM) was adopted against the pod borer in pigeon pea. The data in Table 8.4 on yields of different cropping systems shows that agronomic productivity can be greatly enhanced in ecosystems with a bimodal rainfall pattern by growing two sequential crops under rainfed conditions.

Table 8.4 Agronomic yields and economic returns of a range of crops and cropping systems.

First crop (July–Sept.)	*Second crop (Oct.–Feb.)*	*Yield of first crop (kg/hg)*	*Yield of second crop (kg/ha)*	*Net income (Rs./ha)*
Pigeon pea (LD) (pure crop)	Contd.	263	–	1315
Pigeon pea (LD) + black gram	Red gram (contd.)	198	107	4158
Pigeon pea (LD) + cowpea	Red gram (contd.)	109	123	1780
Pigeon pea (LD) + groundnut	Red gram (contd.)	109	313	4145
Pigeon pea (LD) + lablab	Red gram (contd.)	198	248	4712
Pigeon pea (MD)	Red gram (contd.)	140	–	1120
Black gram	Horse gram	180	63	3620
Green gram	–	188	–	3478
Cowpea	Bengal gram	27	64	1457
Groundnut	Cluster bean & Radish	244	257	3643
Sesame	Horse gram	121	165	1638
Varagu (Kodo Millet)	Varagu (contd.)	634	–	2219
Finger millet	–	594	–	2673

Note: LD = long duration; MD = medium duration.

Pigeon pea + lablab (*Lablab purpureus* L.) and pigeon pea + groundnut intercropping systems produced the highest grain yields, even under conditions of uneven rainfall distribution. Among the double-cropping systems, groundnut followed by cluster bean (*Cyamopsis tetragonolobo* L. Taubert) and radish (*Raphanus sativus* L.) produced the highest monetary returns. In comparison with the monoculture of pigeon pea, the minor millets varagu or Kodo millet (*Paspalum scrobiculatum* L.) and finger millet (*Eleusine coracana* L. Gaertn) also performed well, indicating the potential prospect of including these nutritious minor millets in cropping systems for advancing crop diversification, food and nutrition security.

8.8 SUMMER PLOWING

Summer plowing to ~30 cm depth enhanced soil-water reserves during the S-W monsoon of 2004, primarily by breaking the crust and improving water infiltration rate. With intense rains received in July' 04, farmers included groundnut in the cropping system. Although there was a continuous dry spell for 52 days, the crop survived because of the moisture conserved through summer plowing, resulting in an average yield of 1.67 Mg/ha of pod. In some plots where summer plowing was not done, the groundnut crop sown in July suffered from severe drought stress at the critical stages of flowering, peg formation, and pod filling. The average groundnut yield was 600 kg pods/ha, or 36% of the yield obtained with summer plowing. Vittal, Vijayalakshmi, and Rao (1983) reported the beneficial effects of deep tillage on dryland crop production in red soils of India. The importance of summer plowing is now widely accepted by the farmers of the district, and it forms an important component of the best management practices (BMPs) for dry farming in the region.

8.9 GREEN MANURING

Sunnhemp (*Crotalaria juncea* L.) was raised as a green manure crop, and green biomass of 8.5 Mg/ha was incorporated at 45 days after sowing. Horsegram (*Macrotyloma uniflorum* L. Verde; Var. Paiyur 1) was sown after green manuring. While the yield of horsegram was low due to the prolonged dry spell (Table 8.5), which occurred during the grain-formation stage, the dry-matter production and post-harvest soil-fertility parameters indicate its numerous advantages and the possibility of introducing green manuring within the cropping cycle, particularly in seasons with normal rains. Thus, intercropping, double cropping, and green manuring are important BMPs for production of additional biomass from the same land. Biomass return to the soil is essential to enhancing SOC sequestration (Lal, 2004) and reversing the degradation trends.

8.10 SOIL FERTILITY MANAGEMENT BY USING COMPOST

Farmers, in particular women, participated in the training program for the production of enriched compost using locally available crop residues of cotton (*Gossypium hirsutum* L.), pigeon pea, sugarcane (*Saccharum officinarum* L.), and raw pressmud obtained as a byproduct from the nearby sugar-refining factory. These materials were wetted and mixed with animal dung. Rock phosphate at 25 kg/Mg of the material, zinc sulfate at 2.5 kg/Mg, *Trichoderma viride* at 5.0 kg/Mg, and phosphobacteria at 200 g/Mg were added to the residues, mixed and formed into heaps of 3 × 1.5 × 1 m dimensions. The moisture content of the heap was maintained at about 60% by sprinkling water periodically. The heap was turned over twice at 30 and 45 days, and well-decomposed compost was ready for field application by 60 days after the start of the composting process.

The enriched compost was mixed with the appropriate inoculum of rhizobium or azospirillum (depending on the crop) and applied at the rate of 2.5 Mg/ha. Increase in

Table 8.5 Effect of using sunnhemp as green manure on horsegram yield and soil fertility measured after the crop harvest.

Parameter	*With green manuring*	*Without green manuring*
Dry matter production of horse gram (kg/ha)	1100	1000
Grain yield of horse gram (kg/ha)	75	62.5
Post-harvest soil nitrogen (kg/ha)	165	133
Post-harvest soil phosphorus (kg/ha)	22.5	21.5
Post-harvest soil potassium (kg/ha)	206	175
Post-harvest soil organic carbon (%)	0.41	0.30

Initial soil analysis: Av. N – 93 kg/ha; Av. P – 18 kg/ha; Av. K – 218 kg/ha; Organic carbon – 0.28%.

crop yield by compost was 10% in pigeon pea, 58% in groundnut, 500% in onions (*Allium cepa* L.), and 12% in okra (*Abelmoschus esculentus* L. Moench; Table 8.6). The strategy of using integrated nutrient management practices (INMP) is important for these depleted and degraded Alfisols of southern India. Farmers are now practicing the production of enriched compost as a component of BMPs for INMP strategies.

8.11 SOIL AND WATER CONSERVATION AND WATER HARVESTING

Four techniques of soil conservation and mulching and three of INMP were laid out in a strip-plot design with cowpea as a test crop in 2002 and 2003 seasons. Total amount of rainfall received during winter 2002 and 2003 crop period was 162 and 132 mm, respectively, with distribution of seven rainy days in both years. Ridges, furrows, and tied ridges were formed immediately after the first monsoon rains (Figure 8.5).

Table 8.6 Effect of enriched compost application on yield crops grown at two sites (kg/ha).

Ariyamuthupetti				*Kannivadi*			
Crop	*Without compost*	*With compost*	*% increase*	*Crop*	*Without compost*	*With compost*	*% increase*
Pigeon pea	83	93	10.0	Onion	500	3000	500
Groundnut	302	478	58.0	Okra	15000	17620	12.0

Figure 8.5 Tied ridges.

Crop-residue mulch was applied at the rate of 2.5 Mg/ha about 15 days after sowing. The crop yield figures and the derived resource indices are given in Tables 8.7 and 8.8.

The efficiency indices, such as production efficiency (PE) and economic efficiency (EE), were computed by the following formula (Singh et al., 2005):

$$PE(\%) = \{(Y^{IS} - Y^{TS})/Y^{TS}\} \times 100$$

where Y^{IS} is grain yield in intervention system and management, and Y^{TS} is grain yield in traditional system and management.

$$EE\ (\%) = \{(NR^{IS} - NR^{TS})/NR^{TS}\} \times 100$$

where NR^{IS} is net returns in intervention system and management and NR^{TS} is net returns in traditional system and management.

Energy efficiency was computed by evaluating the input and output of energy for each treatment (Dazhong & Pimentel, 1984). Rainfall-use efficiency (RUE) was calculated by dividing the yield by total quantity of rainfall obtained during the cropping period. Solar radiation-use efficiency (SRUE) for the cropping period was computed as per the procedure of Hayashi (1966) and expressed in g cal^{-1}

The data in Table 8.7 and Table 8.8 show that tied ridges significantly improved cowpea performance through increasing soil-water reserves in the root zone. In addition, mulching with groundnut crop residues reduced soil evaporation and increased crop yield. Among the INMP treatments, application of enriched compost produced the highest yield. During the second year, tied ridges replaced the compartmental

Table 8.7 Effect of *in situ* rainwater harvesting on sustainable indicators of rainfed cowpea.

Moisture conservation	*Total biomass (kg ha⁻¹)*	*Grain yield (kg ha⁻¹)*	*Total factor productivity*	*Production efficiency (%)*	*Energy productivity (kg MJ⁻¹)*	*Energy efficiency*
Winter, 2002						
Farmer's practice	2652.5^{b}	572.2^{c}	0.127	–	0.77	12.60
Ridges & furrows (R&F)	2897.8^{ab}	662.8^{ab}	0.134	15.83	0.78	12.75
Compartmental bunding (CB)	2775.5^{b}	627.8^{b}	0.132	9.71	0.78	12.78
R&F + mulching	3008.8^{a}	715.9^{a}	0.143	25.11	0.81	13.20
CB + mulching	2917.3^{a}	688.9^{b}	0.143	20.39	0.81	13.18
Winter, 2003						
Farmer's practice	956.3^{c}	169.7^{c}	0.066	–	0.37	5.96
Ridges & furrows (R&F)	1117.0^{b}	208.5^{b}	0.072	22.86	0.40	6.53
Tied ridge (TR)	1151.8^{b}	223.2^{b}	0.072	31.52	0.40	6.58
R&F + mulching	1345.2^{a}	279.0^{a}	0.088	64.40	0.46	7.48
TR + mulching	1406.6^{a}	297.4^{a}	0.089	75.25	0.47	7.70

Means within column followed by the same letter are not significantly different (l.s.d at P = 0.05).

Table 8.8 Effect of land and nutrient management on cowpea yield and efficiency indices (2003).

Treatments	*Grain yield (kg/ha)*	*Production efficiency (%)*	*Economic efficiency (%)*	*Rainfall use efficiency (kg/ha/mm)*	*Solar radiation use efficiency (g/cal)*	*Energy efficiency*
Moisture (M)						
Farmers' practice	169.7	–	–	1.3	0.51	6.0
Ridges & furrows (R&F)	208.5	22.86	924	1.6	0.59	6.5
Tied ridge (TR)	223.2	31.52	1024	1.7	0.61	6.6
R&F + mulching	279.0	64.40	4024	2.1	0.71	7.5
TR + mulching	297.4	75.25	4468	2.3	0.74	7.7
CD (P = 0.05)	29.1	–	–	–	–	–
Nutrient (N)						
Farmers' practice	156.6	–	–	1.2	0.46	7.1
Rec. NPK	237.9	51.91	1526	1.8	0.65	6.5
50% NPK + 50% Enriched compost	284.5	81.67	4665	2.2	0.72	7.1
Enriched compost	263.2	68.07	2921	2.0	0.69	6.7
CD (P = 0.05)	20.3	–	–	–	–	–

bunding treatment, and ordinary compost application was replaced by application of 50% of inorganic fertilizer and 50% of enriched compost.

Significantly more grain yield of cowpea was obtained under ridges and furrows with mulching (715.9 kg/ha) and tied ridges with mulching (297.4 kg/ha) during 2002 and 2003, respectively, over farmers' practice. Ridges and furrows with mulching had higher PE (25.11%), EE (13.2), RUE (4.4 kg/ha/mm), and SRUE (1.61 g/cal) during 2002 than other treatments. Tied ridges with mulching were superior to farmers' practice and ridges and furrows with mulching in PE (75.25%), EE (7.7), RUE (2.3 kg/ha/mm), and SRUE (0.74 g/cal) during 2003. Among nutrient-management practices, application of enriched compost during 2002 and integration of 50% inorganic fertilizers and 50% enriched compost during 2003 produced significantly higher grain yield (730.7 and 284.5 kg/ha) and at higher resource-use efficiencies. Application of enriched compost improved PE (30.46 and 68.07%), EE (12.5 and 6.7), RUE (4.5 and 2.0 kg/ha/mm) and SRUE (1.67 and 0.69 g/cal) during 2002 and 2003, respectively, more than inorganic fertilizer application and farmers' practice of no-nutrient application. The advantage of tied ridges over farmer's practice in producing higher yield, even during seasons of sub-normal rainfall (2003), is significant (Ramesh and Devasenapathy, 2005, 2007a).

Field observations during the cropping seasons showed (Ramesh, Devasenapathy & Sabarinathan, 2006) that root growth (length, volume, and dry weight) and nodulation characteristics (numbers and dry weight) increased with the practices of soil moisture conservation (e.g., ridges and furrows, and tied ridges with mulching).

The data in Table 8.9 indicate that the highest yield of pigeon pea was obtained with tied ridges and mulching among soil-conservation treatments, and with application of enriched compost among the INMP treatments. Soil moisture content (0–30 cm

Table 8.9 Effect of land and nutrient management on pigeon pea yield and efficiency indices (2003).

Treatments	*Grain productivity (kg/ha)*	*Water productivity (kg/ha/cm)*	*Energy efficiency*
Moisture (M)			
Farmers' practice	292.3	5.5	11.38
Ridges & furrows (R&F)	353.6	6.6	12.23
Tied ridges (TR)	393.7	7.4	13.45
R&F + mulching	454.5	8.5	15.18
TR + mulching	475.4	8.9	15.63
CD (P = 0.05)	42.6		
Nutrient (N)			
Farmers' practice	271.2	5.1	14.0
Rec. NPK	391.2	7.4	12.5
50% NPK + 50% enriched compost	442.0	8.3	13.6
Enriched compost	471.2	8.9	14.2
CD (P = 0.05)	21.0		

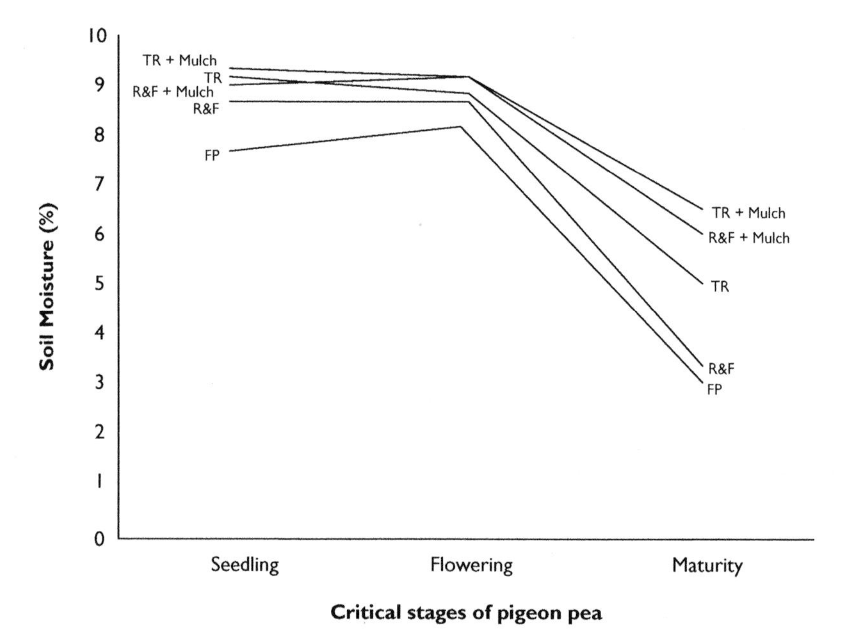

Figure 8.6 Effects of soil management treatments on soil moisture content at different phenological stages of crop growth.

depth) was the highest throughout the growing season under tied ridges with mulching (Figure 8.6). Improved grain yield (475.5 kg/ha), PE (8.9 kg/ha-cm), and EE (15.63) were obtained with tied ridges with mulching. In soil-management treatments, application of enriched compost significantly influenced grain yield (471.2 kg/ha) and

Table 8.10 Effect of *in situ* soil moisture conservation and nutrient management practices on post-harvest available npk (kg ha^{-1}) in cowpea and pigeon pea.

	Cowpea (2002)					*Cowpea (2003)*				*Pigeon pea (2003)*		
Treatments	*Org. C*	*N*	*P*	*K*	*Treatments*	*Org. C*	*N*	*P*	*K*	*N*	*P*	*K*
Moisture (M)					Moisture (M)							
Farmers' Practice	0.28	184	13.8	28.9	Farmers' Practice	0.20	170	12.8	36.4	155	16.9	44.8
Ridges & Furrows (R&F)	0.28	193	16.2	32.7	Ridges & Furrows (R&F)	0.22	181	14.8	50.4	162	18.9	52.2
Compart. Bunding (CB)	0.29	206	17.3	31.7	Tied Ridge (TR)	0.23	173	13.3	40.5	178	21.4	56.8
R&F + Mulching	0.32	235	21.2	49.4	R&F + Mulching	0.26	192	17.5	64.0	205	25.2	69.9
CB + Mulching	0.30	221	22.9	57.8	TR + Mulching	0.27	193	15.9	68.0	219	26.8	78.3
SEd	0.02	12.1	2.3	7.4	SEd	0.02	9.20	0.90	3.40	17.7	1.2	6.4
CD (P = 0.05)	0.04	28.1	5.2	17.1	CD (P = 0.05)	0.03	21.2	2.00	7.80	40.8	2.8	14.8
Nutrient (N)					Nutrient (N)							
Farmers' Practice	0.25	182	9.9	20.9	Farmers' Practice	0.22	158.9	9.6	36.6	145.5	14.5	35.1
Rec. NPK	0.27	199	13.7	29.9	Rec. NPK	0.23	182.0	11.3	45.0	162.4	21.1	44.8
Raw compost	0.33	210	20.1	44.1	50% NPK+50% Enriched compost	0.24	189.6	17.6	57.7	207.1	24.2	73.8
Enriched compost	0.32	242	29.3	65.6	Enriched compost	0.25	198.2	20.8	68.0	221.1	27.6	87.8
SEd	0.01	5.9	1.0	3.6	SEd	0.01	9.75	0.76	2.90	14.5	1.1	2.8
CD (P = 0.05)	0.03	14.5	2.4	8.7	CD (P = 0.05)	0.02	23.90	1.86	7.00	35.4	2.7	6.8

nutrient uptake. The highest net returns of Rs. 5731/ha from pigeon pea + green gram intercropping was obtained with the BMPs comprising tied ridges and mulching. In accordance with the increase in agronomic yields, improvements in the post-harvest available soil NPK and SOC concentration were also observed with mulching in ridges and furrows and tied-ridges treatments as given in Table 8.10 (Ramesh & Devasenapathy, 2007b). Selvaraju and colleagues (1999) reported that tied ridging and application of manures in combination with N and P fertilizer improved soil-water storage and yield of crops compared with sowing on the flat bed in rainfed Alfisols and related soils of the semi-arid tropics.

8.12 GRAVITATIONAL DRIP IRRIGATION SYSTEM USING LOCAL MATERIALS

A ground-level, zero-energy, drip-irrigation system was installed in a farmer's field at the Ariyamuthupatti and Kannivadi sites (Figure 8.7). The system involved installation of a PVC tank of 800-liter capacity at an elevation of about 1 m. The tank was connected with a drip irrigation system. Drip line was laid in alternate rows of the crop, with 1 lph dripper spaced at 30 cm. The tank was filled from the nearby borewell once a day, and tomatoes were transplanted during the dry season. Yield of fresh tomatoes of 23.2 Mg/ha was obtained with 73.8% saving in water.

At Kannivadi site, the plots receiving drip irrigation yielded 15.2 Mg green chillies (*Capsicum annum* L.) per ha compared with the production of 9.4 Mg/ha for plots with surface irrigation, an increase of 61%. The green chillies from the drip irrigation plots were larger, longer, and shinier than those from surface-irrigated treatments, and fetched a higher price in the market.

8.13 ARID HORTICULTURE

Adopting rainfed horticulture is an important strategy for providing another income stream for farmers. In cooperation with the state-funded wasteland development project, seedlings of amla (*Phyllanthus emblic* L.), sapota (*Manilkara zapota* L.),

Figure 8.7 Tomato grown with gravitational drip, okra under gravitational drip.

and mango (*Mangifera indica* L.) were planted on 35 acres (14 ha) of uncultivated wasteland in the village (Figure 8.8). The farmers were advised to use pitcher-pot irrigation (Figure 8.9), and mulching with coconut husk and crop residues (Figure 8 and Figure 8.10) around the saplings for initial establishment and better growth of

Figure 8.8 Coconut husk for moisture conservation—crop residues of cowpea on the background.

Figure 8.9 Pitcher irrigation to sapota.

Figure 8.10 Mulching with crop residue (groundnut) to a sapota sapling.

the seedlings during the summer period. This has been a successful undertaking, and a popular practice with the community.

8.14 SOIL TESTING

A soil-testing campaign was organized in the project area with the assistance of a soil-testing laboratory in Kudumianmalai. The services of the mobile soil-testing laboratory were also availed. Based on the soil-test analyses of individual farms, 'soil-health cards' were printed and distributed to the farmers (Figure 8.11) Specific training sessions were organized for the method of soil-sample collection, soil-test interpretation, and the use of soil-health card. Participating farmers appreciated the usefulness of these cards for periodic monitoring of soil health with the adoption of BMPs and recommended cropping systems.

8.15 SOCIAL MOBILIZATION AND INCOME-GENERATING ACTIVITIES

Self-help groups were formed to create other livelihood opportunities and to generate off-season and off-farm income. The women self-help group formed comprised Mangayi Amman Self Help Group with 16 founding members and Akilandeswari Self Help Group with 12 founding members. Members were invited to the MSSRF

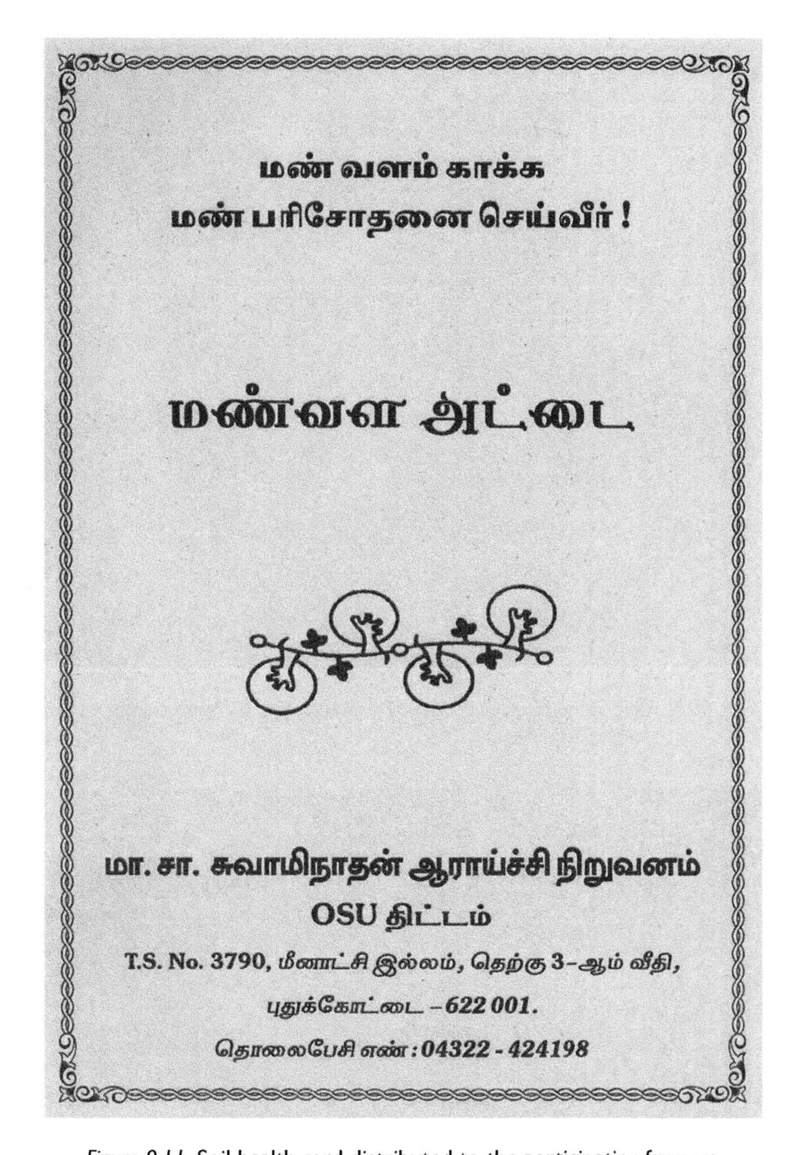

Figure 8.11 Soil health card distributed to the participating farmers.

centers at Kannivadi and Pondicherry to observe the functioning of the self-help groups in these centers and to interact with them in starting micro-finance-assisted micro-enterprise. The self-help group members mastered the technique of producing *Trichogramma parasitoid* and supplied the cards to sugarcane cultivators through a buy-back arrangement with the EID Parry Sugar Factory, located at Aranthangi, 30 km from project site (Figure 8.12). This arrangement was extremely effective in the biological control of sugarcane internode borer. This micro-enterprise and preparation of enriched compost (Figure 8.13) has considerably improved the employment and income-generating capacity of the participating women.

Figure 8.12 Sub-collector (training) keenly watching the *Trichogramma parasitoid* production process.

Figure 8.13 Compost enrichment.

Arrangements were also made for the construction of a "smokeless stove" in 26 households, through the available government scheme. Because of this, the accumulation of smoke inside the house is averted and respiratory and wheezing problems are much reduced among women in rural households.

8.16 CONCLUSIONS

A community-based participatory operational research project on Alfisols (red soils) of Tamil Nadu in the semi-arid region of India benefited the farming community by adopting rainfed agriculture and best management practices of dry-farming technologies. The technologies adopted comprised: 1) scheduling suitable crops and varieties, and standardizing intercropping and double-cropping systems based on rainfall patterns in a region normally used for monocropping of coarse cereals; 2) land configuration for water harvesting and soil- and water-conservation practices for alleviating drought stress; 3) INMP to improve soil fertility and IPM practices; and 4) introduction of arid horticulture and drip irrigation.

These on-farm demonstrations indicated the beneficial impacts on soil quality and agronomic production of the BMPs comprising the following components: 1) disc plowing during summer; 2) establishing tied ridges with mulch; 3) application of pressmud compost enriched with rock phosphate; 4) intercropping of pigeon pea as main crop and other pulses and groundnut as intercrop; 5) double cropping with sunnhemp and pulses; 6) diversification of cropping with nutritious minor millets; 7) drip irrigation for high-value vegetable crops; 8) agriculture-centered, income-generating activities for self-help groups of farm women; 9) utility of soil testing and use of a 'soil-health card' for monitoring soil productivity and sustainable agriculture; and 10) introduction of arid horticulture. The approach used in the project provides a community-based interactive participatory operational research model, with an intrinsic value of large-scale social mobilization and extension domain for adoption of improved dry-farming technology.

ACKNOWLEDGEMENTS

This project was implemented from a generous grant provided by Sir Dorabji Tata Trust and Tata Education Trust of the Tata House, Mumbai. We are grateful for the financial and monitoring support provided by them and to Dr. M.S. Swaminathan for guidance during the execution of the project.

REFERENCES

Aune, J. and R. Lal, 1998. Agricultural productivity in the tropics and critical limits of properties of Oxisols, Ultisols and Alfisols. *Trop. Agric.* 74: 96–103.

Dazhong, W. and D. Pimentel, 1984. Energy flow through an organic agro-ecosystem in China. *Agric. Ecosystem Environ.* 17: 145–160.

Hayashi, K. 1966. Efficiency of solar energy conversion in rice varieties as affected by planting density. *Proc. Crop Sci. Soc. Japan.* 35(304): 205–209.

Jackson, M.L. 1975. *Soil chemical analysis*. New Delhi, India: Prentice Hall.

Lal, R. 2004. Soil carbon sequestration in India. *Climatic Change* 65: 277–296.

MSSRF. 2006. *Sustainable management of natural resources for food security and environmental quality. Project achievements*. Chennai, India: M.S. Swaminathan Research Foundation.

Murthy, R.S., L.R. Hirekarur, S.B. Deshpande, B. Venkat Rao, and K.S. Sankaranarayana, 1981. *Benchmark soils of India*. Nagpur, India: ICAR, National Bureau of Soil Survey and Land Use Planning.

Natarajan, A., P.S.A. Reddy, J. Sehgal, and M. Velayutham, 1997. Soil resources of Tamil Nadu for land use planning. National Bureau of Soil Survey and Land Use Planning, Pub. No. 46b, Nagpur, India: ICAR.

Pretty, J.N. 1994. Alternative systems of inquiry for sustainable agriculture. *IDS Bull.* 25(2): 34–48.

Ramesh, T. and P. Devasenapathy, 2005. Productivity and resource use efficiency of rainfed cowpea (Vigna unguiculata (L.) Walp) as influenced by in situ soil moisture conservation and nutrient management practices. *J. Farming Sys. Res. & Dev.* 11(2): 135–140.

Ramesh, T. and P. Devasenapathy, 2007a. Physical indicators of sustainability in rainfed cowpea (*Vigna unguiculata* (L.) Walp) as influenced by in situ rainwater harvesting. *Legume Res.* 30(4): 256–260.

Ramesh, T. and P. Devasenapathy, 2007b. Natural resource management on sustainable productivity of rainfed pigeonpea (Cajanus cajan L.) *Res. J. Agri. Bio. Sci.* 3(3): 124–128.

Ramesh, T., P. Devasenapathy, and R. Sabarinathan, 2006. Root growth and nodulation characteristics of cowpea (Vigna unguiculata (L.) Walp) as influenced by in situ soil moisture conservation and nutrient management practices under rainfed Alfisols ecosystem. *Crop Res.* 31(1): 37–42.

Selvaraju, R., P. Subbian, A. Balasubramanian, and R. Lal, 1999. Land configuration and soil nutrient management options for sustainable crop production on Alfisols and Vertisols of southern peninsular India. *Soil & Tillage Res.* 52: 203–216.

Singh, J.P., A. Salaria, K. Singh, and B. Gangwar, 2005. Diversification of rice-wheat cropping system through inclusion of basmati rice, potato and sunflower in Trans-Gangetic Plains *J. Farming Systems Res. Develop.* 11(1): 12–18.

Swindale, L.D. 1982. Distribution and use of arable soils in the semi-arid tropics. Proc. 12th Intl. Congress of Soil Science, New Delhi, India.

Velayutham, M. 1999. Crop and land use planning in dryland agriculture. In *Fifty years of dryland agricultural research in India*, eds. H.P. Singh, Y.S. Ramakrishna, K.L. Sharma, and B. Venkateswarlu. Hyderabad, India: Central Research Institute for Dryland Agriculture, SCAR.

Velayutham, M., D.K. Mandal, M. Champa, and J. Sehgal, 1999. Agro-ecological sub-regions of India for planning and development National Bureau of Soil Survey and Land Use Planning. Pub. No. 25, Nagpur, India: ICAR.

Velayutham, M., V. Ramamuthy, and M.V. Venugopalan, 2002. Agricultural land use planning: From theoretical perspectives to participatory action in the Indian context. *The Land* 6(14): 45–60.

Venkateswarlu, J. 1987. Efficient resource management systems for drylands of India. *Adv.Soil Sci.* 7: 165–221.

Vittal, K.P.R., K. Vijayalakshmi, and U.M.B. Rao, 1983. Effect of deep tillage on dryland crop production in red soils of India. *Soil & Tillage Res.* 3: 377–384.

Section IV

Soil and water quality management

Chapter 9

Soil and water quality: Integral components of watershed management

K.R. Reddy & J.W. Jawitz
Soil and Water Science Department, University of Florida, Gainesville, Florida, USA

ABSTRACT

Management of agricultural, forested, range, wetland, and urban land plays an integral part in influencing soil and water quality within a watershed. Non-point-source pollution of blue waters, streams, rivers, groundwater, lakes, wetlands, and estuaries is now linked to the management practices used in these ecosystems. In this article, we present a brief overview of the role of soil and water quality in sustainable management of watersheds. Sustainable watershed management will require integration of information from diverse domains (e.g., physical, biogeochemical, economic, social, cultural, and demographic) at multiple spatial and temporal scales, and development of predictive tools across environmental, hydrologic, economic, and social gradients. Soils in the watershed play a unique role in regulating air and water quality, plant productivity, carbon sequestration, production and consumption of greenhouse gases, and climate. Management strategies used to reduce pollutant loads from watersheds should seek to improve soil quality as a first step to improve water quality. Changes in practices by the user are needed to implement new solutions to improve quantity and quality of water resources. Current educational programs in India and other developing countries are not adequate for watershed technology transfer to the user. Linkage between research and outreach must be improved for effective implementation of watershed management practices. Watershed-management plans should be viewed as the starting point, but not the end of management cycle, thus following the concept of adaptive management. Science and policy must function together for watershed management to be successful.

Keywords: agricultural ecosystems, nutrients, water quality, biomass productivity

Address correspondence to K.R. Reddy at Soil and Water Science Department, University of Florida, Gainesville, FL 32611, USA. E-mail: krr@ufl.edu

Reprinted from: *Journal of Crop Improvement*, Vol. 24, Issue 1 (January 2010), pp. 60–69, DOI: 10.1080/15427520903307726

9.1 INTRODUCTION

Global land and water resources are threatened as land use is converted from natural to urban and agriculture environments; population grows, particularly in water-short regions; demographics change as a large number of people move from rural to urban environments; demands increase for food security and socio-economic well-being; and industrial, municipal, and agricultural pollution contaminates the environment (UNESCO 2006; Yeston et al. 2006). The major challenges for global agriculture are: (1) meeting the food and fiber needs of a world population projected to exceed 7.5 billion by the year 2020; (2) decreasing the rate of soil degradation and ameliorating degraded soils; and (3) protecting the quality of natural resources. In addition, the quality and security of land and water resources are threatened by extreme natural events, such as hurricanes, monsoonal flooding, and droughts intensified by climate change, and more recently by potential human impacts from bioterrorism. Changes in land-use practices resulting from economic and population pressures directly influence the quantity and quality of water resources. At the current rate of global water consumption, changes in land use may result in severe water shortages for more than 2.7 billion people by the year 2025 and another 2.5 billion people will live in areas where it will be difficult to find sufficient freshwater to meet their needs (UNESCO 2006).

The magnitude of global land and water-resource issues and associated infrastructure problems pose the following questions as outlined by the National Research Council (2004): (1) Will drinking water be safe?; (2) will there be sufficient water to support both the environment and future economic growth?; (3) can effective water policy be made?; (4) can water quality be maintained and enhanced?; and (5) will our water-management systems adapt to climate change? These questions highlight the importance of a functional understanding of global watersheds, including urban, agricultural, and forested ecosystems and natural systems, such as wetlands, river basins, estuaries, and aquifers. This is necessary so that we may be able to predict how these systems will respond to socioeconomic shifts that induce changes in land use, water distribution, and water quality, and therefore devise sustainable and resilient policies and management strategies to meet current demands while protecting the needs of future generations.

Water resources may be broadly categorized as blue water and green water. Blue water is stored in lakes, ponds, reservoirs, rivers, and aquifers, while green water is derived from precipitation, stored in soils, and subsequently returned to the atmosphere via evapotranspiration (Falkenmark & Rockstrom 2006). Globally, it is estimated that approximately 56% of the green water (61,600 km^3) from rainfall is evaporated back to the atmosphere from various land uses, and approximately 36% of the rain contributes to blue water sources (42,900 km^3) (Comprehensive Assessment of Water Management in Agriculture 2007). Total annual global freshwater (blue water) withdrawals are estimated to be 3,800 km^3, of which 70% is used for agriculture, 20% by industry, and 10% by municipalities (Comprehensive Assessment of Water Management in Agriculture 2007). For maintenance of ecosystem productivity and food production, conservation of both blue and green waters is important. In water-use assessments, both blue and green waters must be considered (Liu, Zehnder & Yang 2009). Both blue and green waters connect ecosystems across the landscape. Often, this connection is broken by changes in land-use activity and inefficient use of water resources.

Current options to meet the land and water-resource demands in any given country vary and depend on the socio-economic conditions, geographic region, and landscape properties, and intensity of land- and water-use practices. For example, in the most economically developed areas of the world (such as North America and Western Europe), some of the options utilized to meet the water demands require energy-intensive technologies. Such technologies may include a mix of traditional use of ground and surface water, reservoir storage of surface water, aquifer storage and recovery technology, desalination, reuse of reclaimed water, and advanced water-conservation technologies. In rapidly growing countries, such as China and India, more emphasis is sometimes placed on economic development rather than environmental protection, which creates enormous demands and pressures on land and water resources and results in their degradation and depletion. In underdeveloped countries, such as some areas in Sub-Saharan Africa, there is a long history of exploitation of land and water resources to meet minimum living conditions.

Increased population, rapid urban development, large agricultural water demands, and the need to protect natural resources have led to water-resource problems in many parts of the world. For example, the total demand for freshwater in India was estimated to be 634 billion cubic meters (bcm) in the year 2000 (approximately 460 billions of gallons per day [bgd]) and is estimated to be 1447 bcm in the year 2050 (approximately 1050 bgd). The vast majority of water currently used in India is for agriculture (approximately 80%), and irrigation efficiency is extremely low (from 17% to 48%). Furthermore, rapid industrialization and urbanization have also led to increased demand for public water supply, industrial use, and thermal power. All of these competing sectors use good-quality freshwater and have the potential to release poor-quality waters into the natural system. Uneven spatial and temporal distribution of rainfall and the likelihood of rainfall redistribution, caused by climate change, exacerbate problems associated with floods, droughts, soil erosion, poor nutrient-use efficiency by crops, and nutrient leaching. These water-related issues are not unique to the Indian subcontinent, but are global issues. Similar problems of increasing population, rapid urban development, large agricultural water demands, and the need to protect natural resources exist in many parts of the United States. For example, because of the rapidly increasing population, freshwater withdrawals in the state of Florida alone are expected to grow from approximately 8.2 bgd in 2000 to more than 9.3 bgd in 2020, with the obvious potential to produce significant conflict between urban, agricultural, industrial, and natural-system water users. Like in India, agriculture constitutes the majority of Florida's current freshwater usage, but public water supply needs are growing rapidly and are projected to surpass agriculture water needs in the near future.

9.2 WATERSHEDS AND WATER QUALITY

Management of agricultural, forested, range, wetland, and urban land plays an integral part in influencing soil and water quality within a watershed (Figure 9.1). Non-point-source pollution of blue waters, streams, rivers, groundwater, lakes, wetlands,

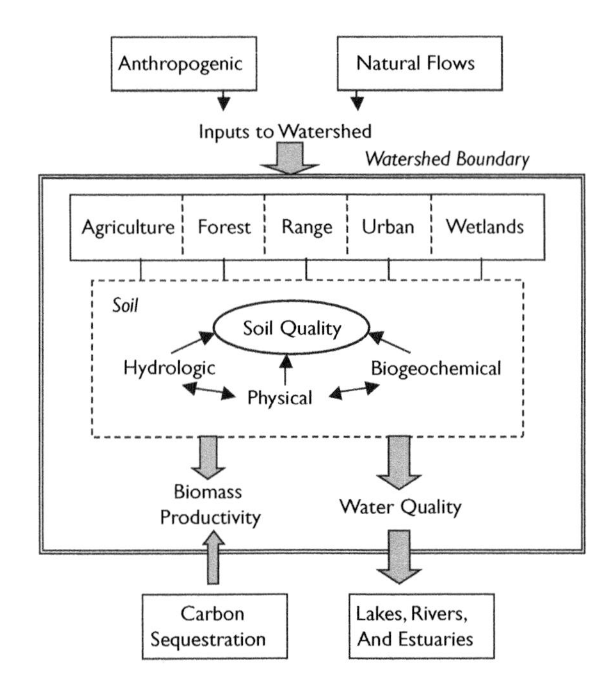

Figure 9.1 Schematic showing the linkage between soil and water quality as influenced by anthropogenic inputs and natural flows.

and estuaries is now linked to the management practices used in these ecosystems. The question of immediate concern is: *Are the current watershed-management practices compatible or adequate to sustain or improve the quality and quantity of our water resources?* Many current practices are compatible, but not all are adequate to sustain water-resource quality. Society demands that the water-resource quality and quantity be protected, placing a greater demand on producers to deliver environmentally sound goods. Many watershed-management practices currently used around the world are clearly insufficient to effectively deal with stresses placed on resources from burgeoning populations. Sustainable watershed management will require integration of information from diverse domains (e.g., physical, biogeochemical, economic, social, cultural, and demographic) at multiple spatial and temporal scales, and development of predictive tools across environmental, hydrologic, economic, and social gradients.

The U.S. National Research Council committee on long-range soil and water conservation defined four broad issues that show promise of maintaining sustainable agriculture, while protecting water quality (1993). These include: (1) conserve and enhance soil quality as a fundamental step to environmental improvement; (2) increase nutrient, pesticide, and irrigation-use efficiencies in farming systems; (3) increase the resistance of farming systems to erosion and runoff; and (4) make greater use of field and landscape buffer zones.

9.2.1 Soil quality: An integral component of watershed management

Soil is a fragile and finite resource. It plays a unique role in regulating air and water quality, plant productivity, carbon sequestration, production and consumption of greenhouse gases, and climate. Use and management of native, agricultural, forested, range, and urban lands play an integral part in influencing soil quality within a watershed. Further, soil management is directly linked to non-point-source pollution of aquatic ecosystems. Understanding the nature of soil quality, which we define as the ability of the soil resource to produce and maintain ecosystem production of plant, animal, and microbial biomass and to buffer or improve water quality, is fundamental to meeting this challenge.

Soils function as long-term storage units for various contaminants, including nutrients, trace metals, human and animal pathogens, pesticides, and other toxic organic compounds of agricultural and industrial origin. Protecting soil resources and maintaining their productivity in a sustainable manner is fundamental to meeting this challenge. Land-use changes and alterations in management practices have significant impacts on soil quality, which ultimately affect the quality of groundwater, adjacent streams, wetlands, lakes, and estuaries. A holistic, integrated approach is required to develop alternate practices that will maintain environmentally sound management of these ecosystems.

The task of defining soil quality and its linkages to sustainable biological productivity and water quality is complex (Doran et al. 1995; Sojka & Upchurch 1999). Unfortunately, there is often insufficient time and/or resources to accomplish such an effort, particularly given the rapid rate of anthropogenically driven changes occurring in many watersheds.

Soil quality is the capacity of a soil to sustain biological productivity, maintain environmental quality, and promote plant and animal health (Doran & Parkin 1994). Historically, soil fertility has been the major soil-quality concern of farmers, especially in developing countries. Soil-quality decline, resulting from erosion and over-cropping, has been compensated through the application of fertilizers and soil amendments, use of genetically improved varieties, and improvement of farming practices (Crosson et al. 1985). Soil-quality decline has been directly linked to water-quality decline. Even though a soil may be considered of high quality for production, it may have a deleterious influence on water quality. A variety of soil physical, chemical, and microbial parameters have been proposed as potential indicators of soil quality (Doran & Parkin 1994; Karlen et al. 1997). The use of ecological indicators in the evaluation of ecosystem health has received widespread attention during the past few decades (Suter 1993; Bortollo 1998; Jackson, Kurtz & Fisher 2000). In watershed management and restoration, there is a pressing need for sensitive, reliable, rapid, and inexpensive soil-quality indicators that can be used to assess soil-nutrient bioavailability and capacity to maintain water quality.

Soils are integral players in biogeochemical cycles that regulate ecosystem functions. Thus, it is appropriate to define soil quality as the functions that soil plays in these ecosystems (Johnson et al. 1992). Policies that protect soil resources should protect the capacity of soils to serve several functions simultaneously, including the production of food, fiber, fuel, nutrient storage, carbon sequestration, waste storage, and the maintenance of ecosystem stability and resiliency. Soil quality is best defined by a combination of physical, chemical, and biological properties. These attributes should form the basis for selection of soil-quality indicators (National Resource Council 1993).

Long-term goals of watershed management should include conservation and enhancement of soil quality. Governmental policies to reduce pollutant loads from watersheds should seek to improve soil quality as a first step to improve water quality. Soils have long-term memory of pollutant accumulation and the legacy of these pollutants should be taken into consideration as part of watershed-management strategies. Soil management in a watershed should also promote carbon sequestration by conserving plant residues and protecting soil organic matter. Appropriate agricultural production systems and buffer zones should be implemented to reduce soil erosion. Protecting soil quality, like protecting air and water quality, should be a fundamental goal of national environmental policy.

9.3 WATER QUALITY – AN INTEGRAL PART OF WATERSHED MANAGEMENT

In the United States, approximately 218 million people live within 16 km of an impaired water body. States have identified about 21,000 polluted lakes and estuaries. Management of agricultural, forested, range, and urban lands has been identified as a major source of pollutants impacting the quality of rivers, lakes, wetlands, and estuaries (EPA, 1992; 1994; 1996; 1997a,b). In the USA, agricultural and forest management alone contributes 70% of the pollution in rivers and 60% of the pollution in lakes (Carey 1991). Excess nutrients in aquatic systems have detrimental effects on designated or existing uses, including drinking water supply, recreational use, aquatic life, and fisheries. In the southeastern United States, approximately 4 million ha of lakes and 560,000 km of rivers are impacted by pollutants, sediment and nutrients being major types (EPA 2000). New regulations place a burden and demand on farmers to change current practices or develop new practices to protect water quality while maintaining a profitable production system. To meet this demand, it is critical for holistic management of watersheds for sustainable productivity while protecting water resources.

Nutrients and sediments are major sources of pollution in watersheds. These pollutants enter or are distributed in the watershed via surface and subsurface flows. A major portion of these pollutant loads is from non-point sources. The U.S. Environmental Protection Agency defines an impaired water body as one that does not meet water-quality criteria that support its designated use. The criteria might be numeric and specify concentration, duration, and recurrence intervals for various parameters, or they might be narrative and describe required conditions, such as the absence of scum, sludge, odors, or toxic substances (EPA 2008). If the water body is impaired, it is placed on Section 303(d) list. For each pollutant listed, the state must develop a restoration target called a total maximum daily load (TMDL). The EPA provides guidance for the watershed planning process and highlights common features of typical watershed-planning processes (EPA 2008). Key steps for this watershed-planning process are listed below:

- *Build partnerships.* Initial activities of planning process should involve interested parties, e.g., identifying stakeholders, integrating other key programs, and conducting outreach.
- *Define scope of watershed planning effort.* Developing information on defining issues of concern, developing preliminary goals, and identifying indicators to assess current conditions.
- *Gather existing data and create an inventory.* Available information should be collected and a data inventory for the watershed created.
- *Identify data gaps and collect additional data.* Data gaps should be identified and, if needed, additional data should be collected.
- *Analyze data to characterize the watershed and pollutant sources.* Primary data analyses are needed to identify sources and loads of pollutants and causes of pollution.
- *Estimate pollutant loads.* Identify and use watershed models to estimate pollutant loads. Select appropriate models for use to estimate pollutant loads.
- *Set goals and identify load reductions.* Set management and water-quality goals, develop management objectives, and determine load reductions needed to meet the goals.
- *Identify possible management strategies.* Identify cost-effective management strategies, including best-management practices, to abate pollutant loads.
- *Evaluate options and select final management strategies.* Develop strategies to screen and research candidate management options, evaluate possible scenarios, and select the final management measures to be included in the watershed-management plan.
- *Design implementation program and assemble watershed plan.* Establish milestones and implementation schedules and identify the technical and financial resources needed to implement the plan, including information/education activities and monitoring, and evaluation components.
- *Implement watershed plan and measure progress.* Use adaptive management techniques to make changes to the watershed plan, including analysis of data to determine whether milestones are being met.

9.4 CONCLUSIONS AND FUTURE RESEARCH AND EDUCATION NEEDS

Sustainable economic agricultural productivity requires systematic implementation of watershed-management plans. These plans must address protection of soil and water quality. Current practices used in rapidly developing countries, such as India and China, are inadequate to protect soil and water quality. Research and collecting more data alone will not address current watershed-management issues, unless this research is transferred to the user and implemented effectively. Current educational programs in India and other developing countries are not adequate for technology transfer to the user. Changes in practices by the user are needed to implement new solutions to improve quantity and quality of water resources. Technology transfer with effective communication and readily available tools at all levels is critical in addressing issues related to efficient watershed management. Non-traditional approaches, such as e-technologies, may be more effective in reaching broader group of users than traditional ones.

Some examples of future research and education needs for watershed management are given below:

- Comprehensive nutrient budgets and nutrient-management plans for site-specific land use (crop, animal operations) within the watershed. Alternate nutrient/contaminant control technologies to assist farmers in meeting future environmental demands. Practical, inexpensive indicators that can be used to determine watershed condition/health. Process-oriented research is essential because it leads to enhanced predictive capabilities, better understanding of cause-effect relationships, and a stronger foundation for planning and management.
- Quantitative methods of data integration (models) can be used to evaluate watershed-management practices (models that link all components and attributes of watershed and models that can be operated by managers).
- Methods to effectively assess risk and uncertainty analysis in making management decisions.
- Watersheds are connected on the landscape and should be managed together [uplands (agricultural, forest, and range lands), wetlands, streams and rivers, floodplains, groundwater, and estuaries].
- Watershed-management plans should be viewed as the starting point, but not the end point of management cycle – adaptive management.
- Integration across disciplines (physical, chemical, biological, social, and economic).
- Nationwide commitment to aggressive educational programs to transfer technology to the user.
- Science and policy must function together for watershed management to be successful.

REFERENCES

Bortollo, P. 1998. Assessing ecosystem health in governed landscapes: A framework for developing core indicators. *Ecosystem Health* 4(1): 33–41.

Carey, A.E. 1991. Agriculture, agricultural chemicals, and water quality. In *Agriculture and the environment: The 1991 yearbook of agriculture,* pp. 78–85. Washington, DC: U.S. Govt. Printing Office.

Comprehensive Assessment of Water Management in Agriculture. 2007. *Water for food, water for life: A comprehensive assessment of water management in agriculture.* London: Earthscan and Colombo, International Water Management Institute.

Crosson, P.R., P. Dyke, J. Miranowski, and D. Walker. 1985. A framework for analyzing the productivity costs of soil erosion in the United States. In *Soil erosion and productivity.* eds. R.F. Follett, B.A. Stewart, pp. 481–503. Madison, WI: American Society of Agronomy.

Doran, J.W. and T.B. Parkin. 1994. Defining and assessing soil quality. In *Defining soil quality for a sustainable environment, SSSA Spec. Publ. 35.* eds. J.W. Doran et al., pp. 3–21. Madison, WI: SSSA.

Falkenmark, M. and J. Rockstorm. 2006. The new blue and green water paradigm: Breaking new ground for water resources planning and management. *J. Water Res. Plann. Manage.* 139: 129–132.

Jackson, L.E., J.C. Kurtz, and W.S. Fisher, eds. 2000. *Evaluation guidelines for ecological indicators,* EPA/620/R-99/005. Research Triangle Park, NC: EPA Office of Research and Development.

Johnson, M.G., D.A. Lammers, C.P. Andersen, P.T. Rygiewcz, and J.S. Kern. 1992. Sustaining soil quality by protecting the soil resource. In *Proc. soil quality standards symposium,* San Antonio, Texas, October 1990. Watershed and Air Management Report No. WO-WSA-2. Washington, DC: USDA, U.S. Forest Service.

Karlen, D.L., M.J. Mausbach, J.W. Doran, R.G. Cline, R.F. Harris, and G.E. Schuman. 1997. Soil quality: A concept, definition, and framework for evaluation. *Soil Sci. Soc. Am. J.* 61: 4–10.

Liu, J., A.J.B. Zehnder, and H. Yang. 2009. Global consumptive water use for crop production: The importance of green water and virtual water. *Am. Geophys. Union Water Resources Res.* 45:W5428.

National Research Council (NRC). 1993. *Soil and water quality: An agenda for agriculture.* Washington, DC: National Academy Press.

National Research Council (NRC). 2004. *Confronting the nation's water problems: The role of research.* Washington, DC: National Academy Press.

Sojka, R.E. and D.R. Upchurch. 1999. Reservations regarding the soil quality concept. *Soil. Sci. Soc. Am. J.* 63: 1039–1054.

Suter, G.W. 1993. A critique of ecosystem health concepts and indices. *Environ. Toxicol. Chem.* 12(9): 1533–1539.

UNESCO. 2006. *Water: A shared responsibility.* United Nations world water development report.

U.S. Environmental Protection Agency. 1992. *Framework for ecological risk assessment, EPA 630-R-92-001.* Washington, DC: EPA, Risk Assessment Forum.

U.S. Environmental Protection Agency. 1994. *National water quality inventory, 1992 report to Congress, 841-R-94-001.* Washington, DC: EPA.

U.S. Environmental Protection Agency. 1996. *Drinking water regulations and health advisories,* EPA 822-B-96-002. Washington, DC Office of Water, EPA.

U.S. Environmental Protection Agency. 1997a. *Pesticides industry sales and usage. 1994 and 1995 market estimates, 733-R-97-002.* Washington, DC: EPA/Office of Prevention, Pesticides and Toxic Substances.

U.S. Environmental Protection Agency. 1997b. *Environmental monitoring and assessment program (EMAP) research strategy.* Washington, DC: EPA Office of Research and Development.

U.S. Environmental Protection Agency. 2000. *Atlas of America's polluted waters, EPA 840-B-00-002.* Washington, DC: Office of Water (4503F), EPA.

U.S. Environmental Protection Agency. 2008. *Handbook for developing watershed plans to restore and protect our water bodies, EPA 841-B-08-002.* Washington, DC: EPA, Office of Water Nonpoint Source Control Branch.

Yeston, J., R. Coontz, J. Smith, and C. Ash, eds. 2006. Freshwater resources. Special issue, *Science* 313: 1005–1184.

Chapter 10

Water pollution related to agricultural, industrial, and urban activities, and its effects on the food chain: Case studies from Punjab

Milkha S. Aulakh, Mohinder Paul S. Khurana & Dhanwinder Singh
Department of Soils, Punjab Agricultural University, Ludhiana, Punjab, India

ABSTRACT

In Punjab, a northwestern state of India, groundwater is depleting at a fast rate because of its excessive use and mismanagement. Contamination of groundwater and water bodies from geogenic and anthropogenic sources is also becoming a serious problem. Selenium (Se) toxicity is prevalent in 1,000 ha in Hoshiarpur and Nawanshahar districts where about 11 and 4% of groundwater samples were found unfit for drinking and irrigation purposes, respectively. About 9 and 66% groundwater samples of the Ludhiana and Bathinda districts, respectively, had fluoride concentration more than the safe limit of 1 mg l^{-1}. In the Bathinda district, 15–44% of groundwater samples had boron more than the maximum permissible limit of 2 mg l^{-1}. Arsenic concentration in alluvial aquifers of Punjab ranged from 4 to 688 µg l^{-1}, and the majority of them were found unfit for human consumption considering maximum permissible limit of 10 µg l^{-1}. Excessive applications of fertilizers, manures, and agrichemicals to field and vegetable crops lead to nitrate and phosphate leaching and contamination of groundwater and water bodies. In certain situations, nitrates exceed the dangerous level of 10 mg l^{-1}. Industrial effluents, released without any treatment to sewage drains, contain potentially toxic elements in concentrations that are several folds higher than those in domestic sewage water and exceed the maximum permissible limits for their disposal onto agricultural lands. The mean concentrations of Pb, Cr, Cd and Ni in sewage water were, respectively, 21, 133, 700, and 2200 times higher than those in tubewell water. In one study, water of several shallow hand-pumps installed in vicinity of a sewage-water drain had 18, 80, 88, and 210 times higher concentration of Pb,

Address correspondence to Milkha S. Aulakh, Department of Soils, Punjab Agricultural University, Ludhiana 141 004, Punjab, India. E-mail: msaulakh2004@yahoo.co.in

Reprinted from: *Journal of New Seeds*, Vol. 10, Issue 2 (April 2009) , pp. 112–137,
DOI: 10.1080/15228860902929620

Cr, Cd and Ni, respectively, than in deep tubewell water. A large number of pathogens were observed in tubewells installed for domestic water supply. Possible remediation options for such deplorable situations are discussed in this article.

Keywords: toxic metals, geogenic contaminants, anthropogenic contamination, sewage effluents, selenium, arsenic, cadmium, lead, chromium, nickel, nitrate, phosphate

10.1 INTRODUCTION

One of the most important challenges facing humanity today is to conserve and sustain natural resources, including water, for increasing food production while protecting the environment. As the world population grows, stress on natural resources increases, making it difficult to maintain food security. In recent years, efforts have increasingly been focused on environmental pollution and its ill effects on humans and animals. Public concern over the effects of environmental pollution continues to increase because of the industrial revolution and an enhanced understanding of the risk to human and animal health.

Water is one of the important and precious natural resources. In 1995, when the world population was ~5.7 billion, 92% people had sufficient water supply; whereas 5% had a supply considered to be "under stress," and 3% had scarce supplies. Projections are that in 2025, when the world population is expected to reach 9.4 billion, water supplies would be sufficient for only 58%, under stress for 24%, and scarce for 18% of Indian population. Of all the water available on the earth, 97.5% is present in oceans and 2.5% is fresh water, of which only 20% is groundwater. The latter is the main source for drinking, irrigation, and industrial purposes. Among these, the agricultural sector is the major consumer of water. In India, agriculture accounts for ~89% of total water use, versus 8% by domestic sector and 3% by industrial sector. Predictions are that by 2025, agriculture's share of the water will be reduced to 73%. Rapid industrialization and urbanization during the past few decades have increased the demand for available water and put stress on the already dwindling water resources.

Punjab, a northwestern state of India, has a total area of 5.03 million ha and is situated between 29°33′–32°31′ N latitude and 73°53′–76°55′ E longitude. About 84% of the state's geographical area is cultivated, which is a record in the country. The area under irrigation is 4.04 m ha, constituting about 95% of the net area sown. About 72% of the area is irrigated by tubewells, 23% is canal irrigated, and the remaining 5% is rainfed. The size of operational holdings is distributed as follows: 19% landholders have less than 1 ha, 46% have 1 to 4 ha, 28% have 4 to 10 ha, and 7% have more than 10 ha. While Punjab state occupies only 1.53% of the total geographical area of the country, it produces about 20% of the wheat, 11% of the rice, and 13% of the cotton for all of India. Punjab contributes about 65% of the wheat and 42% of the rice to India's foodgrain pool and is thus regarded as the "food basket of the country." The production of rice has increased from 1.2 t ha^{-1} in 1960–1961 to 3.9 t ha^{-1} in 2006–2007, and that of wheat from 1.2 t ha^{-1} to 4.2 t ha^{-1} in the corresponding period. The problem of the falling groundwater table in central Punjab, where rice is a predominant crop, is because of the overdraft of water. The rate of groundwater usage is more than water recharge; of a total of 137 blocks, 103 blocks are over-exploited for water use. Recently, Kang, Aulakh, and Dhiman (2008) described a holistic strategy required to overcome the water crisis, which includes crop diversification, delayed transplanting of rice, adoption of water-saving agronomic practices, etc. Agriculture in the state would have to be carefully reoriented to ensure sustainable development with the least disturbance to the ecosystem, including water resources.

Groundwater quality depends upon both geogenic factors and anthropogenic factors. There are naturally occurring minerals in aquitards in different regions of Punjab which control the concentration of geogenic pollutants, such as selenium (Se), fluoride (F), boron (B) and arsenic (As) in alluvial aquifers. Agricultural activities that influence water quality include the application of fertilizers and chemicals. The application of nutrients in excess of crop requirements can increase both the cost of production and the risk of adverse environmental effects. Nitrate leaching into groundwater, P movement into surface water and groundwater, soil acidification, and heavy-metal accumulation in soil can be associated with inefficient or excessive application of fertilizers and manures. Similarly, overexploitation of water and the excessive use of agrochemicals reduce the quality of groundwater.

Urban and industrial sectors not only utilize the available water but also discharge a considerable amount of wastewater. In Punjab and the rest of India, the most important anthropogenic factor responsible for groundwater pollution is urban and industrial wastewater. This wastewater is often not treated before its release into sewerage drains. The most common disposal of wastewater is its use as crop irrigation. Direct release of untreated effluents to land and water bodies can potentially contaminate air, surface, groundwater as well as soils, and eventually the crops grown on these soils would have a bearing on the quality of the food produced.

The main aims of this article were to ascertain the state of knowledge and to devise research, extension and education strategies relative to the following issues: (a) compile available information on water pollution caused by geogenic sources in Punjab; (b) synthesize information on the impact of agricultural, urban and industrial activities on water pollution, which leads to contamination of soil-plant-animal-human food chain; and (c) explore possible options for mitigating water pollution. An additional aim was to pinpoint the gaps in knowledge relative to controlling or minimizing water pollution.

10.2 GEOGENIC POLLUTANTS

As high water-quality standards for drinking and food processing are an important consideration, groundwater quality is also an important consideration for irrigation. It varies with rainfall pattern, and the depth and geology of aquifers. Salinity (electrical conductivity) and sodicity (residual sodium carbonate) hazards are important elements of quality of groundwater for irrigation purposes. While the groundwater of sub-mountainous Punjab is of very good quality, groundwater quality is poor in the southwestern region. High residual sodium carbonate, coupled with medium to high salinity, is dominant in this region. However, naturally occurring pollutants, such as Se, F, B, and As, in different regions of Punjab do affect water quality for human and animal consumption.

10.2.1 Selenium

In the Nawanshahar and Hoshiarpur districts, an area of about 1,000 ha is affected by Se toxicity. In this seleniferous region, the symptoms of Se are visible in plants, animals, and humans (Dhillon and Dhillon 1991). Snow-white chlorosis appears on young leaves and sheaths, some of which turn light pink or develop purple-white tips while mid-veins remain green. Typical symptoms of Se toxicity in animals are overgrowth and cracks, followed by gradual detachment of the hoof, shedding of horn corium, loss of hair, necrosis of tip of tail, loss of body condition, reluctance to move, and stiff gait. Selenium toxicity is known to cause hair loss and nail drop in humans.

Dhillon and Dhillon (2003) determined the Se quality of groundwater drawn from 90 tubewells located in the seleniferous region of Punjab, which ranged between 0.25 and 69.5 $\mu g\ l^{-1}$ with a mean of 4.7 $\mu g\ l^{-1}$. The maximum permissible limit of 10 $\mu g\ l^{-1}$ for drinking purposes was exceeded by 11% of the tubewell water samples, and the maximum permissible level (MPL) of 20 $\mu g\ l^{-1}$ for irrigation purposes was exceeded by 4.4% of water samples. In water samples with >1 $\mu g\ l^{-1}$, a significant positive relationship ($r = 0.92$) between Se and pH was observed. Groundwater pumped from relatively shallow tubewells (24–36 m depth) contained two to three times more Se than that pumped from deep tubewells. Selenium build up in the soil (Se added through irrigation water minus Se removed by crops) was observed in the case of rice and sunflower (Table 10.1; Dhillon and Dhillon 2003). In contrast, mustard, Egyptian clover, pearl millet, and wheat crops removed more Se from the soil than that added through irrigation water. In rice-based cropping sequences except rice-egyptian clover Se balances were positive, suggesting that the cultivation of rice could be discouraged to reduce Se accumulation in soil. Even irrigation with water containing Se at MPL could result in accumulation of Se in the soil under rice, sunflower, sugarcane, maize, or oat.

10.2.2 Fluoride

High concentrations of fluoride, often significantly above the safe limit of 1 mg F l^{-1}, constitute a severe problem in some semi-arid areas of Punjab. The use of groundwater for drinking in these areas has resulted in the onset of widespread fluorosis

Table 10.1 Selenium added by groundwater irrigation, removed by different crop rotations, and accumulated in soil (g ha^{-1} $Year^{-1}$).

Crop rotation	*Se added by irrigation water*	*Se removed by crop*	*Balance of Se in soil*
Rice-wheat	498	379	119
Rice-Egyptian clover	627	657	–30
Maize-wheat	184	309	–125
Sunflower-rice	590	392	198
Sugarcane	221	226	–5

Note: Adapted from Dhillon and Dhillon (2003).

symptoms, from mild forms of dental fluorosis to crippling skeleton fluorosis. Singh, Rana, and Bajwa (1977) surveyed groundwater samples of five blocks of Ludhiana district and found that the F concentration varied from 0 to 10.09 mg l^{-1}. On the basis of maximum permissible limit, 3, 5, 9, and 38% of the total samples collected were high in F concentration in Sidhwan bet, Ludhiana-I, Pakhowal, and Jagraon blocks, respectively (Table 10.2; Singh et al. 1977). While all the water samples collected from Sudhar block of Ludhiana were below the maximal permissible limit, F concentration in irrigation water samples of 100 villages in five development blocks of Bathinda varied from nil to 9 mg l^{-1} (Table 10.2). About 9% samples from Ludhiana and 66% samples from Bathinda had F concentration of more than 1 mg l^{-1}.

10.2.3 Boron

Singh et al. (1977) reported that the B concentration in irrigation water samples of 100 villages in five development blocks of Bathinda district ranged from negligible to 5.75 mg l^{-1} (Table 10.3; Singh et al. 1977). There was a highly significant positive correlation of B concentration with soluble salts, as determined by electrical conductivity ($r = 0.74$; $p = 0.01$). Considering the maximal permissible limit of 2 mg B l^{-1}, 37% samples in Rampura, 15% in Nathana, 29% in Mansa, 44% in Budhlada, and 22% in Phul block were high in B.

10.2.4 Arsenic

Arsenic is a deadly poison, especially when present in high concentrations. People are exposed to As most of the time through drinking groundwater, fortunately at much

Table 10.2 Concentration of fluoride in groundwater of different blocks of Ludhiana and Bathinda districts.

Name of block	*No. of water samples*	*No. of villages*	*Range (mg F l^{-1})*
District Ludhiana			
Sudhar	288	39	0.00–0.88 (0%)[†]
Sidhwan Bet	488	88	0.00–2.25 (3%)
Ludhiana-I	391	73	0.00–10.09 (5%)
Pakhowal	484	61	0.12–1.95 (9%)
Jagraon	224	53	0.31–4.97 (38%)
Overall	1888	314	0.00–10.09 (9%)
District Bathinda			
Rampura	181	20	0.15–4.00 (84%)
Nathana	375	23	0.00–6.50 (71%)
Mansa	191	37	0.15–5.00 (45%)
Budhlada	73	16	0.50–9.00 (74%)
Phul	14	4	0.50–3.60 (44%)
Overall	834	100	0.00–9.00 (66%)

Note: Adapted from Singh, Rana, and Bajwa (1977).
[†]Values in parentheses indicate % of total samples having F greater than permissible limit of 1 mg l^{-1}.

Table 10.3 Concentration of boron in groundwater of different blocks of Bathinda district.

Name of block	*No. of water samples*	*No. of villages*	*Range (mg l^{-1})*
Rampura	181	20	0.00–4.95 (37%)[†]
Nathana	375	23	0.00–3.40 (15%)
Mansa	191	37	0.00–5.75 (29%)
Budhlada	73	16	0.25–5.75 (44%)
Phul	14	4	0.65–2.40 (22%)
Overall	834	100	0.00–5.75 (27%)

Note: Adapted from Singh, Rana, and Bajwa (1977).
[†]Values in parentheses indicate % of total samples having B greater than the permissible limit of 2 mg l^{-1}.

Table 10.4 Concentration of arsenic in groundwater of different zones of Punjab.

Zone	*No. of water samples*	*Range (μg As l^{-1})*	*% samples > 10 μg As l^{-1}*
Zone I	20	4–42	97
Zone II	58	10–43	99
Zone III	90	11–688	100

Note: Adapted from Hundal et al. (2007).

lower concentrations than the deadly levels, and usually unknowingly. Under natural conditions, oxi-hydroxides of iron control the concentrations of metals in aquifers or surface waters. Under oxidized conditions, iron precipitates as hematite and gets deposited on the surface of particulate suspensions or the surface of the reservoirs. Concurrent depositions of metal ions on the rust particles of hematite minimize their toxicity in water (Levy et al. 1999).

Arsenic concentration in alluvial aquifers of Punjab varied from 4 to 688 μg l^{-1} (Table 10.4; Hundal et al. 2007). In groundwater of Zone I, comprising Gurdaspur, Hoshiarpur, Nawanshahr, and Ropar districts, As varied from 4 to 42 μg l^{-1}. The concentration of As in groundwater of Zone II, comprising Amritsar, Tarn Taran, Jalandhar, Kapurthala, Ludhiana, Patiala, Mohali, Barnala, and Moga districts, varied from 10 to 43 μg l^{-1}. In the arid southwestern zone (Zone III), comprising Sangrur, Mansa, Faridkot, Muktsar, Bathinda, and Ferozepur, As concentrations varied from 11 to 688 μg l^{-1}. According to the safe limit of 10 μg As l^{-1}, only 3% groundwater samples from Zone I and 1% from Zone II were fit for dinking purposes with respect to As concentration. In Zone III, all water samples had As concentrations of greater than the safe limits and thus were not suitable for drinking purposes. The presence of elevated As concentrations in groundwater is generally due to the natural occurrences of As in the aquifer materials. Geochemical conditions, such as pH, oxidation-reduction, associated or competing ions, and evaporative environments, have significant effects on As concentration in groundwater (Hundal et al. 2007).

A recent investigation (Hundal et al. 2008) of deep-water tubewells located in Amritsar city, which are used for domestic supply, revealed that As ranged from 3.80 to 19.1 μg l^{-1}, with a mean of 9.8 μg l^{-1}, which implied that 54% water samples were unfit for human consumption. In contrast, As concentration in canal water varied from 0.30 to 8.80 $\mu g\ l^{-1}$, with a mean of 2.89 $\mu g\ l^{-1}$, presumably because of the higher oxidation potential of canal water than deep-tubewell water. These studies suggest that regular monitoring of As concentration in deep-tubewell waters by water-testing laboratories should be done. Consumption of water with As concentrations above the safe limit must be discouraged. In the southwestern districts of Punjab, use of canal water for drinking purposes and domestic use by rural and urban populations over groundwater is recommended.

10.3 AGRICULTURAL ACTIVITIES

The increased use of fertilizers and pesticides in farming because of large-scale adoption of high-yielding, fertilizer-responsive crops and varieties in Punjab has led to a gradual build up of nutrients and other chemicals in soils and groundwater. Movement of nitrogen (N) and phosphorus (P) below the root zone and leaching into the groundwater can cause human and animal health problems. If the drinking water has more than the safe limit of 10 mg NO_3^--N l^{-1}, ingested nitrate is converted to nitrite that is absorbed in the blood, causing methemoglobinemia, commonly known as "blue baby syndrome," and gastric cancer (Alexander 1986). There are reports of increased levels of NO_3^- in groundwater and eutrophication of water bodies due to both high nitrate and phosphate concentration. The concentrations of P that cause eutrophication range from 0.01 to 0.03 mg l^{-1} (Sharpley et al. 1996). Reports on increasing occurrence and severity of surface blooms of cyanobacteria and algae are reported from different parts of the world (Lennox et al. 1997; Carpenter et al. 1998). Many drinking-water supplies throughout the world experience periodic massive algal blooms, which contribute to a wide range of water-related problems including fish kills, unpalatability of drinking water, and formation of trihalomethane during water chlorination. Consumption of algal blooms or water-soluble neurotoxins and hepatoxins, which are released upon senescence of algae, can kill livestock and pose a serious health hazard to humans.

10.3.1 Nitrogen

Commoner (1968) warned the world that intensive application of fertilizer N could lead to increased eutrophication and a potential hazard from nitrate poisoning. When fertilizer N is applied in excess of required N rate, large amounts of NO_3^- accumulate in the soil profile, which is susceptible to leaching with rain and irrigation water. High rates of leaching and nitrification in permeable or porous soils and relatively high fertilizer N rates combine to make nitrate-leaching a serious problem in many irrigated soils (Aulakh and Malhi 2005). In intensively cultivated semi-arid subtropical region of Punjab, where average fertilizer N consumption increased from 56 to 188 kg N ha^{-1} $year^{-1}$ during 1975–1988, NO_3^--N concentration in the shallow-well waters increased by almost 2 mg l^{-1} (Aulakh and Bijay-Singh 1997). In some central districts of Punjab, fertilizer N levels exceed 300 kg N ha^{-1} $year^{-1}$ and on several farms, fertilizers are poorly managed (Aulakh and Pasricha 1997). The soils in this region are predominantly coarse-textured and about 75% of the total rainfall of more than 600 mm is received during the monsoon period (July–September). A survey of groundwater samples from 21–38 meter-deep tubewells located in cultivated fields in various blocks of Punjab revealed that 78% water samples had less than 5 mg NO_3^--N l^{-1} and 22% samples had 5–10 mg NO_3^--N l^{-1} (Bajwa, Bijay-Singh, and Parminder-Singh 1993). Sixty percent of water samples from shallow-depth (9–18 m) hand-pumps had 5–10 mg NO_3^--N l^{-1} and 2% samples had more than 10 mg NO_3^--N l^{-1}. The amount of NO_3^--N contained in the soil profile to a depth of 210 cm in June correlated significantly with the NO_3^--N concentration in well-water in September, confirming that nitrates tend to reach the groundwater during the rainy season (Bijay-Singh and Sekhon 1976a).

A survey of groundwater of several blocks of Ludhiana and Ferozepur districts revealed that 3 to 26% of the samples in Ludhiana and 4 to 22% of the samples in Ferozepur had more than 10 mg NO_3^--N l^{-1} (Table 10.5). Animal wastes appear to be the major contributors to high NO_3^--N in groundwater under village inhabitations and feedlots. In Punjab, animal wastes are generally dumped near feedlots in the outskirts of villages. The level of NO_3^--N in the water samples of 367 hand-pumps used in several villages of four districts and in 45 water samples collected beneath feedlots,

Table 10.5 Concentration of nitrate-N in groundwater of different blocks of Ludhiana and Ferozepur districts.

Name of block	*No. of water samples*	*E.C. (dS m^{-1})*	*Nitrate-N range (mg l^{-1})*
Ludhiana district			
Sudhar	288	0.34–1.80	0.0–23.0 (7%)†
Ludhiana-I	391	0.38–2.34	0.0–23.7 (18%)
Ludhiana-II	558	0.37–2.04	0.0–25.5 (26%)
Macchiwara	359	0.25–1.31	0.0–10.3 (3%)
Dehlon	327	0.40–1.98	0.0–16.8 (9%)
Samrala	637	0.42–2.14	0.2–24.3 (14%)
Doraha	141	0.31–1.86	0.0–10.7 (11%)
Overall	2701	0.25–2.34	0.0–25.5 (14%)
Ferozepur district			
Mamdot	50	0.46–2.79	0–3.8 (0%)
Guru Harsahai	50	0.41–1.83	0–7.8 (0%)
Jalalabad	50	0.75–6.97	0–13.9 (4%)
Abohar	50	0.40–7.58	0–15.0 (16%)
Khuian Sarwar	50	0.78–9.68	0–15.1 (6%)
Fazilka	50	0.68–6.42	0–15.5 (10%)
Ghall Khurd	50	0.52–3.61	0–4.5 (0.6%)
Ferozepur	50	0.45–3.95	0–20.0 (22%)
Zira	50	0.51–3.23	0–14.1 (4%)
Makhu	50	0.29–1.19	Absent (0%)
Overall	500	0.40–9.68	0–20.0 (6.9%)

Note: Adapted from Singh and Bishnoi (2001).
†Values in parentheses indicate % of total samples having NO_3^--N greater than the safe limit of 10 mg l^{-1}.

was several-fold higher than in 236 water samples of tubewells of adjoining areas, clearly illustrating that animal wastes and feedlots act as a point source of nitrates (Table 10.6). Another study of a 600 ha farm at the Punjab Agricultural University, Ludhiana, revealed that NO_3^--N content varied from 1.3 to 11 mg l^{-1} in deep irrigation tubewells, 0.4 to 11 mg l^{-1} in shallow irrigation tubewells and 0.6 to 28 mg l^{-1} in hand-pumps (Table 10.7). The wide variations in NO_3^--N concentration in groundwater is attributed to variations in land use and management practices and unscientific disposal of animal feces and urine around dairy sheds. Water from hand-pumps, even those with toxic concentrations of NO_3^--N, is being used for drinking purposes. Similar alarming situations in many villages are attributed to unscientific management of animal dung, urine, poultry manure, and night soil.

Vegetation retards NO_3^--N leaching from the root zone by absorbing nitrate and water. Rooting habits/patterns of different plants exert profound influence on NO_3^- mobility in the rooting zone. Maximum leaching of NO_3^--N below the root zone occurs from heavily fertilized shallow-rooted crops, such as potato (*Solanum tuberosum* L.), maize (*Zea mays* L.) and rice (*Oryza sativa* L.), as well as heavily manured vegetable crops. In the predominant rice-wheat cropping system of Punjab, NO_3^- leaching to 60 cm during the rice crop was used by the subsequent wheat crop, which has a deeper and more extensive root system (Figure 10.1). Application of 120 kg fertilizer N ha^{-1} to each crop for four years resulted in 35 kg of residual NO_3^--N ha^{-1} in the 150-cm soil profile; whereas only 17 kg NO_3^--N ha^{-1} remained when 120 kg N ha^{-1} was applied through the consecutive use of 20 t ha^{-1} of fresh sesbania green manure and fertilizer N. This decreased potential for groundwater nitrate contamination. Similarly, integrated and balanced application of N, P, and potassium (K) could significantly reduce the amount of unutilized nitrates in the root zone

Table 10.6 Concentration of nitrate-N in groundwater of tubewells and hand pumps in four districts of Punjab.

	No. of observations	*Nitrate-N (mg N l^{-1})*	
		Mean	*Range*
Tubewells	236	3.62	1.0–6.7
Hand pumps	367	5.72	1.0–11.3
Feedlots	45	4.73	1.2–10.4

Note: Adapted from Bajwa, Bijay-Singh, and Parminder-Singh (1993).

Table 10.7 Concentration of nitrate-N in groundwater of Punjab Agricultural University farm (600 ha area).

Source of groundwater	*No. of samples*	*NO_3^-- N(mg l^{-1})*	*% Samples (>10 mg NO_3^-- N l^{-1})*
Hand pump (10–15 m)	11	0.6 – 28	50
Shallow tubewell (40–60 m)	11	0.4 – 11	18
Deep tubewell (100 m)	20	1.3 – 11	5

Note: Adapted from Thind and Kansal (2002).

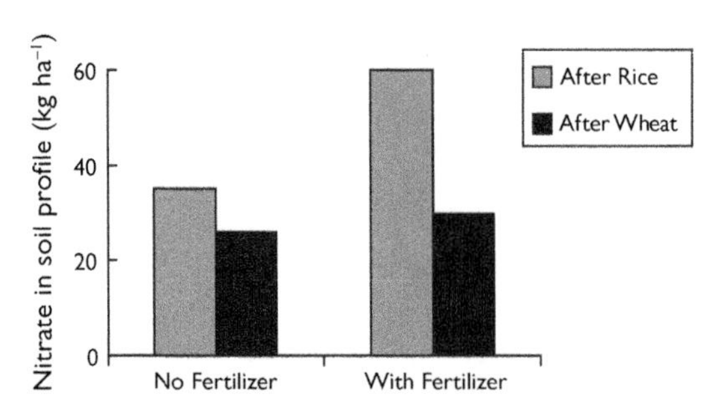

Figure 10.1 Nitrate in 60 cm soil profile after two years of rice and wheat crops. (Drawn with data from Aulakh et al. 2000).

and help enhance nutrient-use efficiency of crops (Bijay-Singh and Sekhon 1976c). Decreased rate/amount, but increased frequency of irrigation water, in conjunction with a split application of fertilizer N, helps in minimizing the movement of NO_3^- N to deeper soil layers (Bijay-Singh and Sekhon 1976b).

10.3.2 Phosphorus

Excessive accumulation of residual P in soil may enhance downward movement of P, which may eventually reach groundwater. Sandy soils have a large number of macropores and thus the resultant bypass flow can lead to greater and deeper leaching of P in such soils. Besides the potential for groundwater contamination, P lost from agricultural soils through leaching may be intercepted by artificial drainage or subsurface flow, accelerating the risk of P transport to water bodies with serious implications for water quality.

Long-term studies, where fertilizer P has been applied at different rates, frequencies and periods, have revealed possibility of P leaching especially in coarse-textured soils (Aulakh, Garg, and Kabba 2007). After 29 years of using groundnut – based cropping systems, 45 to 256 kg of residual fertilizer P accumulated as Olsen-P ha^{-1} in 150-cm soil profile (43–58% below 60 cm depth), illustrating enormous movement of fertilizer P to deeper layers in a coarse-textured soil having low absorption and retention capacity for nutrients (Figure 10.2). Interplay between the fertilizer P management (applied to alternative or all crops, and optimal vis-à-vis excessive rates of fertilizer P in different crop rotations), amount of labile P accumulated in soil profile, and soil characteristics (silt, clay and organic C) largely control the downward movement and resultant potential for P leaching in subtropical soils (Garg and Aulakh 2010).

10.3.3 Agricultural chemicals

The application of pesticides, fungicides, weedicides, and other chemicals in agricultural fields can reach the food chain directly through drinking water, which originates from groundwater and water bodies. There are numerous studies illustrating residues of agricultural chemicals in food grains, vegetables, fruits, and even animal feed, as well as bovine milk. A study across several states of India showed residues of

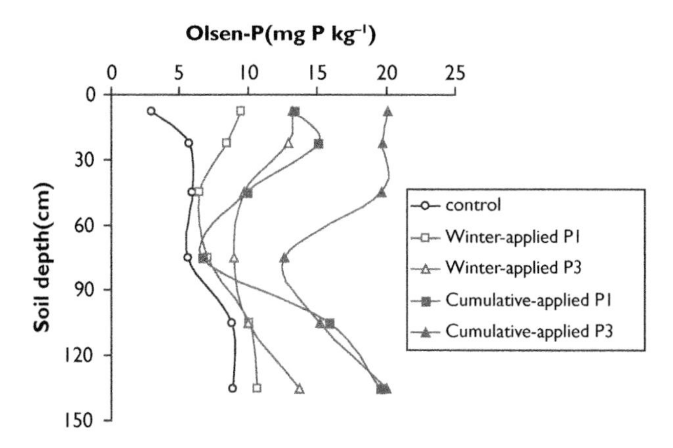

Figure 10.2 Olsen-P accumulation and leaching in soil profiles of no-P control and four fertilizer P treatments after 29 years of groundnut-based cropping systems. (Adapted from Aulakh et al. 2007).

Table 10.8 Residues of hexachlorocyclohexane (HCH) in bovine milk.

Location	*No. of samples analyzed*	*No. of samples contaminated*	*Range of HCH residue (mg l^{-1})*
Hyderabad	127	37	Traces–5.0
Hisar	155	53	Traces–7.0
Himachal Pradesh	100	100	Traces–2.1
Punjab	236	236	0.01–0.6

Note: Adapted from Dhaliwal and Singh (2000).

hexachlorocyclohexane (HCH) ranging from traces to 7 mg l^{-1} (Table 10.8). All of the 236 milk samples collected from Punjab were HCH contaminated, illustrating a grave situation for animal and human health.

10.4 URBAN AND INDUSTRIAL ACTIVITIES

The application of sewage sludge to agricultural soils, and irrigation of field crops with sewage water and untreated industrial effluents alone, or in combination with ground/canal water, are common practices worldwide and are used in Punjab, especially in the vicinity of large cities, as these are considered reusable sources of essential plant nutrients and organic C. It is estimated that more than 15,000 million liters of sewage water is produced every day in India, which approximately contributes 3.2 million t of N, 1.4 million t of P and 1.9 million t of potassium (K) per annum, with an economic value of about Indian Rs. 2,600 million (US$ 520 million). However, some of the elements present in sewage water and untreated industrial effluents, reaching through the soil-plant-animal-human food chain, could be toxic to plants and pose health hazards to animals and humans.

Considering both useful effects (nutritive and irrigation potential) and harmful effects, sewage sludge and water should meet some quality parameters before they are discharged or applied to the agricultural lands for irrigation (Table 10.9).

10.4.1 Chemical composition of sewage waters

Large variations in the composition of sewage waters of industrial and non-industrial cities of Punjab are evident from the data reported in two studies (Table 10.10). The concentration of potentially toxic elements was higher in sewage water of industrial towns of Ludhiana, Jalandhar, and Amritsar, as compared with less or non-industrial towns of Sangrur and Abohar (Singh and Kansal 1985). Further, the composition of sewage water varies within a city depending upon the command area from where the sewage is being collected. The domestic zone sewage contained relatively low amounts of toxic elements whereas the effluents from the electroplating area contained toxic

Table 10.9 Tolerance limits for industrial effluents for their release into public sewer and use for irrigation.

Parameter	*Public sewers*	*Irrigation*
Suspended solids (mg l^{-1})	600	200
Oil and grease (mg l^{-1})	20	10
BOD (mg l^{-1})	350	100
pH	5.5–9	5.5–9
Lead (mg l^{-1})	1	n.a.
Cadmium (mg l^{-1})	1	n.a.
Zinc (mg l^{-1})	15	n.a.
Nickel (mg l^{-1})	3	n.a.
Copper (mg l^{-1})	3	n.a.

Note: Adapted from Environment Protection Act (1986).
n.a. = Not available.

Table 10.10 Concentration (mg l^{-1}) of potentially toxic elements, biological oxygen demand (BOD) and chemical oxygen demand (COD) in sewage water of non-Industrialized and industrialized cities of Punjab.

	Fe	*Zn*	*Cu*	*Pb*	*Cd*	*As*	*Cr*	*BOD*	*COD*	*Source*
Non-industrialized cities										
Sangrur	0.15	0.24	0.26	0.03	–	0.23	0.004	44	175	**
Abohar	1.80	0.08	0.07	0.04	0.010	–	–	–	–	*
Industrialized cities										
Ludhiana	6.2	0.68	0.94	0.05	–	1.38	0.189	130	245	**
Jalandhar	3.5	0.33	0.20	0.05	0.011	–	–	–	–	*
Amritsar	4.5	0.66	0.94	0.08	0.010	–	–	–	–	*

*Adapted from Singh and Kansal (1985).
**Adapted from Khurana et al. (2003).

elements, such as chromium (Cr), Nickel (Ni) and cyanide (CN), in amounts higher than maximal tolerable limits for disposal on agricultural lands (Table 10.11).

Azad et al. (1984) studied the nature of potentially toxic elements and their concentration in the effluents emanating from two groups of industries: the first was involved in the manufacturing of metallic products, such as cycles, spare parts, and electroplating; and the second group comprised the processing of textiles, woolens, and dyes. Analysis of these effluents revealed that, in general, lead (Pb), cadmium (Cd) and Ni were in higher concentration in effluents of industries manufacturing metallic products as compared with textile and woolen industries. These concentrations were, respectively, 2.5, 1.44, and 7.5 times higher in the first type of industries than those in the effluents from the second type of industries.

The chemical analysis of sewage-water samples collected from different locations of an open drain, commonly known as "Budda Nullah," downstream from entry into Ludhiana city, revealed that the concentration of metals in the drain increases many folds as it passes through Ludhiana city (Table 10.12). The concentration of all the toxic elements in sewage water showed increasing trend as the distance downstream increased from the entry point. The mean concentration, in the sewage-water samples collected at the entry point, of iron (Fe) was 0.03 mg l^{-1}, zinc (Zn) 0.04 mg l^{-1}, As 0.005 mg l^{-1}, Pb 0.004 mg l^{-1}, Ni 0.002 mg l^{-1}, and Cr 0.001 mg l^{-1}. These concentrations, increased, respectively, to 10.8, 0.78, 2.10, 0.075, 0.28, and 0.26 mg l^{-1} in the samples collected from about 15 km downstream of the entry point. This is because the number of industries pouring their untreated effluents increased as the distance downstream increased. This implies that the open Budda Nullah drain, which is a natural freshwater stream before its entry into Ludhiana city, turns into a highly polluted sewage channel when it passes through the interior of the city, receiving effluents from various types of industries on its way.

Table 10.11 Concentration of toxic elements (mg l^{-1}) in wastewater of Ludhiana, densely industrialized city of Punjab.

Location	*Cr hexavalent*	*Ni*	*CN*
Electroplating	0.2–2.5	1.0–3.0	0.42–0.97
Domestic	0.1–0.2	0.2– 0.2	0.05–0.07
Maximum limit for disposal on agricultural land	0.1	0.005	0.20

Note: Adapted from Tiwana, Panesar, and Kansal (1987).

Table 10.12 Concentration of toxic elements (mg l^{-1}) and pH of the effluents of open Budda Nullah drains at different locations in Ludhiana city.

Sampling sites	*pH*	*As*	*Pb*	*Ni*	*Cr*	*Fe*	*Zn*
Entry point	8.2	0.05	0.004	0.002	0.001	0.30	0.04
2 km downstream	7.8	1.60	0.060	0.12	0.170	5.8	0.18
15 km downstream	7.2	2.10	0.075	0.28	0.260	10.8	0.78

Note: Adapted from Khurana et al. (2003).

Dheri, Brar, and Malhi (2007) further showed that the concentrations of Pb, Cr, Cd, and Ni were not only significantly higher in water samples of Budda Nullah drain but also in those collected from shallow hand-pumps located within 200 m from the drain, as compared with deep-tubewell water (Table 10.13). The mean concentration of Pb, Cr, Cd, and Ni in drain water was, respectively, 21, 133, and 700, 2,200 times higher than that in tubewell water. Similarly, their concentrations in shallow hand-pumps were 18 (Pb), 80 (Cr), 88 (Cd), and 210 (Ni) times higher than in deep-tubewell water. The concentration of 0.40 mg Cr l^{-1} and 0.44 mg Ni l^{-1} in sewage water was much higher than the maximum permissible limits of 0.1 mg Cr l^{-1} and 0.005 mg Ni l^{-1} for disposal of effluents on agricultural land.

A study of leather complex in Jalandhar city, comprising leather-manufacturing factories, revealed that the concentration of both Cr and Al drastically increased in the sewage water after the disposal of effluents from the leather complex (Table 10.14). The concentration of both the elements at 200 m downstream of the leather complex increased many folds as compared with that from 500 m upstream, indicating that effluents of the leather complex contained a large amount of Cr. However, the concentration of Cr decreased 2 km downstream of the leather complex because of settling of some of the elements at the base of the drain.

Thus, it is evident that the composition of sewage water depends upon the nature of effluents released by different types and density of industrial units. Moreover, open sewage-water drains pollute the shallow groundwater of hand-pumps installed in their vicinity.

Table 10.13 Concentration of toxic metals (mg l^{-1}) in water of Budda Nullah drain, hand-pumps, and tubewells located within 200 m of Nullah.

Source of irrigation	*Pb*	*Cr*	*Cd*	*Ni*
Budda Nullah water	0.35	0.40	0.14	0.44
Hand-pumped water	0.019	0.005	0.002	0.002
Tubewell water	0.017	0.003	0.0002	0.0002

Note: Adapted from Dheri et al. (2007).

Table 10.14 Concentration of potentially toxic elements (mg l^{-1}) in sewage drain near leather complex (LC) of Jalandhar city.

Sampling site	*Cr*	*Al*
500 m upstream of LC	2.71	2.35
200 m down stream of LC	31.6	12.8
2 km downstream of LC	20.4	3.40
Before operation of treatment plant	21.0	3.18
After operation of treatment plant	0.80	1.22
Maximum tolerance limit	0.11	Not available

Note: Adapted from Brar and Khurana (2006).

10.4.2 Effects of polluted water on soil

It has well been documented that irrigation with sewage water increases soil electrical conductivity and organic C, decreases soil pH, and could result in the accumulation of heavy metals in the plow layer of agricultural soils (Singh and Kansal 1985; Brar and Arora 1997; Kansal and Khurana 1999; Aulakh and Singh 2008). The mean concentrations of DTPA-extractable elements in surface soils (0–15 cm) surrounding the densely industrialized city of Ludhiana, irrigated largely with sewage effluents, were 4.2 mg kg^{-1} (Pb), 3.6 mg kg^{-1} (Ni), 0.30 mg kg^{-1} (Cd), 11.9 mg kg^{-1} (Zn), 25.4 mg kg^{-1} (Mn), and 49.2 mg kg^{-1} (Fe) as compared with 2.8, 0.40, 0.12, 2.1, 8.3, and 10.9 mg kg^{-1}, respectively, in the soils around a less industrialized city of Sangrur (Khurana et al. 2003). This was indicative of greater loading of soils of Ludhiana with potentially toxic metals through sewage irrigation. In the industrialized cities of Amritsar and Jalandhar, mean concentrations of these metals, except Pb and Zn in Amritsar, were in-between the values for Ludhiana and Sangrur (Table 10.15). Similarly, increased concentrations of potentially toxic metals in soils with the continuous application of sludge/sewage water have also been reported (Azad, Sekhon, and Arora 1986; Brar, Khurana, and Kansal 2002).

Dheri et al. (2007) showed significantly higher concentrations of both total and DTPA-extractable Pb, Cr, Cd, and Ni in sewage-water-irrigated soil as compared with tubewell-irrigated soil. The concentrations of DTPA-extractable Pb, Cr, Cd and Ni in sewage-irrigated soils were, respectively, 1.8, 35.5, 3.6, and 14.3 times higher than their concentrations in tubewell-irrigated soils. Enormous build-ups of Cr (35.5 times) and Ni (14.3 times) in sewage-irrigated soils could be phytotoxic to crops with continuous application of sewage effluents.

An investigation showed that the mean concentrations of available Cd, Pb and Ni in soils, where Budda Nullah water had been continuously used, were 0.82, 4.71, and 1.94 mg kg^{-1} soil that were, respectively, 530, 361, and 296% higher than those from the adjoining fields where tubewell water had been used for irrigation (Sikka 2003). Earlier, Sharma and Kansal (1986) observed three to five times higher available Cd in Budda Nullah-irrigated than tubewell-irrigated soils. It ranged from 0.119 to 0.253 mg kg^{-1} in Nullah-irrigated soils, as compared with 0.085 to

Table 10.15 Mean DTPA-extractable metals (mg kg^{-1} soil) in soils irrigated with sewage water (SW) and groundwater (GW) in various cities of Punjab.

City	*Source of irrigation*	*Pb*	*Ni*	*Zn*	*Mn*	*Fe*	*Cd*	*Cu*
Ludhiana	SW	4.21	3.58	11.9	25.4	49.2	0.296	–
	GW	1.09	0.78	5.6	5.6	12.8	0.047	–
Jalandhar	SW	3.57	0.47	3.65	7.99	12.86	0.14	5.13
	GW	1.37	0.2	1.25	4.92	6.12	0.08	1.01
Amritsar	SW	5.06	0.65	12.8	9.44	14.7	0.19	14.2
	GW	0.98	0.26	3.78	6.99	13.66	0.02	0.56
Sangrur	SW	2.76	0.4	2.1	8.34	10.88	0.12	1.88
	GW	1.32	0.2	1.45	5.37	6.27	0.02	0.94

Note: Adapted from Khurana et al. (2003).

0.115 mg Cd kg^{-1} in tubewell-irrigated soils. Kansal and Kumar (1994) noted that Ni concentration in the soils was as low as 0.12–0.26 mg kg^{-1} soil with tubewell irrigation, which increased to 0.96 to 1.65 mg kg^{-1} soil in Budda Nullah-irrigated soils around Ludhiana city.

10.4.3 Effects of polluted water on plants

Plant species absorbed higher concentration of potentially toxic metals like Pb, Cu, Co, Cd, Ni, Zn, Mn, and Fe in different plant parts when grown in sewage-irrigated soils, as compared with tubewell-irrigated soils. For example, the concentration of Cd in aboveground parts of maize (*Zea mays* L.), rapeseed (*Brassica juncea* L.) and lady's finger (*Abelmoschus esculentus* L.) grown on polluted soils was 2.0–3.5 times the amount of Cd when grown on non-polluted soils (Table 10.16). The increase in Ni concentration in various crops with waste-water-irrigated crops was 16 to 136% higher than that in tubewell-irrigated crops (Khurana et al. 2003). The roots of all the crops, with a few exceptions, accumulated higher amounts of potentially toxic elements than aboveground parts. Vegetables like spinach (*Spinacea oleracea* L.), cauliflower (*Brassica oleracea* L. *var botrytis*), and cabbage (*Brassica oleracea* L. *var capitata*) tended to accumulate relatively higher concentrations of potentially toxic

Table 10.16 Accumulation of six metals (mg kg^{-1} dry matter) in root and aboveground parts (AGP) of various crops grown in soils of Jalandhar and Sangrur, irrigated with sewage water (SW) and groundwater (GW).

Crop	*Plant part*	*Source of irrigation*	*Cu*	*Fe*	*Mn*	*Zn*	*Ni*	*Cd*
Maize	Root	SW	15.0	392	27.9	68.0	1.97	0.60
		GW	10.0	275	21.8	50.0	1.50	0.36
	AGP	SW	12.6	255	31.3	60.0	1.67	0.42
		GW	8.00	216	19.8	46.0	1.20	0.28
Rapeseed	Root	SW	24.0	550	46.5	82.0	7.00	3.40
		GW	14.0	340	26.4	54.0	6.00	1.62
	AGP	SW	15.4	596	40.6	74.0	6.40	4.00
		GW	10.0	348	28.1	46.0	4.40	1.44
Lady-finger	Root	SW	14.0	440	38.3	70.0	2.48	1.00
		GW	12.0	280	28.0	36.0	2.00	0.50
	AGP	SW	11.0	348	33.7	52.0	1.90	1.00
		GW	8.00	216	26.0	44.0	1.70	0.50
Cauliflower	AGP	SW	8.40	180	48.0	53.6	1.02	0.24
		GW	6.80	84	30.0	32.8	0.50	0.04
Cabbage	AGP	SW	6.80	74	30.0	38.4	0.87	0.048
		GW	5.90	45	21.8	20.0	0.40	0.08
Spinach	AGP	SW	14.2	560	57.0	50.2	3.00	1.98
		GW	10.8	402	32.8	41.8	0.83	0.10
Radish	AGP	SW	11.4	452	60.0	45.0	1.12	0.56
		GW	6.40	270	23.5	29.5	0.90	0.04

Note: Adapted from Khurana et al. (2003).

elements (Table 10.16), as compared with cereal crop like maize. Among the four vegetables, spinach accumulated the highest amount of all the metals (Table 10.16). Aboveground green plants of rapeseed, which are also consumed as a leafy vegetable, had the highest tendency to accumulate potentially toxic metals as compared with other crops, when grown on contaminated soils.

Sikka (2003) indicated that invariably at all the locations around the Budda Nullah drain at Ludhiana, Indian mustard (*Brassica juncea* L.) contained higher concentrations of potentially toxic metals like Cd, Pb, and Ni when grown on soils receiving wastewater, as compared with tubewell-irrigated soils (Table 10.17). The mean concentrations of Cd, Pb, and Ni in plants grown on polluted soils were 7.79, 8.96, and 11.0 mg kg^{-1} as compared with 0.91, 2.05, and 3.60 mg kg^{-1} when grown on non-polluted soils.

More recently, Dheri et al. (2007) found that the mean concentrations of Pb, Cr, Cd, and Ni in crops grown on sewage-irrigated soils were, respectively, 4.88, 4.20, 0.29, and 3.99 mg kg^{-1}, which were significantly higher than their concentrations in tubewell-irrigated soil (Table 10.18). The mean concentrations of Pb, Cr, Cd, and Ni in plants grown on sewage-irrigated soils were, respectively, 1.2, 2.1, 5.8, and 1.9 times higher than in tubewell-irrigated soils. The concentration of toxic metals in plants grown on sewage-irrigated soils was highest for Cd, followed by Cr, Ni and Pb. Spinach accumulated the highest amounts of all these metals, followed by Indian clover (*Trifolium alexandrium* L.) and coriander (*Coriandrum sativum* L). While such crops grown on sewage-irrigated soils could be harmful for humans and animals, hyper-accumulation capability of these crops could be exploited for phytoremediation of toxic elements from polluted soils.

Table 10.17 Concentration of toxic metals (mg kg^{-1} dry matter) in Indian mustard grown in sewage and tubewell-irrigated soils at seven locations around Budda Nullah in Ludhiana.

Sampling sites	*Source of irrigation*	*Cd*	*Pb*	*Ni*
Jamal Pura	SW	8.41	8.64	6.20
	GW	0.97	1.06	4.07
Shingar Cinema	SW	6.81	7.60	5.50
	GW	1.16	2.16	2.87
Bindra Colony	SW	6.53	8.15	9.40
	GW	0.98	3.70	3.97
Sabzi Mandi	SW	8.66	8.39	8.30
	GW	0.16	2.12	3.72
Salem Tabri	SW	8.26	6.25	12.9
	GW	1.22	1.20	3.32
Chauni Mohalla	SW	7.59	12.3	16.1
	GW	0.93	2.46	3.02
Haibowal Chandan Nagar	SW	8.30	11.4	18.7
	GW	0.98	1.68	4.22
Mean	SW	7.79	8.96	11.0
	GW	0.91	2.05	3.60

Note: Adapted from Sikka (2003).

Table 10.18 Concentration of toxic metals (mg kg^{-1} dry matter) in various crops grown in soils irrigated with sewage water (SW) and groundwater (GW).

Crop	*Source of irrigation*	*Pb*	*Cr*	*Cd*	*Ni*
Indian clover	SW	5.12	4.41	0.25	3.16
	GW	4.70	1.30	0.03	2.45
Spinach	SW	5.79	6.74	0.45	8.43
	GW	5.43	2.53	0.08	3.11
Coriander	SW	4.42	3.27	0.29	2.48
	GW	2.44	2.22	0.03	0.62

Note: Adapted from Dheri et al. (2007).

10.4.4 Effect of urban and industrial pollution on potable water

The influence of urban and industrial effluents released into sewage drain without treatment was also observed on groundwater pumped out for drinking purposes and food processing. A pathogenic profile of 100 domestic-supply water samples, collected from Municipal Corporations and Councils of Ludhiana, Jalandhar, and Patiala districts, revealed that the drinking water from tubewells installed in different municipalities contained several pathogenic microorganisms (Figure 10.3).

10.5 WATER DEPLETION AND POLLUTION – AN ALARMING SITUATION: GAPS IN KNOWLEDGE AND POSSIBLE MITIGATION OPTIONS

Fast depletion of groundwater resources and contamination of groundwater and water bodies from geogenic sources and agricultural, urban, and industrial activities pose a great threat to Punjab's ecosystem. It is evident from several studies that the dangers of groundwater pollution are genuine, and in some cases, the situation is alarming. Thus, there is an urgent need to draft and implement a road map for minimizing the depletion of groundwater and mitigating water pollution. The available information on the fast depletion of water resources, pollution of groundwater and water bodies leading to contamination of the soil-plant-animal-human food chain, and possible mitigation options, is summarized below:

1 Excessive pumping of groundwater could create a serious problem for the long-term sustainability of agricultural and industrial productivity as well as for supplying drinking water to an ever-increasing population. The problem of falling groundwater table in the central Punjab, where rice is a predominant crop, is because of overdraft of water. The rate of groundwater usage is more than water recharge, which has led to overexploitation of 103 blocks out of a total of 137 blocks. The holistic strategy to overcome water crisis includes crop diversification, delayed transplanting of rice, adoption of water-saving agronomic practices, etc.

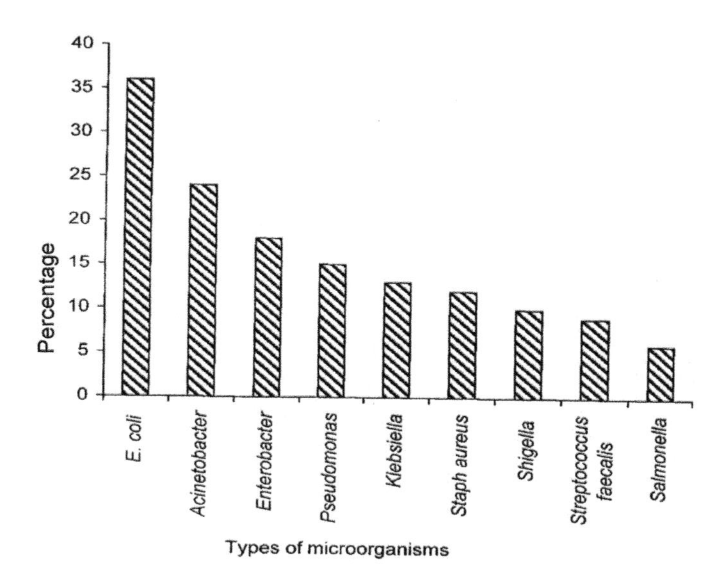

Figure 10.3 Incidence of various pathogens in 100 domestic-supplied water samples collected from Municipal Corporations and Councils of Ludhiana, Jalandhar, and Patiala districts. Used with permission of author (Jairath 2005).

Effectively implemented law for prohibiting the transplanting of rice seedlings during hot and dry period (April to first half of June) in 2008 should hopefully slowdown the fast depletion of groundwater. Similar regulations need to be strictly followed to minimize the wastage of groundwater in cities (e.g., daily washing of floors and vehicles, irrigating lawns, etc.) and industries (releasing water into drains without recycling, etc.). Rainwater harvesting for use in toilets, irrigating lawns, etc., must be encouraged.

2 Selenium, F, B, and As are the geogenic pollutants present in different areas of Punjab.
   a Se toxicity, prevalent in a 1,000-ha area in Nawanshahar and Hoshiarpur districts, causes detachment of the hoof, shedding of horn corium, loss of hair, necrosis of the tip of the tail, loss of body condition, reluctance to move, and stiff gait in animals. In humans, Se toxicity results in hair loss and nail dropping, which can be ameliorated by regulating the entry of Se into the food chain. Gypsum application substantially reduces Se concentration in wheat, maize, oats, and sugarcane. Cultivation of rice in the seleniferous region should be discouraged to reduce the pumping of Se-rich groundwater and the resultant accumulation in surface soil. Mustard crops and tree species, such as poplar, eucalyptus, and mulberry, have a high potential for phytoremediation of excessive Se from the soil.
   b The prevalence of fluoride-rich groundwater and its use for drinking in several blocks of Bathinda and Jagraon block of Ludhiana district has resulted in the onset of widespread fluorosis symptoms, from a mild form of dental fluorosis to crippling skeleton fluorosis.

- c In 100 villages of Bathinda district, 15–44% of groundwater samples exceeded the safe limit of 2 mg B l^{-1}.
- d Arsenic is present in groundwater throughout Punjab; however, its concentration is relatively much higher in the southwestern region. The consumption of water having concentration of As above the safe limit must be discouraged. In the southwestern districts of Punjab, the use of canal water instead of groundwater for drinking and domestic use is recommended. Regular monitoring of As concentration in deep-tubewell waters by water testing laboratories is desirable.

3 Some crops receive large applications of fertilizers, manures, and agrichemicals, which lead to nitrate and phosphate leaching and contamination of groundwater and water bodies. In several villages, groundwater beneath dumps of animal wastes and feedlots is highly NO_3^--N-polluted yet is pumped out for drinking and other domestic purposes. Another deplorable example is the common presence/detection of agrichemicals, such as HCN in water, food, and milk. Ways and means must be found to reduce water pollution caused by nitrate, phosphate, and other agrichemicals.
- a Integrated- and balanced-nutrient managements through inorganic and organic sources have shown immense benefits, e.g., improving soil health and crop productivity, and protecting environment. Formulation and adoption of careful strategies for applying appropriate amounts of fertilizers and manures at proper times, using correct methods, should help synchronize nutrient supply with crop need and avoid excessive use in crops, and, in turn, reduce nitrate and phosphate pollution of groundwater and water bodies.
- b Similarly, a reduced rate/amount of water with an increased frequency of irrigation, in conjunction with a split application of fertilizer N, helps in minimizing the movement of NO_3^--N to deeper soil layers.
- c To minimize N and P leaching into groundwater, rice should not be grown on highly percolating soils, and deep-rooted crops should follow shallow-rooted crops. Various soils and cropping systems, where nitrate and P leaching are possible, should be identified.

4 Application of sewage sludge to agricultural soils, irrigation of field crops with sewage water and untreated industrial effluents alone, or in combination with ground/canal water, are common practices in Punjab, especially in the vicinity of large cities. An estimated 60 to 70% of industrial effluents do not get any treatment before they are released into sewage drains.
- d Water-pollution potential in industrialized cities like Ludhiana (electroplating, bicycles, dying, and other metal industries) and Jalandhar (leather industry) and Amritsar (electroplating, fertilizers, and woolen industries) is many times higher as compared with non- or less-industrialized cities. The biological oxygen demand (BOD) of sewage water of Ludhiana is often found to be more than the maximum limit of 100 mg l^{-1} proposed by the Food and Agriculture Organization of the United Nations for irrigation. Therefore, sewage water of such cities can only be used safely for irrigation after proper treatment.
- e The sustained use of sewage water continues to cause contamination of agricultural land with potentially toxic elements. Several studies have revealed many-folds greater concentration of potentially toxic Pb, Cd, Ni, and

Cr elements in sewage-water-irrigated soils than in soils irrigated with tubewell water.

f Crops grown on soils polluted by sewage water accumulate potentially toxic elements to such an extent as to cause health hazards in animals and humans, depending upon their dietary intake. For example, the concentration of Cd in above-ground parts of maize, rapeseed, pearl millet, and lady's finger grown on such soils was 2.0–3.5 times the amount of Cd in crops grown on non-polluted soils.

g Plant species and varieties differ widely in their ability to absorb, accumulate, and tolerate toxic elements. Leafy vegetables and root crops tend to accumulate a relatively higher concentration of toxic elements than grain crops.

h The threshold levels for Cd and Ni toxicity in tops of vegetable and forage crops, viz., alfalfa, cowpea, Egyptian clover, maize, oats, and pearl millets, have been established. Farmyard manure and Zn are suitable for mitigating Cd toxicity in maize and pearl millet, and Ni toxicity in spinach. Similar information for other crops needs to be generated.

5 In certain situations, environment pollution poses serious health hazards to humans and animals. One glaring example is substantial prevalence of cancer and associated diseases in the southwestern cotton-growing region of Punjab, where groundwater has a high concentration of As; there, the use of pesticides in cotton has been extremely high. Premature graying of hair, degenerated bones, and crippling of humans are prevalent even at the age of 10–15 in this region (Sidhu 2007).

a In areas where groundwater pollution is high, water-works should be developed to remove toxic elements (Se, B, F, As, NO_3^-) from tubewell water, and/or clean canal water before supplying it for human and animal consumption and other domestic purposes. Water from shallow hand-pumps for drinking purposes should be discouraged in polluted areas.

b Fertilizers and agrichemicals should only be marketed and used by farmers under the supervision, recommendation and guidance of agricultural technocrats, as is the practice in developed countries like the U.S. This will ensure the use of correct type and rate of fertilizers and agrichemicals for different crops and soils.

c There is an urgent need to enact and effectively enforce regulations for the release of industrial effluents and their effective enforcement for primary, secondary, and tertiary treatments.

d Educating farmers and public at large about the consequences of dumping animal wastes near feedlots, pumping out shallow polluted water for drinking and domestic purposes, depleting groundwater resources, etc., is desirable.

REFERENCES

Alexander, J. 1986. Health hazards from exposure to nitrate and nitrite. In *Geo-medical consequences of chemical composition of fresh water*, ed. J. Lag, 83–7. Oslo, Norway: The Norwegian Academy of Science and Letters.

Aulakh, M.S. and Bijay-Singh. 1997. Nitrogen losses and fertilizer N use efficiency in irrigated porous soils. *Nut Cycl Agroecosys* 47: 197–212.

Aulakh, M.S., A.K. Garg and B.S. Kabba. 2007. Phosphorus accumulation and leaching, and residual effects on crop yields from long-term applications in subtropics. *Soil Use Manage* 23: 417–27.

Aulakh, M.S., T.S. Khera, J.W. Doran, Kuldip-Singh and Bijay-Singh. 2000. Yields and nitrogen dynamics in a rice – wheat system using green manure and inorganic fertilizer. *Soil Sci Soc Am J* 64: 1867–76.

Aulakh, M.S. and S.S. Malhi. 2005. Interactions of nitrogen with other nutrients and water: Effect on crop yield and quality, nutrient use efficiency, carbon sequestration and environmental pollution. *Adv Agron* 86: 341–409.

Aulakh, M.S. and N.S. Pasricha. 1997. Fertilizer nitrogen management and environmental pollution – Indian scenario. In *Plant nutrient needs, supply, efficiency and policy issues: 2000–2025,* eds. J.S. Kanwar and J.C. Katyal, 296–313. New Delhi, India: National Academy of Agricultural Sciences.

Aulakh, M.S. and G. Singh. 2008. Integrated nutrient management: Experience from South Asia. In *Integrated nutrient management for sustainable crop production*, eds. M.S. Aulakh and C.A. Grant, 285–326. New York: Routledge.

Azad, A.S., B.R. Arora, B. Singh and G.S. Sekhon. 1984. Nature and extent of heavy metal pollution from industrial units in Ludhiana. *Indian J Ecol* 2: 1–5.

Azad, A.S., G.S. Sekhon and B.R. Arora. 1986. Distribution of cadmium, nickel and cobalt in sewage-water irrigated soils. *J Indian Soc Soil Sci* 34: 619–21.

Bajwa, M.S., Bijay-Singh and Parminder-Singh. 1993. Nitrate pollution of groundwater under different systems of land management in the Punjab. In Proc 1st Agricultural Science Congress, National Academy of Agricultural Sciences, New Delhi. pp. 223–230.

Bijay-Singh and G.S. Sekhon. 1976a. Nitrate pollution of groundwater from nitrogen fertilizer and animal waste in the Punjab, India. *Agri Environ* 3: 57–67.

——. 1976b. Some measures of reducing leaching loss of nitrates beyond potential rooting zone I. Proper coordination of nitrogen splitting with water management. *Plant Soil* 44: 193–200.

——. 1976c. Some measures of reducing leaching loss of nitrates beyond potential rooting zone II. Balanced fertilization *Plant Soil* 44: 391–5.

Brar, M.S. and C.L. Arora. 1997. Concentration of micro elements and pollutant elements in cauliflower (*Brassica oleracea var botrytis*). *Indian J Agric Sci* 67: 141–3.

Brar, M.S. and M.P.S. Khurana. 2006. Heavy metal pollution and its impact on environment in Punjab. In All India Seminar on Our Environment: Status and Challenges Ahead, Jan 19–20, Punjab Agricultural University, Ludhiana. pp. 60–70.

Brar, M.S., M.P.S. Khurana and B.D. Kansal. 2002. Effect of irrigation by untreated sewage effluents on the micro and potentially toxic elements in soils and plants. In Proc 17th World Congress of Soil Science, Symposium no 24, Aug. 14–21, Bangkok, Thailand. pp. 198(1)–198(6).

Carpenter, S.R., N.E. Caraco, D.L. Correll, R.W. Howarth, A.N. Sharpley and V.H. Smith. 1998. Non point pollution of surface waters with phosphorus and nitrogen. Ecol Appli 8: 569–8.

Commoner, B. 1968. *The killing of great lake.* World Bank year book. Chicago: Field Enterprises Education Corporation.

Dhaliwal, G.S. and Balwinder-Singh. 2000. Pesticide contamination of fatty food commodities. In *Pesticides and environment,* eds. G.S. Dhaliwal and Balwinder-Singh, 155–90. New Delhi, India: Commonwealth Publishers.

Dheri, G.S., M.S. Brar and S.S. Malhi. 2007. Heavy metal concentration of sewage contaminated water and its impact on underground water, soil and crop plants in alluvial soils of north western India. *Commun Soil Sci Plan* 38: 1353–70.

Dhillon, K.S. and S.K. Dhillon. 1991. Selenium toxicity in soils, plants and animals in some parts of Punjab, India. *Int J Environ Stud* 37: 15–24.

——. 2003. Quality of underground water and its contribution towards selenium enrichment of the soil-plant system for a seleniferous region of northwest India. *J Hydrol* 272: 120–30.

Environment Protection Act. 1986. General standards for the discharge of environmental pollutant. Part A Effluents. Schedule IV of environmental protection rule. http//:scclmines.com/env/Linkfile2.htm (accessed 2 March, 2009).

Garg, A.K. and M.S. Aulakh. 2010. Effect of long term fertilizer management and crop rotations on accumulation and downward movement of phosphorus in semiarid subtropical irrigated soils. *Commun Soil Sci Plan* 41: (in press).

Hundal, H.S., Kuldip-Singh and Dhanwinder-Singh. 2008. Arsenic content in ground and canal waters of Punjab, North-West India. *Environ Monit Assess*, doi: 10.1007/s10661-008-0406-3, http://www.springerlink.com/content/x270834403h412h8/ (accessed 10 December, 2008.)

Hundal, H.S., R. Kumar, Kuldip-Singh and Dhanwinder-Singh. 2007. Occurrence and geochemistry of arsenic in groundwater of Punjab, Northwest India. *Commun Soil Sci Plan* 38: 2257–77.

Jairath, S. 2005. Studies on microbiological quality of water. MSc thesis, Punjab Agricultural University.

Kang, M.S., M.S. Aulakh and J.S. Dhiman. 2008. Agricultural development in Punjab: Problems, possible solution and new initiatives. Alumni Association, College of Agriculture, Punjab Agricultural University, Ludhiana. *Agalumnus* 38: 26–34.

Kansal, B.D. and M.P.S. Khurana. 1999. Extent of contamination of alluvial soils with cadmium. In B.S. Aggarwal, P. Dureja and A.K. Dikshit (Eds.) Proc 2nd International Conference on Contaminants in Soil Environment in Australia–Pacific region. Dec 12–17. Indian network for soil contamination research, New Delhi. pp. 309–310.

Kansal, B.D. and R. Kumar. 1994. Nickel accumulation in plants grown in polluted and non polluted soils. In National Seminar on Development in Soil Science, Diamond Jubilee Convention, Indian Society of Soil Science, New Delhi. pp. 626–627.

Khurana, M.P.S., V.K. Nayyar, R.L. Bansal and M.V. Singh. 2003. Heavy metal pollution in soils and plants through untreated sewage-water. In *Ground water pollution*, eds. V.P. Singh and R.N. Yadava, 487–95. New Delhi, India: Allied Publishers Pvt. Limited.

Lennox, S.D., R.H. Foy, R.V. Smith and C. Jordan. 1997. Estimating the contribution from agriculture to the phosphorus load in surface water. In *Phosphorus loss from soil to water*, eds. H. Tunney, O.T. Carton, P.C. Brookes and A.E. Johnston, 55–75. New York: CAB International.

Levy, D., J. Schramke, J. Esposito, T. Erickson and J. Moore. 1999. The shallow groundwater chemistry of arsenic, fluorine, and major elements: Eastern Owens Lake, California. *Appl Geochem* 14: 53–65.

Sharma, V.K. and B.D. Kansal. 1986. Heavy metal contamination of soils and plants with sewage irrigation. *Pollut Res* 4: 86–91.

Sharpley, A.N., T.C. Daniel, J.T. Sims and D.H. Pote. 1996. Determining environmentally sound soil phosphorus levels. *J Soil Water Conser* 51: 160–6.

Sidhu, H. 2007. Death stalks a generation: Poisoned water, poisoned soil. *Hindustan Times*, September 2.

Sikka, R. 2003. Influence of soil characteristics on available lead and cadmium and their accumulation by *Brassica* species. PhD Diss, Punjab Agricultural University.

Singh, B. and S.R. Bishnoi. 2001. Contribution of underground waters towards nitrogen and sulphur concentration in soils. *J Indian Soc Soil Sci* 49: 188–90.

Singh, J. and B.D. Kansal. 1985. Effect of long term application of municipal waste on some chemical properties of soils. *J Res Punjab Agricultural University* 22: 235–42.

Singh, B., D.S. Rana and M.S. Bajwa. 1977. Salinity and sodium hazards of underground irrigation waters of the Bathinda district (Punjab). *Indian J Ecol* 4: 32–41.

Thind, H.S. and B.D. Kansal. 2002. Nitrate contamination of groundwater under different anthropogenic activities. In Proc 17th World Congress of Soil Science. Symposium no 54, Aug. 14–21, Bangkok, Thailand. pp. 712(1)–712(8).

Tiwana, N.S., R.S. Panesar and B.D. Kansal. 1987. Characterization of waste water of a highly industrialized city of Punjab. In Proc. National Seminar on Impact of Environmental Protection for Future Development of India, Nanital. pp. 119–126.

Section V

Soil and water management in dryland or rainfed agriculture

Chapter 11

Manipulating tillage to increase stored soil water and manipulating plant geometry to increase water-use efficiency in dryland areas

B.A. Stewart
Dryland Agriculture Institute, West Texas A&M University, Canyon, Texas

ABSTRACT

This paper briefly summarizes some of the practices being used in the semiarid U.S. Great Plains to grow crops without irrigation. Fallow periods are commonly used to increase the amount of plant-available water in the soil profile at the time of seeding a crop because growing-season precipitation is nearly always insufficient to produce economic grain yields. Maintaining plant residues on the surface as mulch has been very beneficial for this purpose. The most successful practices depend mostly on herbicides for controlling weeds during the fallow period and eliminating as much tillage as feasible. Reduced plant populations and, more recently, seeding plants in clumps rather than uniformly spacing plants in rows show some promise as useful strategies for reducing early vegetative growth so that more soil water is available during the grain-filling period late in the season. While these practices may not be applicable to other semiarid regions because of differences in soil, climatic, social, and economic conditions, some of the principles may apply and useful technologies can be developed.

Keywords: Fallow, clumps, grain sorghum, evapotranspiration, no-tillage

11.1 INTRODUCTION

There is no single agreed definition of the term drylands. Perhaps the two most widely used definitions are those used by the Food and Agriculture Organization (FAO) of the United Nations and by the United Nations Conference on Desertification (UNCOD).

Address correspondence to B.A. Stewart at the Dryland Agriculture Institute, West Texas A&M University, WTAMU Box 60278, Canyon TX, 79016 USA. E-mail: bstewart@wtamu.edu

Reprinted from: *Journal of Crop Improvement*, Vol. 23, Issue 1 (January 2009), pp. 71–82, DOI: 10.1080/15427520802418319

Table 11.1 Percent of World Land Area (134,900,000 km_2) in Various Regions, Percent of Land in Regions for Different Dryland Areas, Percent of World Population (6,000,000,000) in Various Regions, Percent of Population in Regions Living in Dryland Areas, and Percent of Population in Regions Engaged in Agriculture.

Regions	*World land area*	*Arid*	*Semiarid*	*Dry subhumid*	*Total world population*	*Population in drylands*	*Agricultural population*
	%						
Asia and Pacific	21.5	6	15	17	56	44	59.5
Europe	5.4	12	28	23	12	18	13.5
North Africa and Near East	9.5	4	11	5	5	44	44.3
North America	14.8	12	28	23	5	19	3.0
North Asia and East of Urals	15.6	11	51	33	4	89	17.4
South and Central America	15.4	11	6	10	8	24	23.0
Sub-Saharan Africa	17.7	6	13	19	10	36	65.0
World (Total)	100	7	20	18	100	38	45.8

Source: Stewart, Koohafkan & Ramamoorthy, 2006.

These classification systems are discussed by Stewart, Koohafkan, and Ramamoorthy (2006). The FAO uses the length of the growing period, and areas with less than 179 days are considered drylands. Arid regions have between 1 and 59 growing days, semiarid areas have between 60 and 119 growing days, and dry subhumid regions have between 120 and 179 growing days. The drylands of the world based on this system are presented in Table 11.1 (FAO, 2000). Approximately 38% of the world's land area is classified as semiarid or dry subhumid regions, and an additional 7% is located in arid regions where some form of dryland agriculture occurs. The growing period is defined as the number of days during a year when precipitation exceeds half the potential evapotranspiration (PET), plus a period required to use an assumed 100 mm of water from excess precipitation (or less, if not available) stored in the soil profile.

The UNEP (1992) defined bioclimatic zones based on the climatic aridity index: P/ETP, where P is annual precipitation and ETP is annual potential evapotranspiration calculated by the method of Penman (Doorenbos and Pruitt, 1977), taking into account atmospheric humidity, wind, and solar radiation. According to this classification system, arid lands have P/ETP values >0.03 and <0.20; semiarid lands >0.20 and <0.50; and subhumid lands >0.50 and <0.75.

11.2 INDIA DRYLANDS

Ghosh and Jana (2005) stated that of India's total landmass of 329 million ha with an average annual rainfall of about 1120 mm, 224 million ha are moderately to extremely dry. These areas, however, are home to some 40% of the human population and two-thirds of the cattle population. Most of these areas are generally mono-cropped and dependent entirely on precipitation. Because of these constraints, high-value crops are limited and the major crops are coarse cereals, pulses, oil-seed crops, and cotton. They also reported that 90% of the people in dry areas depend only on agriculture for their livelihood. As a result, poverty is a constant companion of the people living in these areas.

Stewart, Koohafkan, and Ramamoorthy (2006) stated that while dryland farming in India began centuries earlier than in North America, there are some striking similarities between the two regions insofar as the scientific study of dryland farming is concerned. Hegde (1995) reported that in 1917, A.K.Y.N. Aiyer listed the important practices of farmers and found considerable conformity with those that Campbell (1907) proposed during the early 1900s for the U.S. Great Plains. Field bunding, fall plowing, frequent intercultivations, drill sowing, and growing drought-resistant crops, such as finger millet, grain sorghum, and pearl millet, were some of the practices studied.

Scientific study of dryland farming was initiated by the government of India in 1923. Early research focused on improving crop yields. Important practices included bunding to conserve soil and water, deep plowing, use of farmyard manure, use of low seeding rates, and inter-cultivation for weed and evaporation control. These practices gave a 15% to 20% increase in yield over a base yield of 200 to 400 kg ha^{-1} (Hedge, 1995). By the mid-1950s, the emphasis shifted to soil management. Soil conservation research and training centers were established at eight locations focusing on contour bunding. However, negative results were often obtained, particularly on Vertisols, where water accumulations and runoff problems were frequently encountered. Even when yield increases were observed, they were again not more than 15% to 20% above the 200 to 400 kg ha^{-1} base yield.

The importance of shorter duration crops to fit into the soil-water availability period was recognized during the 1960s. It was also during the mid-1960s that high-yielding hybrids and varieties became available that were not only responsive to fertilizers but also to management. An All India Coordinated Research Project for Dryland Agriculture (AICRPDA) was established, and the research emphasis shifted to a multidisciplinary approach to tackle the problems from several viewpoints. Similar efforts were initiated at the International Crops Research Institute for the Semi-Arid Tropics (ICRISAT) at Hyderabad established in 1972.

Although yields of major dryland crops in India have increased, they are still low and total production has actually decreased for sorghum, which was almost equal to wheat in the 1960s. The yield and production of wheat, sorghum, millet, and pulses for various periods from 1961 to 2006 are shown in Table 11.2. These crops are widely grown in the dryland areas, but much of the wheat in India is irrigated so the large increases shown for wheat are largely the result of irrigation, fertilizers, and high-yielding cultivars that were the primary factors associated with the Green Revolution. Most of the sorghum, millet, and pulses in India are grown in the drylands without irrigation. There have been significant increases in yields of all these crops, but at a modest rate. This is not surprising because water is so limited in many of these regions that even significant improvements in water use can only make small improvements in alleviating water stress. The surprising and somewhat alarming information shown in Table 11.2 is that there has been a large decrease in the total production

Table 11.2 Average yield and production of wheat, sorghum, millet, and pulses in India for various time periods.

	1961 to 1969	*1970 to 1979*	*1980 to 1989*	*1990 to 1999*	*2000 to 2006*
Wheat					
kg ha^{-1}	907	1352	1849	2430	2685
1000 mt	12,548	26,607	42,959	61,257	70,674
Sorghum					
kg ha^{-1}	504	597	615	811	771
1000 mt	9,226	9,748	11,121	9,786	7,190
Millet					
kg ha^{-1}	418	510	556	734	1063
1000 mt	7,959	9,588	9,748	9,852	10,555
Pulses					
kg ha^{-1}	420	407	378	534	681
1000 mt	1,223	1,086	860	1,075	1,280

Source: FAOSTAT, 2008.

of sorghum, no increase in the production of pulses, and only a limited increase for millet. The yields of all three crops have shown some increase so the area devoted to these crops has decreased. Although the reasons are probably many, some of the more favorable areas have perhaps been devoted to wheat, and some of the least favorable areas have been removed from cropping. Wheat production since 1961 has increased almost six fold while yields have tripled.

11.3 SIMILARITIES AND DIFFERENCES BETWEEN DRYLAND AGRICULTURE IN UNITED STATES AND INDIA

Although many of the recommended practices for dryland farming in India are in agreement with those for the United States, there are differences. A highly recommended practice for water conservation in India is the use of a dust mulch (Hegde, 1995) similar to that recommended in the U.S. Great Plains in the early 1900s that many scientists have concluded was a major contributor to the infamous Dust Bowl of the 1930s. The Dust Bowl was one of the worst ecological disasters that resulted from human activities. Campbell (1907) erroneously believed "dust mulch" on the surface not only conserved substantial amounts of soil water but also attracted it from the atmosphere. The fact that dust mulching is still recommended and widely practiced in India points out the importance of recognizing and addressing the vast differences that exist between dryland regions, particularly semiarid regions. For example, summer-fallow has played an important role in the U.S. Great Plains because some precipitation occurs every month of the year, but there is no month in which precipitation exceeds PET, and for many cropping areas, there is no monthly precipitation that even reaches one-half of the PET. In contrast, most of the dryland areas in India have more than seven months of the year when there is essentially zero precipitation, and then there is a monsoon season of varying length where the precipitation greatly exceeds the PET for at least a portion of the growing season. Therefore, fallow for storing soil water is not a viable alternative in most cases in India because much of the water saved during the rainy season would be lost during the prolonged dry period and, more importantly, there is usually more than enough precipitation during the monsoon season to fully wet the soil profile. In contrast, successful dryland crop production in the U.S. Great Plains is highly dependent on the amount of water stored in the soil profile at the time the crop is seeded because growing-season precipitation is seldom, if ever, sufficient for economical yields, particularly for cereal crops.

11.4 MANIPULATING TILLAGE TO INCREASE WATER STORAGE

This paper will focus on practices that have been successful for increasing soil and water conservation in the semiarid U.S. Great Plains. While the principles involved are universal, the practices may or may not be applicable to other semiarid regions because of differences in soil, climatic, social, and economic conditions.

11.4.1 Stubble mulch tillage

Crop production, particularly cereal crops, in the U.S. Great Plains is highly dependent on the amount of plant-available water stored in the soil profile at the time of seeding the crop. This is because there is simply not sufficient precipitation during the growing season to result in economic yields in most years. This is true whether the crop is a winter crop, such as wheat, or summer crops like sorghum and cotton. Therefore, fallow is a farming practice where and when no crop is grown and all plant growth is controlled by tillage or herbicides during a season when a crop might normally be grown. The precipitation stored in the soil profile during the fallow period is then available to supplement the growing-season precipitation for the subsequent crop. Such a system results in only one crop every two years, or two crops every three years, or other combinations depending on crop selection. Wheat-fallow was a very common cropping system for the U.S. Great Plains, which meant that only one wheat crop was produced in two years, and the time of fallow between crops ranged from about 16 months when winter wheat was produced to as long as 21 months in northern areas where spring wheat was produced. Prior to its conversion to cropland in the early 1900s, the area was predominantly grassland. Intensive cultivation was widely used during the fallow period and the soil organic matter content of the soil decreased rapidly; when the drought years of the 1930s occurred, wind erosion became rampant and resulted in the infamous Dust Bowl described earlier. This ecological disaster mandated a change if the region was going to survive as an agricultural area.

The use of chisel plows and sweep plows was widely adopted by farmers as a way to retain some of the crop residues on the soil surface. Stubble-mulch tillage was the practice of pulling v-shaped sweeps about 10 cm beneath the soil surface. The sweeps would cut the roots of weeds and volunteer wheat so that water would not be used during the fallow period, and most of the crop residues left on the surface after harvesting the grain would remain on the surface. Pulling a sweep plow through the soil buries only about 15 to 20% of the crop residues, so even using a sweep plow four times during the fallow period would only bury about half of the crop residues. Of course, some of the residues would be decomposed even when they were not buried. Nevertheless, stubble-mulch tillage generally resulted in enough crop residues remaining on the soil surface to control wind erosion and became widely used by farmers. After several years of practice by farmers and research by scientists, it became clear that this practice was not only controlling wind erosion but also increasing the amount of water stored in the soil profile during the fallow period. Consequently, stubble-mulch tillage increased yields as well as controlled wind erosion.

11.4.2 Shortening the fallow period

Although summer fallow was widely practiced in the U.S. Great Plains to increase yields and reduce risk, the precipitation-use efficiency was low. In the central and southern portions of the Great Plains, only about 15% of the precipitation that occurred during the fallow period was actually conserved in the soil for use by the subsequent crop. Research studies showed the efficiency of storage decreased with the length of the fallow period. For example, in the southern portion, winter wheat is grown; when wheat-fallow is practiced, one wheat crop is harvested every two years and the fallow period between wheat crops is about 16 months. Reducing the length of the fallow period to 11 months did not result in much less water in soil profile, so many producers adopted a wheat-sorghum-fallow cropping system. In this system, wheat is seeded in October and harvested the following July. Then, the land is fallowed for 11 months until June of the next year when grain sorghum is seeded and harvested in November. This is followed by another 11-month fallow period when wheat is seeded again in October. Consequently, one crop of wheat and one crop of grain sorghum are harvested from a field every three years. Producers that follow this system usually divide their land into three fields and one-third of the land is in wheat, one-third in grain sorghum, and one-third in fallow in any given year. This system utilizes the total precipitation more efficiently because growing two crops in three years rather than one crop in two years produces more grain. In addition, more crop residues are produced making it easier to control wind and water erosion and to help maintain or enhance the soil organic matter content.

11.4.3 Crop residue management

Crop-residue management is managing the frequency and intensity of soil-disturbing activities related to residue management. The emphasis, particularly in dryland areas, is to maintain as much of the crop residues as feasible on the soil surface. This can include practices often described as conservation tillage, reduced-tillage, ridge-tillage, and no-tillage. Conservation tillage was a widely used term in the U.S. for many years, but the term has been used less in recent years. Conservation tillage was defined by the U.S. Department of Agriculture as "any tillage or planting system in which at least 30% of the soil surface is covered by plant residue after planting to reduce erosion by water, or, where soil erosion by wind is the primary concern, at least 1000 lb/ac

(1120 kg/ha) of flat small grain residue equivalent is on the soil surface during the critical erosion period" (Schertz, 1988). The move away from the term conservation tillage is because of the emphasis on quantity. In dryland regions, there are some years, particularly drought years, in which there are not sufficient amounts of crop residues remaining after harvest to cover 30% of the soil surface after planting the next crop, even if no-tillage is practiced. Therefore, the USDA Natural Resource Conservation Service stresses crop-residue management that focuses on management rather than quantity. Crop-residue management systems include conservation tillage practices, such as no-till, ridge-till, and mulch-till, and other conservation practices that provide sufficient residue cover to protect the soil surface from the erosive effects of wind and water.

A marked change toward reducing tillage in the U.S. Great Plains occurred during the 1970s following the formation and actions of the Organization of Petroleum Exporting Countries (OPEC) that resulted in a significant increase in energy prices. Agriculture in the United States is highly dependent on energy and, prior to OPEC, energy costs had been relatively low and had little or no effect on the frequency and intensity of tillage. The sudden increased cost of tillage resulted in a rapid increase in research and application of practices based on using more herbicides and less tillage for controlling weeds and other vegetation between crops.

Representative studies that showed the use of herbicides in place of tillage to control vegetation during fallow were the work of Unger and Baumhardt (1999; Figure 11.1). The data are from long-term studies at Bushland, Texas, and show the amounts of stored soil water at time of seeding grain sorghum in June, following an 11-month fallow period after harvest of winter wheat. Prior to 1973, vegetation during the fallow period was controlled by sweep tillage and there were generally four tillage operations during the fallow period. Beginning in 1973, there was no

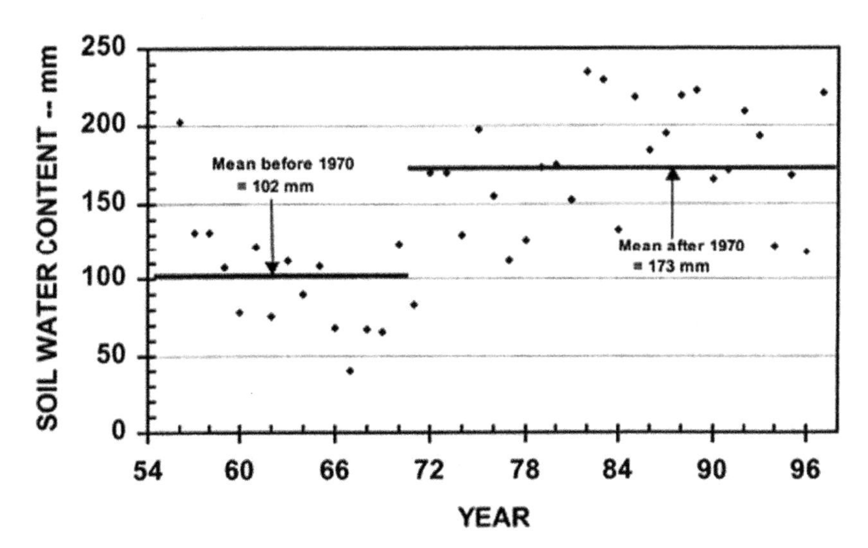

Figure 11.1 Average annual volumetric soil water content at planting time for dryland grain sorghum in studies conducted from 1956 to 1997 at the USDA-ARS Conservation and Production Research Laboratoy, Bushland, Texas (Unger and Baumhardt, 1999).

tillage during the fallow period and herbicides were used to control vegetation. On average, the use of herbicides increased soil-water storage during the fallow period by 71 mm. The precipitation amount that occurs during the 11-month fallow period averages about 415 mm, and there was an average of 102 mm plant-available soil water at time of seeding grain sorghum for the years when sweep-tillage was practiced compared with an average of 173 mm for the years when no tillage was practiced. Therefore, the subsequent grain sorghum crop had considerably more water available during the growing season and the amount of soil water at planting time is a key factor in determining grain yields. Unger and Baumhardt (1999) reported that grain sorghum yields increased 139% during the 1956–1997 period, with 46 of those percentage units resulting from use of improved hybrids, based on results of a uniformly managed 40-year study. The remaining 93 percentage points for that period were attributed to other factors, primarily to soil water at planting. The biggest change in soil water at planting was due to the improved crop-residue management practices that occurred beginning in the 1970s as shown in Figure 11.1.

11.4.4 Water use-grain yield relation

Several studies in the U.S. Great Plains have shown close relations between growing-season water use and grain sorghum yields (Stewart and Steiner, 1990). More recently, Stone and Schlegel (2006) summarized yield-water supply relationships for grain sorghum and winter wheat for studies conducted from 1974 to 2004. In the thirty years of research, various levels of tillage and/or herbicides were used during non-crop periods for weed control, ranging from conventional, stubble-mulch (sweep) tillage with no herbicides to reduced tillage (herbicides and tillage) to no-till (herbicides exclusively). Grain sorghum yields increased an average of 22.1 kg ha^{-1} for each additional mm of plant-available water stored in the soil profile at time of seeding. This is very similar to the 21 kg ha^{-1} increase shown by Unger and Baumhardt (1999) for studies at Bushland. Likewise, Stone and Schlegel showed an increase of 16.6 kg ha^{-1} for each mm of additional evapotranspiration (plant-available soil water at seeding plus growing-season precipitation) above a threshold value of 136 mm when data for all tillage systems were included. This compares very close to the 15.5 kg ha^{-1} value and threshold value of 127 mm cited above that Stewart and Steiner (1990) reported for Bushland studies. Tribune, Kan. is located about 375 km north of Bushland, Texas, in the northernmost part of the southern Great Plains. The mean annual precipitation is 440 mm compared with 475 mm at Bushland, but the potential evapotranspiration is also significantly less.

Stone and Schlegel (2006), however, went further in their analysis in that they separated yield data from no-till, conventional-till, and reduced-till treatments to identify different efficiencies (Figure 11.2). For the no-tillage system, an additional mm of water increased grain yield 18.4 kg ha^{-1} after a threshold of 157 mm, compared with only 12.9 kg ha^{-1} after a threshold value of 102 mm for conventional and reduced-tillage systems. They concluded that mulch increased the amount of precipitation stored in the soil during the fallow period so more plant-available water was present at time of seeding. They also concluded that the mulch resulted in more efficient use of the growing-season precipitation.

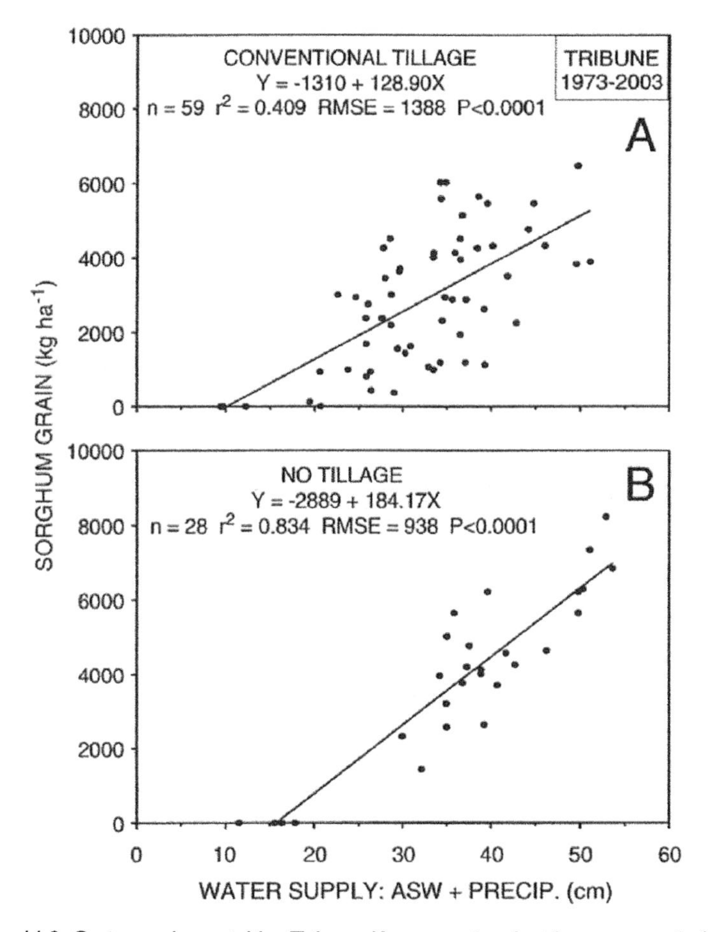

Figure 11.2 Grain sorghum yield at Tribune, Kan., associated with water supply (plant available soil water at emergence plus growing-season precipitation) for dryland conventional tillage (a) and for no-till (b) treatment groups (Stone and Schlegel, 2006).

11.5 CROP MANIPULATIONS TO INCREASE WATER-USE EFFICIENCY

Loomis (1983) discussed ways of manipulating crops to make better use of limited water supplies. He stated that a useful generalization was that when soil resources, such as water, are limiting, non-uniform treatment of the land or the crop can be an advantage. However, where resources are non-limiting, uniform cropping will provide the greatest efficiency in light interception and photosynthesis. In the U.S. Great Plains, reducing the plant population, widening the distance between rows, and using skip-row configurations have been the most commonly used strategies for conserving soil water for use by crops during the grain-filling growth stages when water is generally insufficient. The success of these has been highly variable. For crops such as grain sorghum, reducing the plant population often results in an increased number of tillers that use water and nutrients but often produce little or no grain and negate the expected benefit of having fewer plants.

Using a somewhat different approach, Bandaru et al. (2006) hypothesized that growing plants in clumps would result in fewer tillers and less vegetative growth so that more soil water would be available during the grain-filling period. Results from a three-year study at Bushland, Texas, and a one-year study at Tribune, Kan., showed that planting grain sorghum in clumps of three to six plants reduced tiller formation to about one per plant compared with about three for uniformly spaced plants. Grain yields were increased with clump planting by as much as 100% when yields were in the 1,000 kg ha^{-1} range and 25% to 50% in the 2,000 to 3,000 kg ha^{-1} range, but there was no increase or even a small decrease at yields above 5,000 kg ha^{-1}. There were also marked differences in plant architecture (Bandaru et al., 2006); that is, uniformly spaced plants produced more tillers and the leaves on both the main stalk and tillers grew outward, exposing essentially all of the leaf area to sunlight and wind. In contrast, clumped plants grew upward with the leaves partially shading one another and reducing the effect of wind, thereby reducing water use. These results suggest that, in semiarid environments where plant populations must be kept low to prevent severe water stress, growing plants in clumps rather than in uniform spaces might enhance grain yields. This strategy is somewhat similar to growing plants in skip rows because skip rows also keep plants closer together than uniformly spacing the plants in all rows. In both of these strategies, however, shading of the soil surface is reduced because the plants are concentrated over a smaller area of the soil surface. Therefore, the evaporation portion of evapotranspiration will likely be higher, particularly if there are numerous small precipitation events during the growing season. This effect could be minimized by using these strategies in a no-till cropping system so the mulch would decrease soil evaporation during the growing season.

REFERENCES

Bandaru, V., B.A. Stewart, R.L. Baumhardt, S. Ambati, C.A. Robinson and A. Schlegel. 2006. Growing dryland grain sorghum in clumps to reduce vegetative growth and increase yield. *Agron. J.* 98: 1109–1120.

Campbell, H.W. 1907. *Campbell's 1907 soil culture manual.* Lincoln, NE: H.W. Campbell.

Doorenbos, J. and W.O. Pruitt. 1977. *Crop water requirements.* Food and Agriculture Organization Irrigation and Drainage Paper 24. Rome: FAO.

FAOSTAT. 2008. Food and agriculture organization statistics. Online. http://faostat.fao.org/default.aspx (verified 1 July 2008).

Food and Agriculture Organization. 2000. *Land resource potential and constraints at regional and country levels.* World Soil Resources Report 90. Rome: FAO.

Ghosh, A. and P.K. Jana. 2005. *Dryland farming: A concept of future agriculture.* Ludhiana, India: Kalyani Publishers.

Hegde, B.R. 1995. Dryland farming: Past progress and future prospects. In *Sustainable development of dryland agriculture in India,* ed. R.P. Singh, 7–12. Jodhpur, India: Scientific Pub.

Loomis, R.S. 1983. Crop manipulation for efficient use of water: An overview. p. 345–374. In Limitations to Efficient Use of Water Use in Crop Production, eds. H.M. Taylor, W.R. Jordan, and T.R. Sinclair, ASA, CSSA, SSSA. Madison, WI.

Schertz, D.L. 1988. Conservation tillage: An analysis of acreage projections in the United States *J. Soil Water Conser.* 43: 256–258.

Stewart, B.A., P. Koohafkan and K. Ramamoorthy. 2006. Dryland agriculture defined and its importance to the world. *Agronomy monograph No. 23, Dryland Agriculture* 2nd, ed., eds. G.A. Peterson, P.W. Unger, W.A. Payne, 1–26. Madison, WI. American Society Agronomy, Crop Science Society America, Soil Science Society America.

Stewart, B.A. and J.L. Steiner. 1990. Water-use efficiency. In *Dryland Agriculture: Strategies for Sustainability, Volume* 13: *Advances in Soil Science, eds.* R.P. Singh, J.F. Parr and B.A. Stewart, 151–173. New York, Springer-Verlag.

Stone, L.R. and A.J. Schlegel. 1996. Yield-water supply relationships of grain sorghum and winter wheat. *Agron. J.* 98: 1359–1366.

Unger, P.W. and R.L. Baumhardt. 1999. Factors related to dryland grain sorghum yield increases. *Agron. J.* 91: 870–875.

United Nations Environmental Program (UNEP). 1992. World Atlas of Desertification, UNEP, Nairobi, Kenya.

Chapter 12

Integrated watershed management for increasing productivity and water-use efficiency in semi-arid tropical India

Piara Singh, P. Pathak, S.P. Wani & K.L. Sahrawat
International Crops Research Institute for the Semi-Arid Tropics (ICRISAT), Patancheru, Andhra Pradesh, India

ABSTRACT

Poverty, food insecurity, and malnutrition are pervasive in the semi-arid tropics (SAT) of South Asia, including India. In rural areas, most of the poor make their livelihoods on the use of natural resources, which are degraded and inefficiently used. This is because of the inadequate traditional management practices of managing agriculture as well as the fact that resulting crop yields are much below the expected potential yields. ICRISAT in the early 1970s initiated research on watersheds for integrated use of land, water, and crop management technologies for increasing crop production through efficient use of natural resources, especially rainfall that is highly variable in the SAT and is the main cause of year-to-year variation in crop production in India. Improved watershed management on Vertisols more than doubled crop productivity, and rainfall-use efficiency increased from 35% to 70% when compared with traditional technology. After many years of implementing and evaluating these improved technologies in on-farm situations, many lessons were learned and they formed part of the integrated watershed management model currently being pursued by ICRISAT in community watersheds in rural settings. This watershed model is more holistic and puts rural communities and their collective actions at center stage for implementing improved watershed technologies with technical backstopping and convergence by consortium partners. We describe here the achievements made in enhancing crop productivity and rainfall-use efficiency by implementing improved technologies in on-farm community watersheds in India.

Address correspondence to Piara Singh, ICRISAT, Patancheru 502 324, Andhra Pradesh, India. E-mail: p.singh@cgiar.org

Reprinted from: *Journal of Crop Improvement*, Vol. 23, Issue 4 (October 2009), pp. 402 – 429, DOI: 10.1080/15427520903013423

Keywords: rainfed agriculture, community watersheds, integrated genetic and natural resource management, technology exchange model, food security, rural livelihoods

12.1 INTRODUCTION

South Asia alone accounts for almost all (236 of 237 million) of rural poor living in the semi-arid tropics (SAT) of Asia and about 63% of the rural poor in the SAT worldwide. This also indicates that about 50% of abject poverty in South Asia is concentrated in the SAT (ICRISAT, 2006). Along with pervasive poverty, degradation of agro-ecosystems and declining sustainability are the major concerns of agricultural development in many poor regions of the world where livelihoods depend on exploitation of natural resources. This is especially the case in arid and semi-arid areas where water scarcity, frequent droughts, soil degradation, and other biotic and abiotic constraints lower agricultural productivity and resilience of the system (Shiferaw & Bantilan, 2004). This is further complicated by a policy environment often biased toward high potential regions and incentive systems that discourage adoption of water-saving crops and technologies adapted to dryland areas (Shiferaw, Wani & Nageswara Rao, 2003).

Water is the inherently limiting resource in the SAT for agricultural production on which the human and animal populations are dependent. Erratic rainfall results in widely fluctuating production, leading to production deficit and causing land degradation through soil erosion and reduced groundwater recharge. Population growth accompanied by the heightened demand for natural resources to produce food and to meet needs of the other sectors of the economy further exacerbates the existing problems. Thus, a process of progressive degradation of resources sets in, which intensifies with every drought and the period following it. If not checked timely and effectively, it leads to permanent damage manifested as loss of biodiversity and degradation of natural resources (Wani et al., 2006). Unless the nexus between drought, land degradation, and poverty is addressed, improving the livelihoods dependent mainly upon natural resources can be difficult. Water is the key factor and, through efficient and sustainable management of water resources, entry could be made to beak and not to break the nexus (Wani et al., 2003). In rainfed regions, this would mean enhancing the supply of water through soil and water conservation, water harvesting in ponds, and recharging the groundwater and on the demand side, enhancing its efficient use by adopting integrated soil-, water-, crop-, nutrient- and pest-management practices.

In this paper, we describe the ICRISAT approach of integrated watershed management in rainfed areas of India to enhance the goals of increasing crop production and improving rural livelihoods through sustainable and efficient use of land and water resources.

12.2 OPPORTUNITIES FOR ENHANCING CROP PRODUCTIVITY IN RAINFED REGIONS OF INDIA

The dominant rainfed crops in India are sorghum, pearl millet, pigeonpea, chickpea, soybean, and groundnut. Some area is also under rainfed rice, rainfed wheat, mustard, rapeseed and cotton. Substantial yield gaps exist between current (farmers') and experimental or simulated potential yields (Figure 12.1). The farmers' average yield is 970 kg ha^{-1} for *kharif* sorghum, 590 kg ha^{-1} for *rabi* sorghum, and 990 kg ha^{-1} for pearl millet. Simulated rainfed potential yield in different production zones ranged from 3,210 to 3,410 kg ha^{-1} for *kharif* sorghum, 1,000 to 1,360 kg ha^{-1} for *rabi* sorghum, and 1,430 to 2,090 kg ha^{-1} for pearl millet. Total yield gap (simulated rainfed potential yield – farmers' yield) in production zones ranged from 2,130 to 2,560 kg ha^{-1} for *kharif* sorghum, 280 to 830 kg ha^{-1} for *rabi* sorghum, and 680 to 1040 kg ha^{-1} for pearl millet. These gaps indicate that productivity of *kharif* sorghum can be increased 3.0 to 4.0 times, of *rabi* sorghum 1.4 to 2.7 times, and of pearl millet 1.8 to 2.3 times from their current levels of productivity (Murty et al., 2007).

For legumes, the farmers' average yield is 1,040 kg ha^{-1} for soybean, 1,150 kg ha^{-1} for groundnut, 690 kg ha^{-1} for pigeonpea, and 800 kg ha^{-1} for chickpea. Large spatial and temporal variation in yield gap was observed for the four legumes. The yield gaps for the production zones ranged from 850 to 1,320 kg ha^{-1} for soybean, 1,180 to 2,010 kg ha^{-1} for groundnut, 550 to 770 kg ha^{-1} for pigeonpea, and 610 to 1,150 kg ha^{-1} for chickpea. The results showed that on average the productivity of legumes and oilseeds can be increased 2.3 to 2.5 times their current levels of productivity under rainfed situations. Supplemental irrigation would further increase these yields (Bhatia et al., 2006). Similarly, the national average yield gap relative to simulated rainfed potential yields was 2,560 kg ha^{-1} for rainfed rice, 1,120 kg ha^{-1} for cotton, and 860 kg ha^{-1} for mustard. Such large yield gaps could not be estimated for rainfed wheat because of large percentage of irrigated area in all states (Aggarwal et al., 2008). Whether these biophysical estimates of yield gaps can be abridged economically remains to be quantified, but it gives the scope of increasing crop productivity to meet the future food needs of the country.

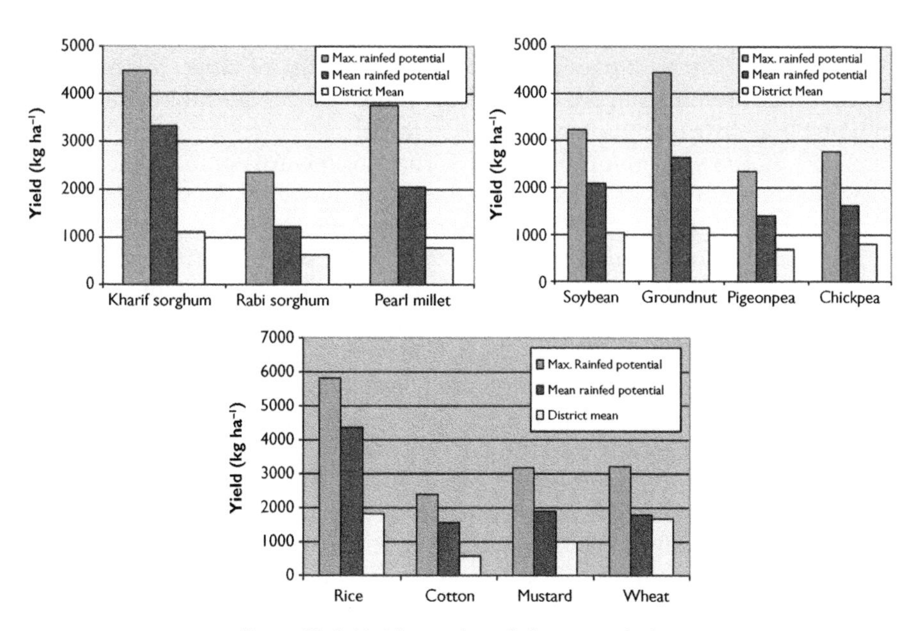

Figure 12.1 Yield gap of rainfed crops in India.

12.3 THE IGNRM APPROACH OF ICRISAT FOR RAINFED AGRICULTURE

ICRISAT has adopted an integrated genetic and natural resource management (IGNRM) approach to enhance agricultural productivity in rainfed areas, which is a powerful integrative strategy of enhancing agricultural productivity. While addressing the core issue of rural development, this approach maximizes synergies among the disciplines of natural resource management, crop improvement, and social sciences, along with people's empowerment through capacity-building measures (Twomlow et al., 2008). ICRISAT has learned that converging different agro-technologies at field level showed greater impact on agricultural productivity in the farmers' holdings than did compartmentalized testing of individual technologies. This was achieved through adoption of an integrated watershed management approach that is holistic in nature to achieve the desired goals of enhancing productivity, reducing land degradation and protecting the environment, which ultimately results in increased economic benefit to rural communities and in alleviating poverty. In our on-stations and on-farm research, the integrated package of technologies was evaluated on watershed scale at various sites in India. The contribution of both individual and combined effects of improved technologies on productivity enhancement and water-use efficiency is presented here.

12.4 ON-STATION WATERSHEDS: TECHNOLOGY DEVELOPMENT AND EVALUATION

12.4.1 Enhancing productivity and resource-use efficiency on Vertisols

The Patancheru area, where ICRISAT is located in India, typifies a SAT climatic environment. The rainy season here begins in June and ends in early October, however, the rains during the month of June, at the time of sowing of crops, are unstable and wide variations are noted in the onset of the rainy season. During the rainy season, the amount and distribution of the rains vary widely.

In 1973, ICRISAT developed a set of Vertisol watersheds at its Patancheru farm. This approach to natural resource management was based on the strategy that, if the land were properly graded and the seedbed prepared in such a way that the rainwater were given enough opportunity time to seep into the soil, the crop-growing season would be secure against water deficits. Grassed waterways were laid out in the watershed hydrological units, so that when heavy rains are received, the water does not stagnate in the field but is safely conducted through the furrows and grassed waterways to the dugout tanks or small reservoirs. These water reservoirs were strategically located in a series to capture most of the surface runoff. These agricultural watersheds have been used to gain insight into the climatic variability impacts on rainfed agriculture production options for sustaining crop yields.

Two Vertisol watersheds were used to evaluate the impact of SAT climate on alternate methods of soil and water conservation and crop productivity under a standard set of agronomic management practices. In the improved watershed-based technology (A), the land was cultivated ahead of the season and prepared to graded BBF system. The rainy season crop was sown in the dry seed bed prior to the on-set of the rainy season; two crops were grown per year in a rotation, which consisted of maize (95 days to maturity) followed by chickpea (120 days to maturity) in year 1, and sorghum (105 days to maturity) intercropped with pigeonpea (210 days to maturity) in year 2. The crops were planted on the graded broad-beds (105 cm wide). Each broad-bed was separated by a furrow (45 cm wide). A standard rate of fertilizers to add 60–80 kg N per ha and 40 kg P_2O_5 per ha was applied every year. All other crop management practices were maintained at the same level throughout the course of this long-term experiment. Under the traditional technology (B) treatment, the seed-bed was kept flat, one crop, either sorghum or chickpea, and was grown during the post-rainy season on conserved soil moisture on the land maintained as fallow during the rainy season. No chemical fertilizers were applied; only farmyard manure was incorporated at 10 t/ha every two years. Farmers in the Patancheru area follow this traditional farming system on the Vertisols. The study has been conducted across the past 30 years, and the following salient results have been obtained.

In the improved system, the two-crop yields were consistently more than 4.7 t/ha, whereas in the traditional system, the average yield across the years was 0.9 t/ha (Figure 12.2). Because of progressive improvements in the soil quality in the improved system, the two-crop yields increased at the rate of 82 kg/ha/year, whereas in the traditional system, the yields increased at 23 kg/ha/year.

The improved system utilized, on average, 67% of seasonal rainfall received as compared with 30% in the traditional management system and endured the climatic risks (seasonal rainfall variability); furthermore, the soil erosion was 1.5 t/ha as against the 6.4 t/ha observed in the traditional system (Table 12.1).

The carrying capacity of the improved system rates at 21 persons/ha/year, compared with 4.6 persons/ha/year under the traditional system. The improved system had no deleterious effects on soil quality, addressed the issues of building up water resources, and is labor intensive.

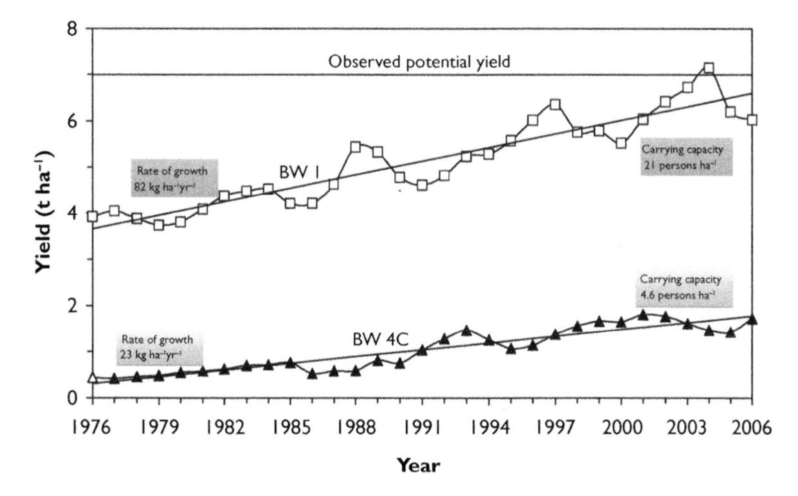

Figure 12.2 Three year moving average of sorghum and pigeonpea grain yield under improved management and on farmers' fields in a deep Vertisol catchment, Patancheru, India.

Table 12.1 Annual water balance and soil loss for traditional and improved technologies in Vertisol watersheds, ICRISAT center, 1976/77 to 1983/84.

	Water balance component (mm)				
Farming systems technology	*Annual rainfall*	*Water used by crops*	*Water lost as surface runoff*	*Water lost as bare soil evaporation and deep percolation*	*Soil loss (t/ha)*
Improved system					
Double cropping on †BBFs	904	602 (67)*	130 (14)	172 (19)	1.5
Traditional system					
Single crop in post-rainy season and cultivation on flat	904	271 (30)	227 (25)	406 (45)	6.4

Note: *Figures in parentheses are amounts of water used or lost expressed as percentage of total rainfall.
†BBF = Broad-bed and furrow system.

Significant positive changes were observed in the soil physical properties under improved management compared with the traditional technology. Organic C content, which is a good indicator of soil quality, was significantly higher for improved management than for traditional management. The total soil organic C (SOC) content in the top 60 cm soil depth contained 6 t more SOC per ha for the improved management watershed than in the traditional management (27.4 vs. 21.4 t/ha) watershed and this difference was reduced to 1.4 t/ha for the 60–120 cm soil depth for these fields. Similar differences were observed with respect to total N and available P contents, especially for the 60 cm soil depth for these management treatments.

12.4.2 Enhancing productivity and resource use efficiency on Vertic Inceptisols

Out of 72 M ha of black soils in India, Vertic Inceptisols occupy about 60 million hectare (M ha) mostly in the states of Madhya Pradesh, Maharashtra, and Andhra Pradesh (Sehgal & Lal, 1988). These soils have similar physical and chemical properties as the Vertisols, except that these are shallower (depth of black soil material), lighter in texture, and have low to medium available water-holding capacity (100–200 mm plant extractable water). Annual rainfall in Central India, where these soils predominate, varies from 750 to 1300 mm with almost 80% received from June until September (Singh, 1997). Because of their location in toposequence, major constraints for crop production on these soils are a high run-off of rainwater and associated soil erosion and depletion of nutrients and beneficial organisms, all leading to a decline in crop productivity. Under such biophysical conditions, farmers find soybean as the most suitable crop in the region. During past two decades, soybean has been introduced on these soils, and currently some 6.0 million ha area is under this crop. However, the productivity of the crop has stagnated at less than 1.0 t/ha. Therefore, for sustainable increase in crop yields of soybean-based systems, it is essential that suitable land management and agronomic practices be introduced to minimize land degradation and to use natural resources efficiently for crop production. Because the Vertic Inceptisols are relatively shallow in depth, the focus of the technology for these soils was to recharge the groundwater through land management and harvesting of excess rainfall in percolation tanks. Integrated nutrient management practices (legumes in the system, bio-fertilizers, and chemical fertilizers) were followed to meet the nutrient needs of the crops and to minimize the pollution of surface and ground waters.

At ICRISAT, Patancheru, we evaluated the long-term effects of improved vis-à-vis traditional management of a Vertic Inceptisol on the productivity of two soybean-based cropping systems, surface runoff and soil erosion, and balance of organic carbon and other nutrients in the soil. Improved management comprised sowing on BBF landform with *Gliricidia* [*Gliricidia sepium* (L.)] grown on graded field bunds between plots; the pruned materials were added to the soil during the cropping season along with the addition of composted crop residues of soybean, chickpea, and pigeonpea at the start of the season (Figure 12.3). Traditional management comprised of flat landform with sowing along the graded field bunds. No organic matter was added to this treatment except the additions through natural leaf senescence and roots. The two cropping systems evaluated from 1996/97 to 2003/04 were soybean-chickpea sequential (SB-CP) and soybean/pigeonpea intercrop (SB/PP) systems, except during the first year (1995/96 season) when only SB-CP was grown in all the four hydrological units.

Improved management of the Vertic Inceptisol decreased surface runoff by 24%–27% and soil loss by 44%–47% as compared with the traditional management (Table 12.2). Water use by the two crops ranged from 50% to 100% of seasonal rainfall across years. Surface runoff and deep drainage water was captured in surface tanks, which resulted in an increase in water level in dug wells – this water was used to provide irrigation to horticultural crops (Figure 12.4). Overall rainfall-use efficiency on watershed basis was greater than 50% in most years.

Figure 12.3 Vertic Inceptisol watershed (BW7 watershed) at ICRISAT center, Patancheru.

Table 12.2 Mean seasonal rainfall, surface runoff, and soil loss in the improved and traditional management treatments during 1996/97 to 2003/04 cropping seasons.

Stat.	*Rainfall (mm)*	*Surface runoff (mm)*		*Water use (mm)**		*Soil loss (t ha^{-1})*	
		Imp.	*Trad.*	*Imp.*	*Trad.*	*Imp.*	*Trad.*
Medium depth							
Mean	751	131	179	515	509	1.9	3.6
Range	401–1062	0–477	0–641	395–563	400–565	0–6.5	0–12.0
Shallow depth							
Mean	751	126	166	482	481	1.9	3.4
Range	401–1062	0–489	0–588	400–565	388–558	0–6.7	0–110.1

*From 1995 to 2001. Imp. = Improved; Trad. = Traditional.

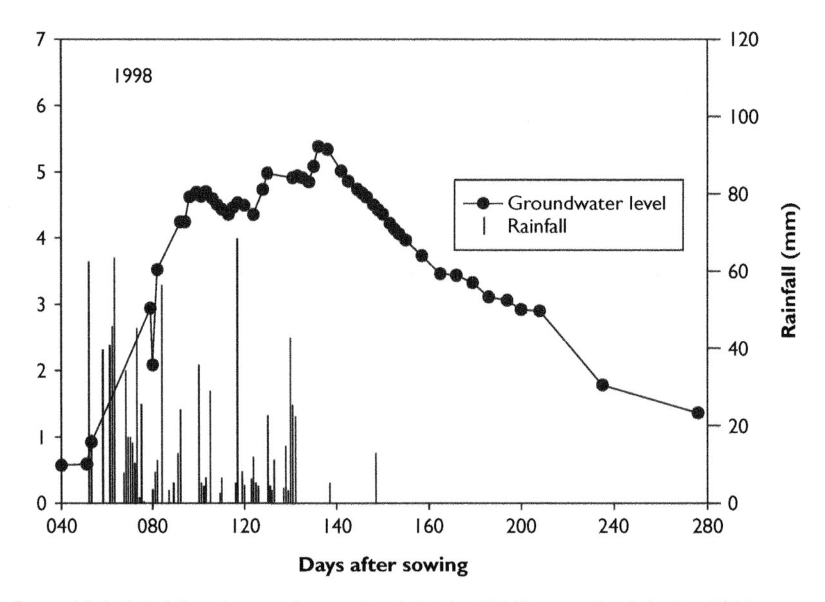

Figure 12.4 Rainfall and groundwater levels in the BW7 watershed during 1998 season.

As the watershed area was cleared of vegetation at the start of the watershed experiment, it had high levels of soil organic carbon (SOC) in the soil profile. Because of rapid decline in SOC in the initial years of the experiment, the productivity of both the cropping systems for biomass and grain yield declined across time; however, the decrease in productivity of the SB/PP system was less than that of the SB-CP system in both the management treatments. In the last three years of the study, total grain yield productivity of SB-CP system was 21% higher on the medium-deep soil and 9% higher on the shallow soil with improved management than with traditional management (Table 12.3). Total grain yield of the SB/PP inter-crop system was marginally higher by 4%–7% with improved management on both the soil types. Across

Table 12.3 Total grain yield of soybean-chickpea sequential and soybean/pigeonpea intercrop systems in the improved and traditional management treatments during the first three and the last three years of study. Each data point is mean of three years.

	Soybean + Chickpea (kg/ha)		*Soybean + Pigeonpea (kg/ha)*	
Soil depth	*Traditional*	*Traditional*	*Improved*	*Improved*
	1995/96 to 1997/98 season		1996/97 to 1998/99 season	
Medium depth	2750	2710	2000	2180
Shallow depth	2520	2360	2100	2030
SEd	111.7		123.3	
	2001/02 to 2003/04 season			
Medium depth	2170	1780	1780	1710
Shallow depth	1400	1290	1690	1590
SEd	129.7		143.5	

SEd = Standard error of the difference between means.

eight years of cropping, organic carbon (OC), total N, and available P content of the soil declined in all the treatments; and in 2002, the extractable sulfur (S), zinc (Zn), and boron (B) in the soil were found to be deficient. After eight years of cropping, improved management retained 2.0 t ha^{-1} more SOC compared with the traditional system, and SB/PP system retained 2.4 to 6.4 t ha^{-1} more OC compared with the SB-CP system in the medium-deep soil, but not in the shallow soil. The conclusion was that for improving and stabilizing productivity of the soybean-based cropping systems on Vertic Inceptisols – in addition to improved land and water management – more balanced nutrition of crops through additions of organics and inorganics, including application of micronutrients, would be required than the low input practice evaluated in this study.

12.4.3 Efficient use of supplemental irrigation water

In the SAT regions, water is a scarce resource, and the amount of water available for supplemental irrigation is generally limited. Once the surplus water has been harvested in surface ponds or the groundwater is recharged, its efficient use is important for increasing crop productivity in a sustainable manner. Efficient use of water involves both the timing of irrigation to the crop and efficient water application methods. Broadly, the methods used for application of irrigation water can be divided into two types, viz., surface irrigation systems (border, basin, and furrow) and pressurized irrigation systems (sprinkler and drip). In the surface irrigation system, the application of irrigation water can be divided into two parts: first, the conveyance of water from its source to the field and, second, application of water in the field.

12.4.4 Conveyance of water to the field

In most SAT areas, the water is carried to cultivated fields by open channel, which is usually unlined and therefore a large amount of water is lost through seepage.

On SAT Vertisols, generally there is no need to line the open-field channels, as the seepage losses in these soils are low mainly due to very low saturated hydraulic conductivity in the range of 0.3 to 1.2 mm hr^{-1} (El-Swaify et al., 1985). On Alfisols and other sandy soils having more than 75% sand, the lining of open-field channel or use of irrigation pipes is necessary to reduce the high seepage water losses. The uses of closed conduits (plastic, rubber, metallic, and cement pipes) are getting popular especially with farmers growing high-value crops, viz., vegetables and horticultural crops.

12.4.5 Efficient application of supplemental water on SAT Vertisols

Formation of deep and wide cracks during soil drying is a common feature of SAT Vertisols. The abundance of cracks is responsible for high initial infiltration rates (as high as 100 mm hr^{-1}) in dry Vertisols (El-Swaify et al., 1985). This specific feature of Vertisols makes efficient application of limited supplemental water to the entire field a difficult task. Among the various systems studied at ICRISAT, the BBF system was found to be most appropriate for applying irrigation water on Vertisols. As compared with narrow ridge and furrow, the BBF saved 45% of the water without affecting crop yields. Compared with narrow ridge and furrow and flat systems, the BBF system had higher water application efficiency, water distribution uniformity, and better soil-wetting pattern. Studies conducted to evaluate the effect of shallow cultivation in furrow on efficiency of water application showed that the rate of water advance was substantially higher in cultivated furrows as compared with uncultivated furrows. Shallow cultivation, in moderately cracked furrows before the application of irrigation water, reduced the water required by about 27% with no significant difference in chickpea yields (Table 12.4).

12.4.6 Efficient application of supplemental water on SAT Alfisols

On Alfisols, surface irrigation on flat cultivated fields results in very poor distribution of water and high water loss. At ICRISAT research station, Patancheru, India, experiments were conducted to determine the most appropriate land surface

Table 12.4 Grain yield of chickpea in different treatments, Vertisols, ICRISAT center.

Treatment	*Mean depth of water application (cm)*	*Grain yield (kg ha^{-1})*
No supplemental irrigation	0	690
One supplemental irrigation on uncultivated furrows	6.3	920
One supplemental irrigation on cultivated furrow	4.6	912
SEM		19
CV%		5.55

configuration for the application of supplemental water. The wave-shaped broad beds and furrows with checks at every 20 m length along the furrows were found to be the most appropriate for efficient application of supplemental water and increasing crop yields. The moisture distribution across the beds was uniform in the case of the wave-shaped broad-beds with checks compared with normal BBF system. The sorghum yield in wave-shaped broad-beds with checks was higher at every length of run compared with normal BBF (Table 12.5). We found that when irrigation water was applied in normal BBF system on Alfisols, the center of the broad-bed remained dry. The center row crop did not get sufficient irrigation water, resulting in poor crop yields. In another experiment on Alfisols, normal BBF system (150 cm wide) was compared with narrow ridge and furrow system (75 cm wide). We found that the narrow ridge and furrow system performed better than BBF system both in uniform water application and higher crop yields. Therefore, for Alfisols, the wave-shaped broad-bed with checks in furrow is the most appropriate land surface configuration for efficient application of supplemental irrigation water, followed by narrow ridge and furrow system.

The improved surge flow irrigation method can also be used for improving the performance of furrow irrigation. This system saves water, uses less energy, and improves water productivity. With proper planning and design, surge flow system can be extensively used for efficiently irrigating high-value crops grown using the ridge and furrow landform (Singh, 2007). The modern irrigation methods, viz., sprinklers and drip irrigation, can play vital roles in improving water productivity. These irrigation systems are highly efficient in water application and have opened up opportunities to cultivate light textured soils with very low water-holding capacity and in irrigating undulating farmlands. The technology has also enabled regions facing limited water supplies to shift from low-value crops with high water requirements, such as cereal, to high value crops with moderate water requirements, such as fruits and vegetables (Sharma & Sharma, 2007). Implementation of these improved irrigation techniques can save water, energy, and increase crop yields. However, currently the use of these improved irrigation methods is limited, primarily because of the high initial cost. Favorable government policies and the availability of credit are essential for popularizing these irrigation methods.

Table 12.5 Sorghum grain yield (t ha^{-1}) as affected by the water distribution in different surface irrigation systems on Alfisols.

Length of run (m)	*Normal BBF*[†]	*Wave-shaped broad beds with checks in furrow*
0	2.07	2.52
20	2.38	3.91
40	2.56	4.42
60	3.06	4.54
80	3.26	4.53
100	3.08	4.42

[†]BBF = Broad bed and furrow system.

12.4.7 Conjunctive use of rainfall and limited irrigation water

Stewart, Musick, and Dusek (1983) developed a limited irrigation dryland system (LID) for the efficient use of limited irrigation water for crop production. The objective of the LID concept was to maximize the combined use of growing-season rainfall, which varies for any given year, with a limited supply of irrigation water. This system was studied at ICRISAT research center, Patancheru, India, for rainy season sorghum on Alfisols. Results demonstrated the usefulness of LID system in the application of limited water under uncertain and erratic rainfall conditions. The LID system increased both the crop yields and water-application efficiency (WAE) during the two years of study (Table 12.6).

12.4.8 Crop responses to supplemental irrigation

Srivastava and colleagues (1985) studied the response of post-rainy season crops to supplemental irrigation grown after maize or mung bean on a Vertisol. The highest WAE was recorded for chickpea (5.5 kg mm^{-1} ha^{-1}), followed by chillies (4.0 kg mm^{-1} ha^{-1}), and safflower (2.0 kg mm^{-1} ha^{-1}) (Table 12.7). They concluded that one pre-sowing irrigation to the sequential crops of chickpea and chillies was profitable on Vertisols. Average additional gross returns due to supplemental irrigation were about Rs 1630 ha^{-1} for safflower, Rs 7900 ha^{-1} for chickpea, and Rs 14600 ha^{-1} for chillies.

Impressive benefits have also been reported from supplemental irrigation to rainy and post-rainy season crops on Alfisols at the ICRISAT center (El-Swaify et al., 1985; Pathak & Laryea, 1991). The average water-application efficiency (WAE) for sorghum (14.9 kg mm^{-1} ha^{-1}) was more than that for pearl millet (8.8 to 10.2 kg mm^{-1} ha^{-1}) (Table 12.8). An intercropped pigeonpea responded less to irrigation and its average WAE ranged from 5.3 to 6.7 kg mm^{-1} ha^{-1} for both sorghum/pigeonpea and pearl millet/pigeonpea intercrop systems. Tomatoes responded very well to water application with an average WAE of 186.3 kg mm^{-1} ha^{-1} (Table 12.8).

For the sorghum/pigeonpea intercrop, two irrigations of 40 mm each gave an additional gross return of Rs 9750 ha^{-1}. The highest additional gross return from supplemental irrigation was obtained by growing tomato (Rs 58300 ha^{-1}). These results

Table 12.6 Effect of irrigation on sorghum (CSH6) yield (kg ha^{-1}) on different sections of the slope, Alfisols, ICRISAT center, 1985–1986.

	Upper section 0–20 m		*Middle section 20–40 m*		*Lower section 40–60 m*		*Average yield*		*WAE*[b] *(kg mm^{-1} ha^{-1})*	
	1985	*1986*	*1985*	*1986*	*1985*	*1986*	*1985*	*1986*	*1985*	*1986*
Rainfed	1058	2220	1618	2110	1710	2140	1659	2150	–	–
Full irrigation[a]	3716	3404	3516	3200	2960	3458	3390	3352	6.9	7.5
LID system	3413	3090	2600	2710	2000	2110	2671	2636	12.1	9.2

[a]5 irrigations totaling 250 mm and 4 irrigations totaling 130 mm were applied during 1985 and 1986, respectively, on full irrigation and LID (upper section) treatments on area basis.
[b]Water-application efficiency (WAE) = increase in yield due to irrigation/Depth of irrigation.

Table 12.7 Response of sequential crops to supplemental irrigation on a Vertisol watershed, ICRISAT center, 1981–85.

	Mean yield (kg ha^{-1})		
Cropping system	*Supplemental irrigation*	*Increase due to irrigation*	*Water application efficiency (mm kg^{-1} ha^{-1})*
1. Maize + chickpea sequential	1540	493	5.6
2. Mung + chillies sequential	1333	325	4.1
3. Maize + safflower sequential	1238	165	2.1

Table 12.8 Grain-yield response (t ha^{-1}) of cropping systems to supplemental irrigation on an Alfisol watershed, ICRISAT, Patancheru, 1981–82.

One irrigation turn of 40 mm	*Increase due to irrigation*	*WAE*[†] *(kg ha^{-1} mm^{-1})*	*Two irrigations 40 mm each*	*Increase due to irrigation*	*WAE (kg ha^{-1} mm^{-1})*	*Combined WAE (kg ha^{-1} mm^{-1})*
Intercropping systems						
Pearl millet			Pigeonpea			
2.353	0.403	10	1.197	0.423	5.3	6.8
	Sorghum			Pigeonpea		
3.155	0.595	14.9	1.22	0.535	6.7	9.4
Sequential cropping systems						
Pearl millet			Cowpea			
2.577	0.407	10.2	0.735	0.425	5.3	6.9
Pearl millet			Tomato			
2.215	0.35	8.8	26.25	14.9	186.3	127.1

[†]WAE = Water-application efficiency.

indicate that on Alfisols, significant returns can be obtained from relatively small quantities of supplemental water.

It is interpreted from the above studies that on Alfisols, the best results from the limited supplemental irrigation were obtained during the rainy season. On Vertisols in medium to high rainfall areas, pre-sowing irrigation for post-rainy season crops was found to be the most beneficial. The best responses to supplemental irrigation were obtained when irrigation water was applied at the critical stages of crop. To get the maximum benefit from the available water, growing high-value crops, viz., vegetables and horticultural crops, is getting popular even with poor farmers.

12.5 INITIAL ON-FARM EVALUATION OF WATERSHED TECHNOLOGIES AND LESSONS LEARNED

Based on impressive successes, with on-station watersheds using new technologies for double cropping on Vertisols, researchers expected that this approach could be 'transferred' to farmers' fields, thereby enhancing the productivity of rainfed systems. The whole process evolved around the 'demonstration' of the technology package and of its possible benefits under farmers' conditions. The Vertisol technology package was demonstrated in the village watersheds. It included land smoothing, drain construction, introduction of the BBF system, use of a bullock-drawn Tropicultor, summer cultivations, dry seeding, and the use of appropriate nutrient and pest-management options along with improved high-yielding crop varieties. Yields in the improved watershed were compared with those in the traditional farmers' system. The trials performed during 1981/82 confirmed that on-farm yields could be similar to those from on-station operational research watersheds. With improved management, sorghum/pigeonpea intercrop system produced higher grain yields (1.9 t ha^{-1}) and net returns of Rs. 3838 ha^{-1} $year^{-1}$ compared with those from the traditional farmers' fields, which recorded 0.55 t ha^{-1} of grain yield and net returns of Rs. 1234 ha^{-1} $year^{-1}$. Similar on-farm evaluations were done at several locations in Maharashtra, Gujarat, Madhya Pradesh, Karnataka, and Andhra Pradesh.

However, subsequent evaluation of these watersheds after 15 years revealed that, in most of them, the farmers went back to their normal practices and that only selected components of the technology package were continued. As part of the watershed evaluation exercise, hundreds of farmers were interviewed and a multidisciplinary team of scientists analyzed the process, farmers' responses and possible reasons for the low adoption of the technology package. Several constraints affected the adoption of technology and higher adoption rates were observed in assured high-rainfall Vertisol areas. From many years of experience of working with watershed technologies, the major lessons learned were: mere on-farm demonstration of technologies by scientists does not guarantee their adoption by farmers; a higher degree of farmers' participation through a consultative to cooperative mode from the planning to evaluation stage is needed; a consortium of organizations is needed for technical guidance, as no single organization can provide support to all the problems of a watershed; a holistic systems approach through the convergence of different activities is needed and it should improve farmers' livelihoods and not merely conserve soil and water in the watershed; efficient technical options are needed to manage natural resources for sustaining systems; appropriate technology applications to address region-specific constraints need to be identified and simple broad recommendations do not help; individual farmers should first realize tangible economic profits from the watersheds; and involvement of women and youth groups is essential, as they play an important role in decision making in the families (Wani et al., 2001).

12.6 CURRENT MODEL OF INTEGRATED WATERSHED MANAGEMENT FOR ENHANCING PRODUCTIVITY AND EFFICIENT USE OF NATURAL RESOURCES

Based on the lessons learned from the extensive watershed-based research and initial on-farm evaluation of watershed technologies, ICRISAT scientists articulated a new model of implementing watershed-based interventions to enhance crop productivity and efficient management of natural resources (Wani et al., 2003). The important components of the new integrated watershed management model are as follows:

- The farmer participatory approach through cooperation and not through the contractual agreement with the farmers. The factors that promote collective action by the community are: the program be demand driven and provide tangible benefits to the individual farmers; initial entry into the watersheds or villages should be knowledge-based to enhance agricultural productivity; and practices that promote equity, equal partnership, shared vision, and trust should be encouraged. Transparency in the use of financial resources, social audit and good local leadership would enhance cooperation and collective action.
- The use of new scientific tools for management and monitoring of watersheds.
- Linking of on-station research watersheds with on-farm community watersheds for technology transfer and for addressing the emerging technical issues.
- A holistic farming systems approach to improve livelihoods of people and not merely conservation of soil and water.
- A consortium of institutions for technical guidance on the on-farm watersheds.
- A micro-watershed within the watershed where farmers conduct strategic research with technical guidance from the scientists.
- Minimize free supply of inputs for undertaking on-farm evaluation of technologies.
- Low-cost soil- and water-conservation measures and structures.
- The amalgamation of traditional knowledge and new knowledge for efficient management of natural resources.
- Emphasis on individual farmer-based conservation measures for increasing productivity of individual farmers along with community-based soil- and water-conservation measures.
- Continuous monitoring and evaluation by stakeholders.
- Empowerment of community of individuals and strengthening of village institutions for managing natural resources.

Since 1999, using the new integrated watershed management model, we have initiated new on-farm watersheds in India. Various interventions made in various community watersheds to enhance productivity and resource-use efficiency are presented.

12.7 ENHANCING PRODUCTIVITY AND WATER USE EFFICIENCY IN ON-FARM COMMUNITY WATERSHEDS

12.7.1 In situ soil and water conservation

Implementation of the type of land- and water-management system depends on the characteristics of the soil, climate, farm size, capital, and availability of human and power resources. Land smoothening and forming field drains are the basic components of land and water management for conservation and safe removal of excess water. Broad-bed and furrow (BBF) system is an improved *in situ* soil and water conservation and drainage technology for the Vertisols. The system consists of relatively flat beds approximately 100 cm wide and shallow furrow about 50 cm wide laid out in the field with a slope of 0.4% to 0.8% (Figure 12.3). The BBF system helps safely dispose of excess water through furrows when there is high intensity rainfall with minimal soil erosion; at the same time, it serves as land surface treatment for *in situ* moisture conservation. Contour farming is practiced on lands having medium slope (0.5%–2%) and permeable soils, where farming operations, such as plowing and sowing, are carried out along the contour. The system helps reduce the velocity of runoff by impounding water in a series of depressions and thus decreasing the chance of developing rills in the fields. Contour bunding is recommended for medium to low rainfall areas (<700 mm) on permeable soils with less than 6% slope. It consists of a series of narrow trapezoidal embankments along the contour to reduce and store runoff in the fields. Conservation furrows is another promising technology in red soils receiving rainfall of 500–600 mm with a moderate slope (0.2%–0.4%). It comprises a series of dead furrows across the slope at 3–5 m intervals, where the size of furrows is about 20 cm wide and 15 cm deep.

On-farm trials on land management of Vertisols of central India revealed that BBF system resulted in a 35% yield increase in soybean during rainy season and yield advantage of 21% in chickpea during post-rainy season when compared with the farmers' practice. Similar yield advantage was recorded in maize and wheat rotation under BBF system (Table 12.9). Yield advantage of 15% to 20% was recorded in maize, soybean, and groundnut with conservation furrows on Alfisols over farmers' practices of Haveri, Dharwad, and Tumkur watersheds in Karnataka (Table 12.10). Yield advantage in rainfall-use efficiency (RUE) were also reflected in cropping systems involving soybean-chickpea, maize-chickpea, soybean/maize-chickpea under improved land management systems. The RUE ranged from 10.9 to 11.6 kg ha^{-1} mm^{-1} under BBF systems across various cropping systems compared with 8.2 to 8.9 kg ha^{-1} mm^{-1} with flat-on-grade system of cultivation on Vertisols (Table 12.11).

Table 12.9 Effect of land configuration on productivity of soybean and maize-based system in the watersheds of Madhya Pradesh, 2001–05 (BBF = Broad Bed and Furrow).

Watershed location	*Crop*	*Farmer's practice*	*BBF system*	*% Increase in yield*
Grain yield (t ha^{-1})				
Vidisha and Guna	Soybean	1.27	1.72	35
	Chickpea	0.80	1.01	21
Bhopal	Maize	2.81	3.65	30
	Wheat	3.30	3.25	16

Table 12.10 Effect of improved land and water management on crop productivity in Sujala watersheds of Karnataka during 2006–07.

Watershed	*Crop*	*Farmers' practice*	*Conservation furrows*	*% Increase in yield*
Grain yield (t ha^{-1})				
Haveri	Maize	3.57	4.10	15
Dharwad	Soybean	1.50	1.80	20
Kolar	Groundnut	1.05	1.22	16
Tumkur	Groundnut	1.29	1.49	15

Table 12.11 Rainfall use efficiency of different cropping systems under improved land management practices in Bhopal, Madhya Pradesh, India.

Cropping system	*Flat-on-grade*	*Broad-bed and furrow*
Rainfall-use efficiency (kg ha^{-1} mm^{-1})		
Soybean-chickpea	8.2	11.6
Maize-chickpea	8.9	11.6
Soybean/maize-chickpea	8.9	10.9

'-' = Sequential system.
'/' = Intercrop system.

12.7.2 Water harvesting and groundwater recharge

In medium to high rainfall areas, despite following the *in situ* moisture conservation practices, rainfall runoff caused by high intensity storms or water surplus after filling up the soil profile does occur. This excess water needs to be harvested in surface ponds for recycling through supplemental irrigation or to recharge the groundwater for later use in the post-rainy season. Various types of water-harvesting structures were built in Adarsha watershed in Kothapally village in Andhra Pradesh with the participation of farmers (Figure 12.5). Water harvesting in these structures resulted in increased groundwater levels (Figure 12.6). Additional water resource thus created was used by the farmers in providing supplemental irrigation to the crops especially to provide come-up irrigation to the post-rainy season crops, such as chickpea, or to grow high-value crops, such as vegetables. Small and well-distributed water-harvesting structures in the watershed area provided equity and benefited more farmers than the large-sized structures that benefit only a few farmers.

12.7.3 Improved crop varieties and cropping systems

The adoption of improved varieties always generates significant field-level impact on crop yield and stability. The yield advantage through the adoption of improved varieties has been recognized undoubtedly in farmer participatory trials across India under rainfed systems. Recent trials during rainy season conducted across Kolar and Tumkur districts of Karnataka, India, revealed that mean yield advantage of 52% in finger millet was achieved with high-yielding varieties like GPU 28, MR 1, HR 911, and L 5 under farmers' management (traditional management and farmers inputs) compared with use of local varieties and farmer management (Table 12.12). These results showed the efficient use of available resources by the improved varieties reflected in grain yields under given situations. However, yield advantage of 103% was reported in finger millet due to improved varieties under best-bet management practices (balanced nutrition including the application of Zn, B, and S, and crop protection). Similarly, use of improved groundnut variety ICGV 91114 resulted in pod yield of 2.32 t ha^{-1} under farmer management compared with local variety with similar inputs. The yields of improved varieties further improved by 83% over the local variety with improved management that included balanced application of nutrients.

Figure 12.5 Water-harvesting structure in Adarsha watershed Kothapally, Andhra Pradesh.

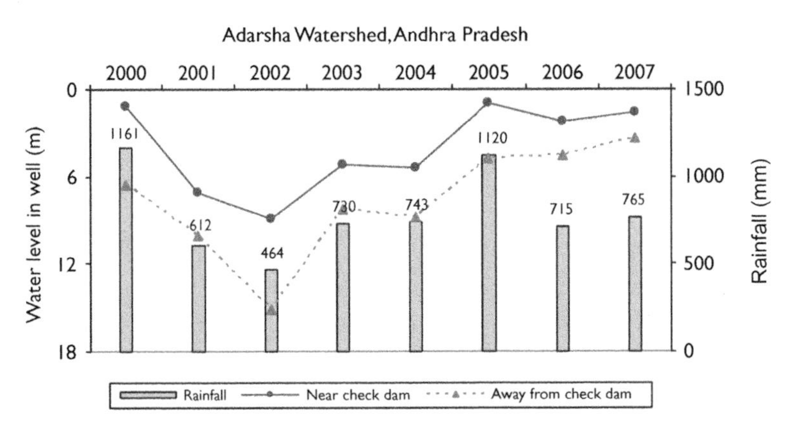

Figure 12.6 Impact of water-harvesting structures on groundwater levels in Adarsha watershed, Kothapally, Andhra Pradesh.

12.7.4 Integrated nutrient management

Low fertility is one of the major constraints responsible for the low productivity under rainfed system besides water scarcity. The deficiency of N and P among the nutrients is regarded as an important issue in soil fertility-management programs. However, ICRISAT-led watershed program across the subcontinent provided the opportunity to diagnose and understand the widespread deficiencies of secondary nutrients, such as S, and micronutrients, such as B and Zn, in the soils of rainfed areas (Sahrawat et al.,

Table 12.12 Effect of improved varieties of finger millet and groundnut under different levels of management in Kolar and Tumkur districts, Karnataka during 2005.

Finger millet yield (t ha^{-1})			*Groundnut yield (t ha^{-1})*		
Variety	*Farmers' practice*	*Improved mgmt.*	*Variety*	*Farmers' practice*	*Improved mgmt.*
Local	1.97	–	TMV 2 (local)	1.38	1.74
GPU 28	3.00	3.68	JL 24	1.92	2.80
MR 1	2.83	3.93	ICGV 91114	2.32	3.03
HR 911	2.90	3.66			
L 5	3.20	4.65			
Mean	3.00	4.00		1.88	2.52
% increase over local variety	52	103		36	83

2007). An on-farm survey across various states revealed that out of 1,926 farmers' fields, 88% to 100% were deficient in available S; 72%–100% in available B and 67%–100% in available Zn (Table 12.13).

On-farm trials evaluated the response of crops to the application of S and micronutrients at the rate of 30 kg ha^{-1} S, 0.5 kg ha^{-1} B, and 10 kg ha^{-1} Zn. The study revealed 79% yield advantage in maize, 61% in castor, 51% in greengram, and 28% in groundnut compared with the yield levels without application of S and micronutrients (Table 12.14). Impressive economic gains due to improved soil-fertility management to the extent of Rs 5948 and Rs 4333 ha^{-1} in maize and groundnut, respectively, were also reported from ICRISAT-led watershed program across Andhra Pradesh (Table 12.15). Addition of micronutrients and S substantially increased productivity of crops and thus resulted in increased rainfall-use efficiency (RUE). RUE of maize for grain yield under farmer inputs of nutrients was 5.2 kg mm^{-1} compared with 9.2 kg mm^{-1} with S, B, and Zn application over and above the farmer nutrient inputs; respective values in the same order of treatment were 1.6 kg mm^{-1} and 2.8 kg mm^{-1} for groundnut and 1.7 kg mm^{-1} and 2.9 kg mm^{-1} in mung bean. However, addition of recommended dose of N and P along with S, B, and Zn in legumes further increased agricultural productivity, RUE, and incomes of the farmers (Table 12.16).

12.7.5 Integrated pest management

Integrated pest management (IPM) is an effective and environmentally sensitive approach to pest management that relies on a combination of available pest suppression techniques to keep the pest populations below the economic thresholds. In other words, IPM is a sustainable approach to managing pests by combining biological, cultural, physical, and chemical tools in a way that minimizes economic, health, and environmental risks. New IPM products and methods are developed and extended to producers to maximize yields. On-farm trials on IPM were evaluated in Bundi watershed, Madhya Pradesh, which clearly demonstrated that IPM encompassing suitable varieties, clean cultivation, scouting through pheromone traps, use of NPV against

Table 12.13 Magnitude of deficiency of micronutrients in the soils of various states in India (Sahrawat et al., 2007).

	% of Deficient fields		
Locations	*Zn*	*B*	*S*
Kurnool (Andhra Pradesh)	81	92	88
Dewas (Madhya Pradesh)	100	96	100
Bundi (Rajasthan)	67	72	72
Bharuach (Gujarat)	85	100	40
Gurgaon (Haryana)	89	93	60
Tirunelveli (Tamil Nadu)	100	100	100

Table 12.14 Effect of sulfur and micronutrient amendments in different field crops.

	Crop yield (kg ha^{-1})		
Crop	*Control*	*Sulfur + micronutrients*	*% Increase over control*
Maize	2800	4560	79
Green gram	770	1110	51
Castor	470	760	61
Groundnut pod	1430	1825	28

Table 12.15 Yield and economic returns in response to application of nutrients in maize and groundnut in Andhra Pradesh.

	Maize		*Groundnut*	
Treatment	*% Yield increase over FP*	*Economic returns (Rs ha^{-1})*	*% Yield increase over FP*	*Economic returns (Rs ha^{-1})*
Farmer's practice (FP)	–	13931	–	12490
FP+S	26	17228	12	13660
FP+Zn	33	17479	27	14780
FP+B	33	18354	20	14850
FP+S+B+Zn	49	19429	48	16830
FP+S+B+Zn+N+P	75	21766	78	19520

Table 12.16 Effect of micronutrient application on rainfall use efficiency in various field crops in Andhra Pradesh and Madhya Pradesh, India.

	Rainwater use efficiency (kg mm^{-1} ha^{-1})	
Crop	*Farmers' practice*	*Farmers' practice + micronutrients*
Andhra Pradesh (Kurnool, Mahabubnagar and Nalgonda districts)		
Maize	5.2	9.2
Groundnut	1.6	2.8
Mung bean	1.7	2.9
Sorghum	1.7	3.7
Madhya Pradesh (Vidisha district)		
Soybean	1.4	2.7

lepidopteron pests, and installing bird perches resulted in yield advantage of 18% and increased net returns by 39% in green peas compared with practice of chemical control alone (Table 12.17).

12.7.6 Crop intensification: A case study from guna watershed, Madhya Pradesh

The practice of fallowing Vertisols and associated soils in Madhya Pradesh has decreased after the introduction of soybean. However, estimates are that about 2.02 M ha of cultivable land is still kept fallow in Central India, where there is a vast potential for having crop during *kharif* season. However, the survey indicated that the introduction of *kharif* crop delays the sowing of post-rainy crop and frequent waterlogging of crops during *kharif* season, which is a major problem forcing farmers to keep the cultivable lands fallow. Under such situations, ICRISAT demonstrated the avoidance of water logging during initial crop growth period on Vertisols by preparing the fields to BBF along with grassed waterways. Simulation studies using SOYGRO model showed that early sowing of soybean in seven out of 10 years was possible by which soybean yields can be increased three-fold along with appropriate nutrient management. Hence, evolving timely sowing with short-duration soybean genotypes would pave the way for successful post-rainy-season crop where the moisture carrying capacity is sufficiently high to support successful post-rainy-season crop. Yield maximization and alternate crops can be focused on post-rainy season, as there is assured moisture availability in Vertisol regions. On-station research was initiated with Indian Institute of Soil Science (IISS), Bhopal, to address issues related to soil, water, and nutrient management practices for sustaining the productivity of soybean-based cropping systems in Madhya Pradesh. Then, the conceptual best-bet options were scaled up in farmers' fields, and yield advantages to the tune of 30% to 40% over the traditional system were recorded.

On-farm soybean trials conducted by ICRISAT involving improved land configuration (BBF) and short-duration soybean varieties along with fertilizer application (including micronutrients) showed a yield increase of 1,300 to 2,070 kg ha^{-1} compared with 790 to 1,150 kg ha^{-1} in Guna, Vidisha and Indore districts of Madhya Pradesh. Similarly, soybean varieties evaluated were Samrat, MAUS 47, NRC 12, Pusa 16, NRC 37, JS 335, and PK 1024, out of which performance of JS 335 was better in Guna watershed of Madhya Pradesh. Increased crop yields (40%–200%) and incomes (up to 100%) were realized with landform treatment, new varieties, and other best-bet management options.

Table 12.17 Effect of IPM on crop productivity and net returns in green peas, Bundi watershed, Rajasthan.

Technology	*Cost of cultivation (Rs ha^{-1})*	*Cost of pest management (Rs ha^{-1})*	*Yield (t ha^{-1})*	*Net returns (Rs ha^{-1})*
Farmers' practice	8520	1800	3.53	10870
IPM	7800	1080	4.16	15070

12.7.7 Crop diversification with supplemental irrigation

The primary constraints to food security in developing countries are low productivity per unit area, shrinking land and water sources available for cropping, and escalating costs of crop production. Under these circumstances, crop diversification can be a useful means to increase crop output under different settings of available resources either through broadening the base of the system by adding more crops coupled with efficient management practices or replacing traditional crops with high-value crops. Crop diversification allows realization of the real value of improved water availability through watershed programs either through growing high-value crops like vegetables or a larger number of crops with supplemental irrigation. However, crop diversification takes place automatically from traditional agriculture to high-value/commercial agriculture at the field level once the water availability is improved. On-farm survey in Ringnodia watershed in Madhya Pradesh revealed the spread of high-value crops like potato, coriander, garlic, etc., and an increase in net income from farming activities once the scope for supplemental irrigation was established in the watershed (Table 12.18).

12.7.8 Crop diversification with chickpea in rice fallows

It is estimated that about 11.4 M ha of rice fallows are available in India. The amount of soil moisture remaining in the dry season after rice crop is usually adequate for raising a short-duration legume crop. Despite low yields, legumes grown after rice due to progressively increasing biophysical stresses, their low-cost of production, and higher market prices often result in greater returns to the farmer. Thus, the twin benefits of income and nutrition could be realized from legumes rather than from rice in spite of moderate yields of legumes. Introduction of early-maturing cool-season chickpea in the rice fallows by addressing the crop establishment constraints should certainly improve cropping intensity and sustainability of the system. Main constraints to the production of legumes in rice fallows are low P in the soil, poor plant establishment, low or absence of native rhizobial population, root rot, and terminal drought. On-farm trials in eastern states of India on early-maturing chickpea in rice fallows with suitable best-bet management practices revealed that chickpea grain yields in the range of 800–850 kg ha^{-1} can be obtained.

Table 12.18 Crop diversification with high-value crops with supplemental irrigation in Ringnodia watershed, Madhya Pradesh.

Crops	*Area covered (ha)*	*Yield (t ha^{-1})*	*Net income (Rs ha^{-1})*
Potato	8.3	17.5	29130
Onion	1.0	25.2	42000
Garlic	1.5	7.6	15750
Hybrid tomato	1.5	66.8	55000
Coriander	2.9	6.1	12700

Table 12.19 Effect of seed priming with sodium molybdate on the performance of chickpea in rice fallows with residual moisture.

	Chickpea yield (kg ha^{-1})		
States	*Control*	*Seed priming with mo*	*Yield advantage (%)*
Madhya Pradesh	814	917	12.7
Uttar Pradesh	2053	2207	7.5
Orissa	284	323	13.7
Jharkhand	664	663	–
West Bengal	309	317	2.6

Table 12.20 Effect of short-duration rice varieties in the rice + rice fallow chickpea system in rainfed uplands.

State	*Varieties*	*Yield (kg ha^{-1})*	*Yield advantage* (%)*
Bihar	Kalinga 3, Vandana, Tulasi	032	144
Eastern Uttar Pradesh	NDR 118, Narendra 97	3613	75
Orissa	Nilagiri, Chandeswari, Vandana	2940	125

*Over check or traditional varieties.

Molybdenum (Mo) deficiency is considered rare in most agricultural cropping areas. However, our on-farm research since 2002 has suggested that in the acid soils of rice fallows, Mo is relatively unavailable, and nodulation, growth, and yield of chickpea can be improved by providing small amounts of Mo (Kumar Rao et al., 2008). The study revealed that seed priming with sodium molybdate resulted in a yield advantage of 2.6% to 13.7% in rice fallow chickpea compared with control (Table 12.19). It is assumed that residual soil moisture after the harvest of rice in target regions could be 100 mm in the soil profile and hence moisture-use efficiency of rice fallow chickpea was calculated to be in the range of 8.0 to 9.0 kg ha^{-1} mm^{-1}.

In a few cases, delayed harvest of rice affects the sowing of chickpea and its grain yield due to terminal drought. Hence to address this issue, on-farm trials were conducted with sowing of short-duration rice varieties during the rainy season. The results showed that sowing of Kalinga 3, Vandana, and Tulasi in Bihar, NDR 118 and Narendra 97 in eastern UP, and Nilagiri, Vandana, and Chandeswari in Orissa resulted in increased rice grain yield and facilitated timely sowing of chickpea in rice fallows (Table 12.20).

12.8 CONCLUSIONS

Rainfed environments in India have great potential to contribute to increasing agricultural production as evidenced by the large yield gaps between potential and actual yields realized by the farmers. ICRISAT and its consortium partners across years of experience have developed the integrated watershed management approach, which has demonstrated promising results in enhancing crop productivity and efficient use of natural resources under both on-station and on-farm watersheds with local community participation. The major contributions to productivity enhancement came from adoption of improved crop varieties and integrated nutrient management and their interaction with soil and water conservation practices. Integrated pest management practices contributed toward reducing the cost of production and protecting the environment. Water harvesting in ponds and recharging of groundwater supported production of high-value crops with supplemental irrigation. Crop diversification and intensification took place automatically at field level once the water availability was established, which in turn enhanced the system productivity and rainfall-use efficiency. The adoption of this new approach to technology development and adoption needs to be promoted to benefit a large number of farmers.

REFERENCES

Aggarwal, P.K., K.B. Hebbar, M.V. Venugopalan, S. Rani, A. Bala, A. Biswal and S.P. Wani. 2008. *Quantification of yield gaps in rainfed rice, wheat, cotton and mustard in India. Global Theme on Agroecosystems Report no. 43. Patancheru 502 324.* Andhra Pradesh, India: International Crops Research Institute for the Semi-Arid Tropics, pp. 36.

Bhatia, V.S., Piara Singh, S.P. Wani, A.V.R. Kesava Rao and K. Srinivas. 2006. *Yield gap analysis of soybean, groundnut, pigeon pea and chickpea in India using simulation modeling. Global Theme on Agroecosystems, Report No. 31, Patancheru 502 324.* Andhra Pradesh, India: International Crops Research Institute for the Semi-Arid Tropics, pp. 156.

El-Swaify, S.A., P. Pathak, T. Rego and S. Singh. 1985. Soil management for optimized productivity under rainfed conditions in the semi-arid tropics. *Adv. Soil Sci.* 1: 1–64.

ICRISAT. 2006. *ICRISAT's vision and strategy to 2015. Patancheru 502 324.* Andhra Pradesh, India: International Crops Research Institute for the Semi-Arid Tropics.

Kumar Rao, JVDK., D. Harris, M. Kankal and B. Gupta. 2008. Extending rabi cropping in rice fallows of eastern India. In *Improving agricultural productivity in rice-based systems of the High Barind Tract of Bangladesh, eds.* C.R. Riches, D. Harris, D.E. Johnson and B. Hardy, 193–200. Los Banos, Philippines: International Rice Research Institute.

Murty, MVR., Piara Singh S.P. Wani, I.S. Khairwal and K. Srinivas. 2007. *Yield gap analysis of sorghum and pearl millet in India using simulation modeling. Global Theme on Agroecosystems, Report No. 37. Patancheru 502 324.* Andhra Pradesh, India: International Crops Research Institute for the Semi-Arid Tropics, pp. 85.

Pathak, P. and K.B. Laryea. 1991. Prospects of water harvesting and its utilization for agriculture in the semi-arid tropics Proceedings of the symposium of the SADCC land and water management research program scientific conference, 8–10 Oct. 1990. Gaborone, Botswana, pp. 253–268.

Sahrawat, K.L., S.P. Wani, T.J. Rego, G. Pardhasaradhi and K.V.S. Murthy. 2007. Widespread deficiencies of sulphur, boron and zinc in dryland soils of the Indian semi-arid tropics. *Current Sci* 93(10): 1428–1432.

Sehgal, J.L. and S. Lal. 1988. *Benchmark swell-shrink soils of India-morphology, characteristics and classification. NBSS publication No. 19.* Nagpur, India: National Bureau of Soil Survey and Land Use Planning, Indian Council of Agricultural Research.

Sharma, K.D. and Anupama Sharma. 2007. Strategies for optimization of groundwater use for irrigation. Ensuring water and environment for prosperity and posterity souvenir. 10th Inter-Regional Conference on Water and Environment (ENVIROWAT 2007), 17–20 October 2007, organized by Indian Society of Water Management in collaboration with Indian Society of Agricultural Engineers and International Commission on Agricultural Engineering. pp. 52–58.

Shiferaw, B. and C. Bantilan. 2004. Rural poverty and natural resource management in less-favored areas: revisiting challenges and conceptual issues. *J Food Agri. Environ.* 2(1): 328–339.

Shiferaw, B., S.P. Wani and G.D. Nageswara Rao. 2003. Irrigation investments and groundwater depletion in the Indian semi-arid villages: The effect of alternative water pricing regimes. Socio-economics and policy working paper 17, ICRISAT Patancheru, India.

Singh, B.P. 1997. Focus on Indian soybean in relation to agroclimatic conditions. *In Workshop on user requirements for agrometeorological services,* Pune, India. 10–14 November 1997, pp. 2–12.

Singh, H.P. 2007. Enhancing water productivity in horticultural crops. In Ensuring water and environment for prosperity and posterity souvenir. 10th Inter-Regional Conference on Water and Environment (ENVIROWAT 2007), 17–20 October 2007. Organized by Indian Society of Water Management in collaboration with Indian Society of Agricultural Engineers and International Commission on Agricultural Engineering. pp. 40–48.

Srivastava, K.L., P. Pathak, J.S. Kanwar and R.P. Singh. 1985. Watershed-based soil and rain water management with special reference to Vertisols and Alfisols. Paper presented at the National Seminar on soil conservation and watershed management, 5–7 September 1985, New Delhi, India.

Stewart, B.A., J.T. Musick and D.A. Dusek. 1983. Yield and water use efficiency of grain sorghum in a limited irrigation-dryland system. *Agron. J.* 75: 629–634.

Twomlow, S., B. Shiferaw, P. Cooper and J.D.H. Keatinge. 2008. Integrating genetics and natural resource management for technology targeting and greater impact of agricultural research in the semi-arid tropics. *Experimental Agric.* 44: 1–22.

Wani, S.P., P. Pathak, H.M. Tam, A. Ramakrishna, P. Singh and T.K. Sreedevi. 2001. Integrated watershed management for minimizing land degradation and sustaining productivity in Asia. Proceedings of a Joint UNU-CAS International Workshop, Beijing, China, 8–13 September 2001, Integrated Land Management in Dry Areas, pp. 207–230.

Wani, S.P., Piara Singh, K.V. Padmaja, R.S. Dwivedi and T.K. Sreedevi. 2006. Assessing impact of integrated natural resource management technologies in watersheds. In *Impact assessment of watershed development – Issues, methods and experiences, eds.* K. Palanisami and D. Suresh Kumar, 38–58. Karol Bagh, New Delhi, India: Associated Publishing Company.

Wani, S.P., T.K. Sreedevi, H.P. Singh, T.J. Rego, P. Pathak and Piara Singh. 2003. A consortium approach for sustainable management of natural resources in watershed. Paper presented at Integrated Watershed Management for Land and Water Conservation and Sustainable Agricultural Production in Asia, ADB-ICRISAT annual meeting and 6th MSEC Assembly, Hanoi, Vietnam, 10–14 December 2001, pp. 218–225.

Chapter 13

Strategies for improving the productivity of rainfed farms in India with special emphasis on soil quality improvement

K.L. Sharma & Y.S. Ramakrishna
Central Research Institute for Dryland Agriculture, Santoshnagar, Hyderabad, India

J.S. Samra & K.D. Sharma
National Rainfed Area Authority, Government of India, New Delhi, India

U.K. Mandal, B. Venkateswarlu, G.R. Korwar & K. Srinivas
Central Research Institute for Dryland Agriculture, Santoshnagar, Hyderabad, India

ABSTRACT

Rainfed agriculture encounters several constraints on account of climatic, edaphic, and social factors. Out of the 97 million farm holdings, about 76% come under marginal and small categories. The productivity levels of these areas have remained lower across years because of frequent droughts occurring due to high variability in the quantum and distribution of rainfall, poor soil health, low fertilizer use, imbalanced fertilization, small farm size and poor mechanization, poor socio-economic conditions and low risk-bearing capacity, low credit availability and infrastructure constraints. Consequently, farmers are distracted from agriculture and tend to migrate to cities to look for alternative jobs. Hence, there is a great need to increase the productivity of rainfed crops and overall net returns to keep the farmers in agriculture. A paradigm shift in rainfed agriculture can be expected through technological thrusts and policy changes. The strategies that need to be emphasized include: (i) land care and soil-quality improvement through conservation agricultural practices, balanced fertilization, harnessing the potential of biofertilizers and microorganisms, and carbon sequestration; (ii) efficient crops, cropping systems, and best plant types; (iii) management of land and water on watershed basis; (iv) adoption of a farming-systems approach by diversifying enterprises with high-income modules; (v) mechanization for timely agricultural operations and precision agricultural approach; (vi) post-harvest, cold-storage, value-addition modules; (vii) assured employment and wage system; (viii) organic farming; (ix) rehabilitation of rainfed wastelands; (x) policy

Address correspondence to K.L. Sharma, Central Research Institute for Dryland Agriculture, Santoshnagar, Hyderabad, Andhra Pradesh, 500059, India. E-mail: klsharma@crida.ernet.in

Reprinted from: *Journal of Crop Improvement*, Vol. 23, Issue 4 (October 2009), pp. 430–450, DOI: 10.1080/15427520903013431 changes and other support system; and (xi) human-resource development, training and consultancy. This paper deals in depth with some of these issues and strategies.

Keywords: rainfed farming, crop productivity, land degradation, land care and soil quality, watershed development strategies

13.1 INTRODUCTION

Rainfed agriculture in India extends across 90.9 m ha that constitutes nearly 64% of the net cultivated area, contributing 40% of the food grains and supporting 40% of the Indian population of >1 billon, and plays a significant role in ensuring food security. There has been some decline in the rainfed area because of development of irrigation facilities since independence, but two-thirds of the net sown area still remains rainfed. Even when the full irrigation potential of the country is realized, 50% of the net sown area is expected to continue to remain rainfed. A host of food, fodder, and industrial crops are grown under rainfed conditions. Two-thirds of livestock population inhabit these areas. Among these, animals, small ruminants, like sheep and goat, predominate. These facts highlight the importance of dry farming in the country. To meet the growing demands for food grains, fodder, and fiber on a sustainable basis, it is inevitable to improve the production and productivity in these areas. Rainfed agriculture prevails in the desert terrain of Rajasthan in the northwest; the plateau region of Central India; the alluvial plains of the Ganga-Yamuna River Basin; the Central Highlands of Gujarat, Maharashtra, and Madhya Pradesh; the rain-shadow region of Deccan in Maharashtra; the Deccan Plateau in Andhra Pradesh; and the Tamilnadu Highlands. About 15 million ha of rainfed area represents the zone that receives <500 mm rainfall; another 15 million ha is in the 500–750 mm rainfall zone, and 42 million ha is in the 750–1,150 mm rainfall zone, with the remaining 25 million ha receiving >1,150 mm rainfall per annum (Kanwar, 1999; Singh et al., 2004). The monsoon season, which starts in June and ends in September, contributes the major share of about 74% of annual rainfall. Drought occurrence is a common feature once in three to five years either because of a deficit in seasonal rainfall during the main cropping season or from inadequate soil-moisture availability during prolonged dry spells between successive rainfall events (Ramakrishna et al., 1999). Many times, very low yields and crop failures lead to food and fodder scarcity, resulting in a near-famine situation. The rainfed regions mostly represent Alfisols, Entisols, Vertisols, and associated soils in the semi-arid tropical (SAT) areas (Virmani, Pathak & Singh, 1991). These soils encounter several constraints on account of physical, chemical, and biological soil-quality constraints. Except Vertisols, the soils are generally coarse textured, highly degraded with low water retentive capacity, and have multiple nutrient deficiencies.

Crops and cropping systems in the semi-arid tropics are diverse depending on soil type and length of growing season. In addition to major livestock production systems, about 87% of coarse cereals and pulses, 55% of upland rice, 77% of oil seeds, and 65% of cotton are grown in these areas (Central Research Institute for Dryland Agriculture (CRIDA), 2007). Of the about 97 million farm holders in these regions, 76% are small (<2 ha) and marginal, cultivating only 29% of the unconsolidated and scattered arable land. This results in greater dependence of farmers on livestock and other sources of income. Further, the small size of holdings, coupled with resource-poor farmers, poor infrastructure, and low investment in technology, are the predominant features of rainfed regions.

13.2 CROP PRODUCTIVITY TRENDS IN RAINFED REGIONS

The productivity levels in the rainfed areas have remained low across years. In general, the rainfed ecosystem suffers from the problems of (i) frequent droughts because of high variability in the quantity and distribution of rainfall; (ii) poor soil health because of continued degradation and inadequate replenishment of nutrients; (iii) low animal productivity because of an acute scarcity of green fodder; and (iv) low risk-bearing capacity of farmers because of poor socio-economic base, low credit availability, and infrastructure.

There has been a slow but steady improvement in the productivity of rainfed crops because of the introduction of modern agronomic practices; the most important one being increased fertilizer use. However, a number of biophysical and socio-economic constraints still limit the average productivity of rainfed crops at 0.7 to 0.8 t ha^{-1} against 2 t ha^{-1} in irrigated areas. Controlled use of irrigation and balanced fertilization significantly contributed to the increased agricultural production in irrigated areas. However, with the exception of sorghum and pearl millet, the use of high-yielding varieties (HYVs) has been insignificant in these areas. This, coupled with the low fertilizer use, has resulted in continued low productivity of rainfed crops. However, the opportunities for further improvement in productivity in irrigated crops being limited, the only option left is to enhance productivity of rainfed crops with the application of modern packages of practices, including the use of HYVs, effective land use, diversification, balanced fertilizer use, and efficient use of rainwater.

By 2025, human population in rainfed regions in India is likely to reach 600 million from the present level of 410 million. To meet the food needs of this increased population, it is essential to increase productivity from the present level of 0.8–1.0 t ha^{-1} to 2.0 t ha^{-1} by 2025. Furthermore, the quality of produce must improve to meet the world-market standards. In addition, the cost of production needs to be reduced not only to improve farmers' net income but also to remain globally competitive. In view of escalating costs of inputs and prevailing low procurement prices for the rainfed agricultural commodities, the task becomes very tough. It means that research and developmental efforts need to be increased. This calls for multidisciplinary and holistic research and developmental activities to maximize crop productivity and profitability, while not endangering the sustainability of rainfed-farming systems in the region (Singh, Sharma & Subba Reddy, 2002).

13.2.1 Major constraints in improving the productivity and net returns of rainfed farms

The major constraints in improving the productivity and returns from rainfed farming in India are as follows: (i) erratic and uncertain rainfall, leading to moisture scarcity, droughts and failure of crops, especially annual crops; (ii) soil degradation and poor soil quality on account of dismally low amount of soil organic C (SOC) because of low returns of residues back to the soil, fast rate of decomposition because of high temperature and frequent inversion tillage, low fertility, excessive nutrient-removal-use gap because of low use of fertilizer inputs, water logging because of subsurface compaction and low infiltration, salinity, sodicity, acidity, compaction, hard setting, etc.; (iii) fragmented and low holding size, leading to constraints in mechanization; (iv) poverty among growers and constraints in availability and purchase of essential inputs, such as seeds and fertilizers, bullock-drawn small seed-cum-fertilizer drills, etc.; (v) lack of assured credit and financial support and marketing; (vi) inadequate infrastructure for post-harvest value-addition and storage of produce; (vi) low procurement prices of agricultural commodities, in general; and (vii) inadequate earnings for livelihood from the farming profession because of low volume of business due to small holding size, low productivity, and low produce prices, etc. The consequences of these constraints are likely to lead the marginal and small-farming communities toward distraction from agriculture, migration to cities to look for alternate assured wages, suicides, etc. To mitigate these constraints and transform the rainfed farming to an attractive option, there is a strong need for strategic planning and policy changes in a phased manner.

13.3 STRATEGIES FOR THE DEVELOPMENT OF RAINFED FARMING

The following multi-component strategies are suggested for improving the productivity and returns in rainfed farming.

13.3.1 Land care and soil quality improvement

The yield curve in most of the crops, such as rice, wheat, and maize, in irrigated areas has plateaued out because of a stagnated response to added inputs. Therefore, it is anticipated that, if at all another green revolution were expected in Indian agriculture, it would come from grey areas only. Even after devoting a good number of years in agricultural research, the gap between possible potential yield of rainfed crops and their actual realized yields could not be narrowed down to a desirable level. Apart from moisture limitations, rainfed areas are also at a disadvantage because of low soil organic matter content and poor soil fertility. The predominant soil orders that represent rainfed agriculture are: Alfisols, Inceptisols, Entisols, Vertisols, Oxisols, and Aridisols. Out of the total geographical area of 328.3 m ha, Entisols constitute 24.4% (80.1 m ha), Inceptisols 29.1% (95.8 m ha), Vertisols 8.02 % (26.3 m ha), Aridisols 4.47% (14.6 m ha), Mollisols 2.43% (8 m ha), Ultisols 0. 24% (0.8 m ha), Alfisols 24.25% (79.7 m ha), Oxisols 0.08% (0.3 m ha), and non-classified soils 7.01% (23.1 m ha). About 187.7 m ha area, which constitutes 57.1% of the total geographical area, is degraded. Of the total degraded area, water erosion constitutes 148.9 m ha (45.3%), wind erosion 13.5 m ha (4.1%), chemical deterioration 13.8 m ha (4.2%), and physical deterioration 11.6 m ha (3.5%). Another 18.2 m ha land (5.5%), which is constrained by ice caps, salt flats, arid mountains, and rockout crops, is not fit for agriculture at all (Sehgal & Abrol, 1994). According to some other estimates, the total degraded area in India accounts for 120.7 m ha, of which 73.3 m ha is affected by water erosion, 12.4 m ha by wind erosion, 6.64 m ha by salinity and a Alkalinity, and 5.7 m ha by soil acidity (Anonymous, 2008). The soils in rainfed areas have been severely affected because of (i) loss of finer fraction of top soils, organic matter, and nutrients because of soil erosion and runoff processes; (ii) virtually no or low recycling back of crop residues to the soil because of competing demand for crop residues as animal fodder; and (iii) temperature-mediated fast oxidation of organic matter because of frequent tillage operations, resulting in breaking up of micro-aggregates and exposure of SOC entrapped in them. As lack of assured moisture does not support higher cropping intensity in these regions, contribution of root biomass to organic matter in the soil is also not very high. Apart from these, low and imbalanced fertilizer use has also resulted in multi-nutrient deficiencies in rainfed regions. Scarcity of moisture caused by an uncertain rainfall pattern does not give desirable return per unit of fertilizer use. Resultantly, soils encounter a diversity of constraints on account of soil quality and ultimately end up with poor functional capacity. Earlier also, it was emphasized that soil degradation is a widespread problem in drylands and largely results from wind and water erosion, organic matter depletion, chemical deterioration, and salinization (Stewart & Koohafkan, 2004).

During the past, some research and developmental efforts have been made toward land care and soil-quality improvement for (i) location-specific, soil-water conservation practices to conserve surface soil and its productive constituents; (ii) adequate supplementation of nutrients and enhancement of organic carbon through integrated-nutrient management, green manuring, residue recycling, biofertilizer application, tree-green leaf-manuring, and by way of capitalization of legume effect; (iii) method of soil-quality assessment; (iv) land cover management and conservation tillage; and v) soil test-based fertilizer recommendation. Some of these past experiences have revealed that these efforts significantly contribute in checking land degradation and improving soil quality and its resilience. Despite these efforts, a considerable part of the research findings and technologies has not gone to the farmers. The farming community is still applying only a few kilograms of farmyard manure and inorganic fertilizers. Soil-water conservation practices have also not been adopted on an appreciable scale for several reasons. Research and extension focus needs to be continued on these aspects to generate adequate data through on-farm and on-station trials. Considering the above, the following strategies are suggested for checking land degradation and improving soil quality.

13.3.1.1 Controlling soil degradation and soil quality

Lands can be protected from being degraded and soil quality can be improved by adopting suitable conservation agricultural-management practices, depending upon climatic and edaphic factors, on a long-term basis. The sparse rainfall and high temperatures in dryland regions are major constraints along with the demand of crop residues for feed and fuel, but adoption of conservation agriculture may be the key to a sustainable crop production in marginal areas (Stewart & Koohafkan, 2004). Conservation practices, apart from protecting the topsoil, also help in enhancing the organic matter in the soil. Organic matter is considered to be panacea to many of the soil-related constraints in rainfed agriculture. However, its buildup is an utmost difficult task. Therefore, it is absolutely necessary to focus research and development efforts on enhancing organic matter in the soil at all costs through conservation tillage, residue recycling, integrated-nutrient management, incorporation of legumes in cropping systems, surface-residue application and stubble retention, and balanced fertilization for higher root biomass. Specifically, the following aspects need to be given more emphasis in research and development programs:

1 Exploring the possibilities of surface-residue management, land cover, and residue recycling in a non-competitive manner-residue quality, decomposition pattern, carbon turnover, build-up of carbon pools, and carbon-turnover models.
2 Conjunctive use of organics and inorganics, off-season generation of biomass, green manuring, and tree-green leaf manuring.
3 Conservation tillage – standardization of methodology of growing crops under reduced or minimum tillage, method of seeding and appropriate seeding devices, weed control mechanism, frequency of tillage, quantification of fluxes of CO_2 emission, water-relation studies, etc.
4 Efficient crops and cropping systems – identification and promotion of best carbon-sequestering systems, quantification of carbon contribution through roots, and fertilizer needs for higher root-biomass contribution.
5 Assessment of soil quality using standard methods – precise methods of computation of soil-quality index using key indicators and other parameters.
6 Quantification of soil fertility with inter/mixed-cropping systems.
7 Amelioration of problematic soils, such as saline and sodic soils and acid soil, using appropriate amendments and correcting waterlogged soils.

Further, the crops and cropping systems of rainfed areas have been poorly nourished through external application of manures/fertilizers owing to failure as a result of unpredictable monsoon and also of poor risk-taking ability of the resource-poor farmers. This practice has across years resulted in multiple nutrient deficiencies. The changing price polices regarding fertilizers have caused the resource-poor farmers of rainfed areas to feed their crops with only certain type of fertilizer, which has resulted in low nutrient-use efficiency and profitability because of deficiency/antagonistic relationships of certain essential nutrients. The fertilizer use in some of the rainfed crops is considerably low. Only 9% of the districts in India use more than 200 kg of $N + P_2O_5 + K_2O$ per hectare (Tiwari, 2006) (Figure 13.1). On the other hand, only 32% of districts used <50 kg $N + P_2O_5 + K_2O$ per hectare. Most of the rainfed regions fall in this category. On an average, in India, the average fertilizer nutrient use is quite low (120 kg ha^{-1}) compared to many other countries of the world, and consequently the yields of some principal crops followed the similar trend (Table 13.1; Fertilizer Association of India, 2008). Hence, apart from many other reasons, low fertilizer use in India is definitely one of the important causes of low yields. Efforts need to be made to redefine fertilizer doses by synchronizing with the anticipated water availability. To ensure adequate and balanced fertilization, the following issues need to be given appropriate attention:

a Mapping spatial variation in soil properties on a watershed scale and designing precise management practices for maximizing land productivity;
b Identification and correction of deficient nutrients in rainfed conditions;

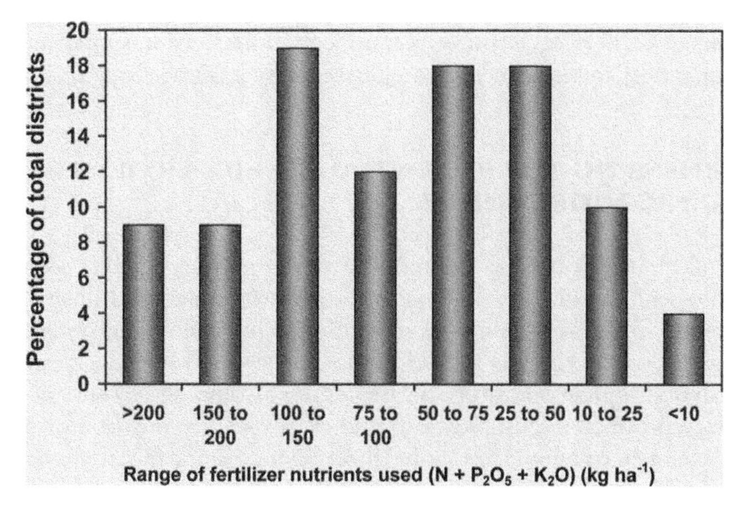

Figure 13.1 Fertilizer use pattern in India.

Note: No. of districts = 490.
Source: Adapted from Tiwari, 2006.

Table 13.1 Fertilizer consumption per hectare of arable land and yield of principal crops in India and other countries during the year 2005.

Country	*Fertilizer consumption ($N + P_2O_5 + K_2O$) (kg ha^{-1})*	*Yield (kg ha^{-1})*			
		Paddy	*Wheat*	*Maize*	*Potato*
India	120	3154	2602	1938	17923
China	301	6253	4275	5288	14521
Japan	373	6648	4097	2500	31634
Korea Rep	474	6568	3206	4841	27323
Pakistan	168	3174	2586	2984	18079
Austria	358	–	5029	10657	34398
Germany	205	–	7465	9214	41980
Netherlands	666	–	8593	12200	43442
Norway	251	–	4820	–	24640
UK	287	–	7961	–	43384
New Zealand	309	–	8092	10931	45331
Egypt	625	9987	6493	8120	25000
World	109	4084	2845	4844	16912

Source: Adapted from FAI, 2008.

c Identification of nutritional constraints in soils and devising balanced fertilization schedules to correct them with special emphasis on cropping systems and promotion of customized fertilizer application;
d Yield maximization by combining all limiting nutrients, and integrated nutrient-management practices;
e Use of precision farming principles to enhance input-use efficiency;

f Integration of database on soil fertility indicators through soil test network laboratories of the country, other organizations and agencies of central and state governments to develop fertilizer recommendation decision support system; and
g Development of soil health cards on a case-study basis as ready reckoners.

13.4 HARNESSING THE POTENTIAL OF BIOFERTILIZERS AND MICROORGANISMS

Biological soil quality is one of the important pillars of soil quality as a whole. To improve biological soil quality, it is essential to restore microbial diversity, which is most threatened in stressed rainfed agro-ecology. To exploit the potential of biofertilizers and microbes, there is a need to isolate new and effective strains that can withstand harsh dryland conditions. Strategies are also needed to enhance their shelf life and reduce production cost. Region- and crop-specific consortia of biofertilizers need to be developed to popularize biofertilizers. The specific initiatives needed in this regard are:

a Identification of unexplored beneficial soil microorganisms and development of protocols for isolation, culturing, and mass multiplication; and
b Identification of soil microorganisms for specific purposes, such as nutrient dissolution, mobilization and delivery, scavenging heavy metals and harmful compounds, degrading plastics, and suppression of pathogens.

13.4.1 Carbon sequestration through afforestation and efficient cropping systems

In accordance with the global mandate, India also needs to focus on reducing global warming by reducing emission of greenhouse gases. With a large land area and diversity of eco-regions, India has considerable potential for terrestrial soil carbon sequestration. Of the total land area of 329 m ha, arable land constitutes 162 m ha, forest and woodland 69 m ha, permanent pasture 11 m ha, permanent crops 8 m ha, and other land uses 58 m ha. The soil organic carbon (SOC) pool is estimated as 21 Pg (1 Pg = 1×10^{15} g (billion ton)) at 30 cm depth and 63 Pg at 150 cm depth. The soil inorganic pool is estimated at 196 Pg at 1 m depth. The SOC concentration in most cultivated soils is less than 5 g kg^{-1} compared with 15 to 20 g kg^{-1} in uncultivated soils. Low SOC concentration is attributed to plowing, removal of crop residue and other bio-solids, and mining of soil fertility. Accelerated soil erosion by water leads to emissions of 6 Tg C yr^{-1}. Important strategies for soil C sequestration include restoration of degraded soils and adoption of recommended management practices (RMPs) on agricultural and forestry soils. Potential of soil C sequestration in India is estimated at 7 to 10 Tg C yr^{-1} for restoration of degraded soils and ecosystems, 5 to 7 Tg C yr^{-1} for erosion control, 6 to 7 Tg C yr^{-1} for adoption of RMPs on agricultural soils, and 22 to 26 Tg C yr^{-1} for secondary carbonates. Thus, potential of soil C sequestration is 39 to 49 (44 ± 5) Tg C yr^{-1} (Lal, 2004). The strategy for improving SOC and SIC pools in these soils involves improving water and nutrient-use efficiencies by decreasing losses and increasing biomass production. Land-use/farming practices to achieve these goals include conservation tillage and mulch farming, as well as cover crops in the rotation cycle, mixed farming/cropping, agroforestry, ley farming (putting the land under pastures and hay after growing grain crops), and adoption of integrated nutrient and pest management practices.

Keeping in view the above, terrestrial ecosystem is one of the greatest sinks for carbon. This sink capacity can be enhanced through afforestation, i.e., by growing long-term perennials of economic importance. The carbon sequestration by perennials/tree species has multiple advantages, such as (i) utilization of CO_2 from the atmosphere, (ii) economical output as produce for income generation, and (iii) entrapping of carbon through extensive root system in the subsurface soil for a longer period. The suitability and choice of tree species for different rainfed regions need to be researched. The quantification of carbon stocks aboveground and belowground and computation of carbon balance in soil using suitable simple models is also a potential theme that needs research focus.

Soil could be another great sink for sequestering carbon by adopting an efficient cropping system having an extensive hard-to-decompose root system. Enhancing SOC by way of conservation agricultural practices, such as residue recycling, reduced tillage, organic manuring, optimum fertilization, etc., also helps in entrapping/sequestering the carbon in soils. Apart from creating a sink for CO_2 in the atmosphere, it will also improve the soil quality. Therefore, carbon sequestration through afforestation (energy plantation, perennials of industrial importance) and through efficient cropping systems under wider climatic and edaphic conditions is a potential subject that needs research thrust with special emphasis on rainfed agriculture. The specific thrust areas in this regard are:

a Management practices for enhancing the productivity of short-rotation forestry/ energy plantations;
b Quantifying belowground biomass accumulation for precise quantification of carbon;
c Management practices that can enhance and maintain soil carbon levels in agroforestry systems/annual cropping systems;
d Identification of potential tree-based systems for various degraded lands for enhancing biomass production and also income;
e Carbon accounting in tree-based systems using tree-growth models; and
f Exploiting carbon sink capacity of soil using efficient crops and cropping systems, their effective management, and appropriate conservation agricultural practices. Some of these practices were mentioned in the foregoing section.

13.4.2 Efficient crops, cropping systems, and best plant types

Identification and recommendation of most efficient crops and cropping systems for rainfed areas are a must for ensuring higher yields. A good number of improved varieties of millets, pulses, and oil seeds have been evaluated for their yield with reference to the local cultivars being grown by the farmers (All India Coordinated Research Project for Dryland Agriculture, 2000). A yield increase between 15% and 50% was reported when local cultivars were replaced by high yielding varieties. During monsoon season, the quantity and distribution of rainfall determines the effective growing season and cropping systems for a given region. For instance, in regions receiving 350–600 mm of rainfall and having a 20 week effective growing season, only single cropping is possible with all soil types, except deep Vertisols. In deep Vertisols, a single post-rainy season crop is possible in areas receiving 350–600 mm rainfall and having a 20 week effective growing season. Intercropping (150% cropping intensity) is possible in regions having a 20–30 week effective growing season and 650–750 mm rainfall. In areas receiving more than 750 mm of rainfall and having an effective growing season of more than 30 weeks, double cropping (200% cropping intensity) is assured (Singh et al., 2004; Singh & Subba Reddy, 1986). There is a need to reexamine recommendation of best crops and cropping systems as per land capability, soil type, rainfall availability, and length of growing period for different agro-climatic zones. Availability of good quality seeds and planting material through seed banks, seed-village programs, etc., needs to be ensured. Specific programs need to be launched to ensure quick seed availability to the growers. Best and timely seed availability has been one of the important inputs in the first green revolution.

The postulates given by Creswell and Martin (1998) for the better plant types universally still hold true. These postulates envisage that better plant types must have the following:

- Short stem with less leaf surface for less transpiration;
- A deep, prolific root system for better moisture utilization; and
- Crops that can mature before dry conditions start.

Further, the plants chosen for rainfed regions must have good adaptability. The requirements of crops for better adaptation are (Shantz, 1956):

- Drought escape: Limited-season, small-stature crops that can respond to limited water supply. Further, short-season crops with fast-growing root system and high grain to straw ratio are also capable to escape drought.
- Drought resistance: Crops with extensive root system that obtain soil moisture during repeated water stress and crops capable of maintaining adequate cell water content when tissue moisture stress occurs.
- Drought tolerance: In the process of drought tolerance, crops prepare themselves to go in to temporary dormancy during water stress and resume normal situation on improvement of conditions.

13.4.3 Management of land and water on watershed basis

While managing the land and water on a watershed basis, the following points need to be considered:

- Emphasis on *in situ* moisture conservation to ensure adequate charge of soil profile and higher response to fertilizer and manure inputs.
- Soil water conservation, water harvesting, and its most efficient use through micro-irrigation techniques (drip and sprinkler irrigation) for high-value enterprises

for maximum returns. The principle that "harvested water is gold" need to be advocated.

- The central focus of soil water conservation should be individual farmer holdings instead of the macro approach being followed at present.
- Efficient use of existing ground water for higher output and most remunerative commodities only.
- Development of additional water resources wherever possible through linkage of canals, rivers, lift irrigation, etc.
- Incentives and 'community movement' for rooftop water harvesting, percolation, and efficient use.
- Development of effective policies for water management and its sharing at community level using Israeli/Californian models.

13.4.4 Adoption of farming systems approach by diversifying enterprises

Farming systems approach by diversifying enterprises is most needed in rainfed agriculture. This will help in increasing the productivity and profitability and reducing poverty and the extent and magnitude of risks in rainfed farming. Different modules could be alternative agricultural practices, such as agri-horticulture, silvi-pasture, agroforestry systems, livestock integration, rainfed horticulture, medicinal and aromatic plants, etc. There is a need to initiate special incentives and programs for providing seeds of forage crops, grasses, seedlings of horticultural plants and top feed-tree species through suitable nurseries. Development of infrastructure for micro-irrigation and post-harvest processing is essential. Assured marketing linkage needs to be established.

13.4.5 Mechanization for timely agricultural operations and precision agricultural approach

Rainfed agricultural operations are most time-bound and hence suitable mechanization for timely and precision agricultural operations is essential. Postponement of sowing date of rainfed crops can lead to a risk of losing 15% to 100% yield of crops. Special drive is needed to ensure the availability of bullock- and tractor-drawn small and medium implements. The state-based agro-industries and promotion of custom-hiring services for implements, etc., in rural areas need to be strengthened by employing educated rural youth. Provision need to be made for subsidized purchase of implements by the farmers. These efforts should help replace the age-old practice of broadcast (spreading) of seed and fertilizer randomly in field. A significant difference in fertilizer- and moisture-use efficiency and yield enhancement can be expected out of these initiatives, if implemented at a mass scale.

13.4.6 Post-harvest, cold storage, value-addition modules

To enhance the value of the produce and enable farmers to obtain higher price, it is essential to strengthen the post-harvest processing units and cold-storage facilities.

These will not only help the farmers in enhancing their income, but also in providing/generating employment for the rainfed-area rural community.

13.4.7 Assured employment or wage system

Recommendation for providing assured employment to at least one member of a farm family and landless laborers and off-season employment to other members of the family could very much support/sustain their livelihood. The salary of the lowest paid government employee in India at present is nearly rupees one lakh year^{-1}, which is approximately equal to US$2000 year^{-1}. The small and marginal farmers, who possess 1–2 hectares of rainfed land holdings, despite using their entire family as labor input in the agricultural enterprise for a whole year (if two crop seasons are available), might not even get 1/5 of the above income as net return under prevailing price policy for agricultural commodities. Even after discounting for the marginal subsidies that a farmer is availing on some of the inputs, the amount he gets as net return is miserably low from the present kind of crops/cropping systems/commodities. Hence, employment support is inevitable to make him stay in this profession. It is worthwhile to mention here that the Government of India has already initiated good efforts in this direction by enacting a program viz. National Rural Employment Guarantee Act (NREGA), through which rural families are getting assured jobs (wages) for at least 100 man-days. Many more improvements are expected in this program in future.

13.4.8 Organic farming

There is a scope to introduce organic farming in some selected rainfed crops, which may be helpful in increasing the income of the farmers as well as improving soil quality. The list of selected crops is given as follows:

Table 13.2 Scope for introducing organic farming in some selected rainfed crops for generating higher income.

Crop/commodity	*Potential part used*	*Scope*
Safflower	Petals	Herbal tea and yellow dye
Finger millet	Grain	Health foods for export
Soybean	Seed	Pesticide free export
Cotton	Lint	Demand for organically grown fiber
Rainfed vegetables	Economic parts	Residue free consumption
Rainfed fruits	Fruits	Residue free consumption
Rainfed rice	Grain	Scented rice for export
Dried beans	Fruits	To meet International demand

13.4.9 Development of rainfed wastelands

To harness the potentials and to make efficient use of the rainfed wastelands, it is essential to rehabilitate them by using selective technological modules. These are:

- Soil and rainwater conservation through terracing, bunding, trenching, water-storage ponds, vegetative barriers, rainwater harvesting, increase in water storage, and recycling;
- Encouraging natural vegetation, planting and sowing of multipurpose trees, shrubs, grasses, legumes, pastures, fruit, timber, and fodder species;
- Growing of biodiesel plants such as *Jatropha, Pongamia*, etc., considering market availability;
- Controlling wind erosion through shelter belt/wind-break plantations;
- Stabilization of sand dunes;
- Integrated soil fertility management;
- Ravine reclamation through mechanical and vegetative means;
- Rehabilitation of mine spoils;
- Management/utilization of saline lands and industrial effluents; and
- Linking employment for landless laborers with wasteland management. In these activities, community participation is a must.

13.4.10 Policy changes and other support required

Besides technological interventions, appropriate changes in policies and provision of other support system are needed for the development of rainfed agriculture and ensuring an adequate income and livelihood of the farming community. Some of the important policy issues are listed below:

- Provision for special subsidies on certain inputs;
- Revision of procurement prices of rainfed agricultural commodities periodically;
- Provision for soft loans to small and marginal rainfed farmers/growers;
- Family health cards and medical and crop insurance to the rainfed farmers;
- Soil health cards and periodical updates;
- Kisan credit cards for purchasing seed, fertilizer, and other inputs instantaneously whenever needed without delay;
- Quick drought monitoring and relief;
- Reservation and priority in providing assured jobs;
- Assured marketing;
- Development of contract farming and cooperative farming modules for improving the performance of rainfed agriculture. A mechanism of "land for mutual working but with individual titles" needs to be created;
- Capacity building and training of farmers in specialized farm activities; knowledge buildup initiatives need to be started for the farming community; and
- Opening of 'information hubs' and agri-clinics for better decisions and technical support.

13.4.11 Human resource development, training and consultancy

The persons associated with the technical functionaries on rainfed agriculture need periodical exposure to the latest advances in rainfed agricultural technologies being developed abroad. If required, it is desirable to establish a network of consultants from suitable countries for executing the mega developmental projects on rainfed agriculture in India. Special exposure trainings of short duration are required for the grassroots level and cutting-edge personnel involved in implementation of watershed and rainfed agricultural development programs in various states. Some of the training aspects are listed below:

- Package of practices of rainfed crops.
- Alternate land-use system.
- Control measures of different pests and diseases, and integrated pest management.
- Concept of integrated watershed management.
- Strategy and approach of watershed management.
- Criteria of site selection.
- Priorities of watershed planning and peoples participation.
- Survey and watershed planning.
- Preparation of base map for watershed and estimates for financial outlay.
- Contingency crop planning.
- Improved farm implements.
- Pasture development.
- Dryland horticulture.
- Agro-climatologic parameters.
- Land capability classification.
- Land leveling.
- Chain survey.
- Gully control measures.
- Water-harvesting structures.
- PRA technique.

13.5 DEVELOPMENT OF COMPREHENSIVE DATABASE ON RAINFED AGRICULTURE

There is a need to develop a comprehensive database on rainfed agriculture for periodic planning. The database could be from the viewpoint of national and international perspective.

13.5.1 National perspective

- Climate, land, and water resources.
- Crops and cropping systems and input-use pattern and their productivity.
- Forest and environment, coverage, status, threats, if any.
- Complete profile of human resources and their socio-economic conditions.
- Credit marketing and R & D organization.
- Agricultural and rural development programs launched on rainfed agriculture in the country by different ministries – their success, functioning, budget, scope for convergence, etc.
- Database on allied activities, such as health and education guarantees, insurances, kisan credit cards, old-age pensions, subsidies on agricultural inputs, such as seed,

fertilizers, insecticides/pesticides, etc., drip irrigation, lift irrigation, fertigation, seed production, mechanization level, small implements, tractors, electricity/energy, post-harvest processing, export, etc.
- Others.

13.5.2 International perspective

- Database on land and water resource for different countries.
- Technological hubs on rainfed farming.
- Data- and information-sharing mechanism.
- Frequent exchange of experts in rainfed farming, consultancies for mega projects, etc.
- Development of international commissions for rainfed farming for coordinating research and technology transfer among different countries.

13.6 SOME LESSONS LEARNED

Some of the lessons learned on various aspects of rainfed farming are briefly enumerated below:

- There has been an increase in rainfed area under oilseeds and cotton (*Gossypium spp.*) to the extent of 130% and 50% increase, respectively. Area under coarse cereals decreased by 25% and no significant change was recorded for pulses (FAI, 2001). Further, an increase of 50%–100% in cotton and corn (*Zea mays*) was observed because of additional irrigation for these crops.
- Timely sowing of rainfed crops makes a significant difference in yields. A delay of 9–14 days in sorghum and upland rice (*Oryza sativa*) resulted in yield losses of 43–137 and 36 kg ha^{-1} day^{-1}, respectively (Singh & Das, 1984). Similarly, sowing castor (*Ricinus communis* L.) during the second half of July reduced bean yield up to 850–250 kg ha^{-1}. A 15 day delay in sowing of sorghum caused a reduction in grain yield of 850 kg ha^{-1}.
- Tillage plays an important role in influencing the conservation of soil and rainwater. Deep tillage (25–30 cm) helps in soil pulverization, increased rainwater infiltration, and better root growth, thereby increasing crop yield (Thyagaraj et al., 1999; Vittal, Vijayalaxmi & Rao, 1983; AICRPDA, 2000). Off-season or pre-monsoon tillage has a significant impact on weed control and rainwater infiltration. Grain yields of sorghum and barley (*Hordeum vulgare*) were 2,600 and 1,570 kg ha^{-1} with off-season tillage compared with 1,870 and 1,370 kg ha^{-1} without off-season tillage (AICRPDA, 1986).
- Studies on reduced-till farming indicated that conventional tillage using recommended fertilizer and weeding, with or without off-season tillage, resulted in higher grain yields of barley, rice, lentil (*Lens culinaris*), wheat (*Triticum aestivum*), soybean, groundnut, fingermillet (*Eleusine coracana* (L.) Gaertn), and pearl millet (AICRPDA, 1999). However, excessive tillage reduces organic carbon and accentuates soil erosion.
- Incorporating sorghum stubbles at 5 t ha^{-1} to cover 69% soil surface resulted in a 0.24 t ha^{-1} soil loss and 25 mm runoff compared with a 1.58 t ha^{-1} soil loss and 83 mm runoff when this treatment was not applied (AICRPDA, 2000).

- Mulching also reduced soil temperature and resulted in 25% greater moisture storage in the 0–30 cm soil profile.
- Cultivation during the vegetative stage enhanced the productivity of castor, sunflower, and pigeon pea (*Cajanus cajan* (L.) Millsp) by 15–20% compared with no cultivation (Subba Reddy et al., 1996).
- In Vertisols, spreading of crop residues at 5 t ha^{-1} enhanced the productivity of post-rainy season sorghum and sunflower by about 25% probably through efficient utilization of stored soil moisture (Indian Council of Agricultural Research–Australian Centre for International Agricultural Research (ICAR-ACIAR), 2001).
- In Alfisols, incorporation of corn residue at 4 t ha^{-1} increased crop yield in a succeeding crop by about 80% (Gajanan et al., 1999).
- Sorghum yield under vertical mulching at 5-m intervals was about 25% higher than that without mulching (Itnal, 1981). Using *Gliricidia spp.* branches/lopping at 5 t ha^{-1} in sorghum + pigeon pea–castor intercropping rotation reduced runoff by 56% and soil loss by 72% and increased castor bean yield from 328 to 984 kg ha^{-1} (ICAR-ACIAR, 2001).
- Application of FYM at 10 t ha^{-1} along with recommended fertilizer doses stabilized the productivity of finger millet at about 3,400 kg ha^{-1}.
- Continuous application of chemical fertilizers resulted in a decline in finger millet grain yield from an average of 2,880 kg ha^{-1} during initial five years of the study to 1,490 kg ha^{-1} by the 19th year (Gajanan et al., 1999).
- In Vertisols, providing 50% of recommended fertilizer dose through crop residues and the remaining 50% through *Leucaena leucocephala* loppings enhanced the sorghum yield by 87%, 31%, and 45%, respectively, compared to application of 25 kg N ha^{-1} and 50 kg N ha^{-1} of chemical fertilizers alone (AICRPDA, 1999).
- Application of FYM in set rows resulted in an additional yield increase of about 20%, 30%, 90%, and 20% for sorghum, sunflower, castor, and pigeon pea, respectively. At the same time, water-holding capacity and organic carbon content increased in the set rows by 8.5% and 5.7%, respectively (CRIDA, 2002).
- Silvipasture systems involving palatable grasses like Buffel or Anjan grass (*Cenchrus ciliaris* L.) and legumes Stylo (*Stylosanthes hamata* (L.) Taubert) with trees such as Subabul (*Leucaena leucocephala* Lam.), Siris (*Albezzia lebbeck* (L.) Benth), Anjan (*Hardwickia binata* Roxb.), and Sisso (*Dalbergia sisso* Roxb.) were found to be more productive and profitable in the drylands (Singh, 2002).
- Conservation furrows at a 3.6 m horizontal interval supplemented with glyricidia (*Gliricidia maculata* (Jack.) Walp.) mulch at 5 t ha^{-1} increased sorghum grain yield from 1,821 kg ha^{-1} under control to 4,003 kg ha^{-1}, reduced runoff by 73%, and decreased soil loss by about 50%.
- Conjunctive use of gliricidia loppings at 2 t ha^{-1} (fresh) with 60 and 90 kg N ha^{-1} under conventional tillage maintained soil quality index as high as 1.19 and 1.27, respectively, in long-term castor-sorghum system under semi-arid tropical Alfisol. Predominant soil quality indicators, which contributed considerably towards SQI, were available N (32%), microbial biomass carbon (MBC) (31%), available K (17%), hydraulic conductivity (HC) (16%), and available S (4%) in rainfed Alfisols (Sharma et al., 2005).

- Based on a five-year study on an Alfisols, it was found that conventional tillage maintained 14.5% higher sorghum grain yield compared with reduced tillage. The two INM treatments, 2 t gliricidia loppings + 20 kg N and 4 t compost + 20 kg N, were found to be most effective in increasing the sorghum grain yield by 84.62% and 77.7% over control, respectively. For green gram, 2 t compost + 1 t gliricidia loppings followed by 2 t compost + 10 kg N, were most promising in increasing grain yield by 51.6% and 50.8% over control, respectively. Highest amount (0.82%) of organic carbon content was recorded in 4 t compost + 2 t Gliricidia loppings treatment (Sharma et al., 2004). Based on the soil quality indices computed in the same experiment using key indicators after eight years of the study, it was observed that purely organic treatments maintained highest soil quality index. The predominant key soil quality indicators emerged in this study were microbial biomass C (MBC), available N, DTPA-Zn, and Cu, hydraulic conductivity (HC), and mean weight diameter of the soil aggregates (MWD) (Sharma et al., 2008).

13.7 CONCLUSION

Rainfed agriculture in India needs to be re-examined as far as technological and policy interventions are concerned. The suggestions made in the foregoing sections, if implemented systematically, would definitely help in increasing the productivity and net income of the rainfed-area farmers in the country. Ultimately, the objective is to make rainfed farming a viable livelihood option on a sustainable basis while also protecting the environment, and to help the farming community to stay in agriculture and distract farmers from migration to cities for alternative jobs.

REFERENCES

All India Coordinated Research Project for Dryland Agriculture (AICRPDA). 1986. *Annual report of the All India Coordinated Research Project for Dryland Agriculture.* Hyderabad, India: Central Research Institute for Dryland Agriculture.

All India Coordinated Research Project for Dryland Agriculture (AICRPDA). 1999. *Annual report of the All India Coordinated Research Project for Dryland Agriculture.* Hyderabad, India: Central Research Institute for Dryland Agriculture.

AICRPDA, 2000. Annual report of the All India Coordinated Research Project for Dryland Agriculture. Hyderabad, India: Central Research Institute for Dryland Agriculture.

Anonymous, 2008. Harmonization of wastelands/degraded lands datasets of India. National Rainfed Area Authority (NRAA), Ministry of Agriculture, Government of India, NASC Complex, DP Sastri Marg, New Delhi-110012, 5p.

Creswell, R., and F.W. Martin. 1998. Dryland farming: Crops and techniques for arid regions. Educational Concerns for Hunger Organisation (ECHO) Staff. Online. Available at http://www.echonet.org//tropicalag/technotes/drylandF.pdf

Central Research Institute for Dryland Agriculture (CRIDA). 2002. *Annual progress report.* Hyderabad, India: Central Research Institute for Dryland Agriculture (CRIDA).

Central Research Institute for Dryland Agriculture (CRIDA). 2007. *Annual progress report.* Hyderabad, India: Central Research Institute for Dryland Agriculture.

Fertilizer Association of India (FAI). 2001. *Fertilizer statistics 2000–2001*. New Delhi: Fertilizer Association of India, New Delhi.

Fertilizer Association of India (FAI). 2008. *Fertilizer statistics 2007–08*. New Delhi: Fertilizer Association of India, New Delhi.

Gajanan, G.N., B.R. Hegde, Ganapathi, Panduranga, and K. Somashekhar. 1999. *Organic manure for stabilizing productivity: Experience with dryland fingermillet*. Bangalore, India: All India Coordinated Research Project on Dryland Agriculture, University of Agricultural Sciences.

ICAR-ACIAR. 2001. *Progress report of Indian Council of Agricultural Research–Australian Centre for International Agricultural Research (ICAR–ACIAR) project on tools and indicators for planning sustainable soil management on semi arid farms and watersheds in India*. Hyderabad, India: Central Research Institute for Dryland Agriculture.

Itnal, C.J. 1981. *Water conservation measures for increased production in black soils of Bijapur. Report on All India Coordinated Research Project on Dryland Agriculture*. Bangalore, India: University of Agricultural Sciences.

Kanwar, J.S. 1999. Need for a future outlook and mandate for dryland agriculture in India. In *Fifty years of dryland agricultural research in India,* eds. H.P. Singh, Y.S. Ramakrishna, K.L. Sharma, and B. Venkateswarlu, 11–19. Hyderabad, India: Central Research Institute for Dryland Agriculture.

Lal, R. 2004. Soil carbon sequestration in India. *Climate Change* 65: 277–296.

Ramakrishna, Y.S., G.G.S.N. Rao, B.V. Ramana Rao, and P. Vijay Kumar. 1999. Agrometeorology. In *Fifty years of natural resource management research,* eds. G.B. Singh and B.R. Sharma, 32–60. New Delhi, India: Indian Council of Agricultural Research.

Sehgal, J., and I.P. Abrol. 1994. *Soil degradation in India: Status and impact*. New Delhi: Oxford/IBH Publishing Co.

Shantz, H.L. 1956. History and problems of arid lands development. In *The future of arid lands,* ed. G.F. White, 3–25. Publ. 43. Washington, DC: Am. Assoc. Adv. Sci.

Sharma, K.L., U.K. Mandal, K. Srinivas, K.P.R. Vittal, Biswapati Mandal, J. Kusuma Grace, and V. Ramesh. 2005. Long term soil management effects on crop yields and soil quality in dryland Alfisols. *Soil Tillage Res.* 83: 246–259.

Sharma, K.L., K. Srinivas, U.K. Mandal, K.P.R. Vittal, J. Kusuma Grace, and G. Maruthi Sankar. 2004. Integrated nutrient management strategies for sorghum and green gram in semi arid tropical Alfisols. *Indian J. Dryland Agricultural Res. Develop.* 19(1): 13–23.

Sharma, K.L., J. Kusuma Grace, Uttam Kumar Mandal, N. Pravin Gajbhiye, K. Srinivas, G.R. Korwar et al. 2008. Evaluation of long-term soil management practices using key indicators and soil quality indices in a semi-arid tropical Alfisol. *Austral. J Soil Res.* 46: 368–377.

Singh, H.P. 2002. Farming systems and best practices for drought prone areas of India. In *Farming systems and best practices for drought prone areas in Asia and the Pacific*. Hyderabad, India: Central Research Institute for Dryland Agriculture.

Singh, H.P., K.D. Sharma, and G. Subba Reddy. 2002. Rainfed agriculture: Challenges of the 21st century – National perspectives. Proceedings of the 89th Indian Science Congress – Part IV. Lucknow, India, pp. 3–4.

Singh, H.P., K.D. Sharma, G. Subba Reddy, and K.L. Sharma. 2004. Dryland agriculture in India. In *Challenges and strategies of dryland agriculture,* eds. L.K. Al-Amoodi, K.A. Barbarick, C.A. Roberts, Srinvas C. Rao, and John Ryan, 67–92. CSSA Special Publication 32, Chapter 6. Madison, WI: Crop Science Society of America, American Society of Agronomy.

Singh, R.P., and S.K. Das. 1984. Nitrogen management in cropping systems with particular reference to rainfed lands of India. In *Nutrient management in drylands with special reference to cropping systems and semi-arid red soils,* 1–56. Hyderabad, India: All India Coordinated Research Project for Dryland Agriculture.

Singh, R.P., and G. Subba Reddy. 1986. *Research on drought problems in arid and semi-arid tropics*. Patancheru, India: International Crops Research Institute for Semi-Arid Tropics.
Stewart, B.A., and P. Koohafkan. 2004. Dryland agriculture: Long neglected but of worldwide importance. In *Challenges and strategies of dryland agriculture, eds.* L.K. Al-Amoodi, K.A. Barbarick, C.A. Roberts, Srinvas C. Rao, and John Ryan, 11–24. CSSA Special Publication 32, Chapter 2. Madison, WI: Crop Science Society of America, American Society of Agronomy.
Subba Reddy, G., D. Gangadhar Rao, S. Venkateswarlu, and V. Maruthi. 1996. Drought management options for rainfed castor grown in Alfisols. *J. Oilseeds Res.* 13(2): 200–207.
Thyagaraj, C.R., K.P.R. Vittal, V.M. Mayande, and K.L. Sharma. 1999. Tillage and soil management for higher productivity in drylands. In *Fifty years of dryland agricultural research in India*, eds. H.P. Singh, Y.S. Ramakrishna, K.L. Sharma and B. Venkateswarlu. 329–344. Hyderabad, India: Central Research Institute for Dryland Agriculture.
Tiwari, K.N. 2006. Getting agriculture moving-reverting stagnation in fertilizer consumption and Foodgrain Production. *Fertilizer Marketing News,* 37(3): 1–14, 29.
Virmani, S.M., P. Pathak, and R. Singh. 1991. Soil related constraints in dryland crop production in Vertisols, Alfisols and Entisols of India. In *Soil related constraints in crop production,* eds. T.D. Biswas, G. Narayanasamy, N.N. Goswami, G.S. Sekhon, and T.G. Sastry. Bulletin No. 15, Indian Society of Soil Science. 80–95. New Delhi: Indian Society of Soil Science.
Vittal, K.P.R., K. Vijayalaxmi, and U.M.B. Rao. 1983. Effect of deep tillage on dryland crop production in red soils of India. *Soil Tillage Res.* 3: 77–384.

Section VI

Biotechnological and agronomic strategies

Chapter 14

Biotechnology and drought tolerance

Satbir S. Gosal, Shabir H. Wani & Manjit S. Kang
Punjab Agricultural University, Ludhiana, India

ABSTRACT

Abiotic stresses present a major challenge in our quest for sustainable food production as these may reduce the potential yields by 70% in crop plants. Of all abiotic stresses, drought is regarded as the most damaging. The complex nature of drought tolerance limits its management through conventional breeding methods. Innovative biotechnological approaches have enhanced our understanding of the processes underlying plant responses to drought at the molecular and whole plant levels. Hundreds of drought stress-induced genes have been identified and some of these have been cloned. Plant genetic engineering and molecular-marker approaches allow development of drought-tolerant germplasm. Transgenic plants carrying genes for abiotic stress tolerance are being developed for water-stress management. Structural genes (key enzymes for osmolyte biosynthesis, such as proline, glycinebetaine, mannitol and trehalose, redox proteins and detoxifying enzymes, stress-induced LEA proteins) and regulatory genes, including dehydration–responsive, element-binding (DREB) factors, Zinc finger proteins, and NAC transcription factor genes, are being used. Using *Agrobacterium* and particle gun methods, transgenics carrying different genes relating to drought tolerance have been developed in rice, wheat, maize, sugarcane, tobacco, *Arabidopsis*, groundnut, tomato, and potato. In general, the drought stress-tolerant transgenics are either under pot experiments or under contained field evaluation. Drought-tolerant genetically modified (GM) cotton and maize are under final field evaluations in the United States. Molecular markers are being used to identify drought-related

Address correspondence to Satbir S. Gosal at the School of Agricultural Biotechnology, Punjab Agricultural University, Ludhiana, India 141 004. E-mail: ssgosal@rediffmail.com

Reprinted from: *Journal of Crop Improvement*, Vol. 23, Issue 1 (January 2009), pp. 19–54,
DOI: 10.1080/15427520802418251

quantitative trait loci (QTL) and their efficient transfer into commercially grown crop varieties of rice, wheat, maize, pearl millet, and barley.

Keywords: drought tolerance, abiotic stress, transgenics, molecular markers, crop improvement

14.1 INTRODUCTION

Coping with plant environmental stress is the foundation of sustainable agriculture. Stress is a phenomenon that limits crop productivity or destroys biomass. Stress can be biotic, caused by insects and diseases, or abiotic, which may include drought, flooding, salinity, metal toxicity, mineral deficiency, adverse pH, adverse temperature, and air pollution. Among the abiotic stresses affecting crop productivity, drought is regarded as most damaging (Borlaug and Dowswell, 2005). Drought and salinity are widespread in many regions and are expected to cause, by 2050, serious salinization of more than 50% of all arable lands (Vinocur and Altman, 2005). The world food grain production needs to be doubled by the year 2050 to meet the food demands of the ever-growing population (Tilman et al., 2002), which is going to reach 9 billion by that time (Virmani and Ilyas-Ahmed, 2007). Abiotic stresses present a major challenge in our quest for sustainable food production, as these may reduce the potential yields by 70% in crop plants (Katiyar-Agarwal et al., 2006). There is an increasing scarcity of fresh water, and plants account for about 65% of global fresh-water use (Postel, Daily & Ehrlich, 1996). Soil salinity limits crop production in about 20% of irrigated lands (Flowers & Yeo, 1995). Drought and salinity stresses also limit crop production even under irrigated conditions (Chinnusamy, Xiong & Zhu, 2006). Drought is an extended dry period that results in crop stress and reduction in harvest. Different plant species, or even different varieties of a species, exhibit variable responses to drought tolerance, which may be attributed to escape, avoidance, or resistance. Development of genetic resistance is the best approach to mitigate drought effects.

14.2 DEVELOPMENT OF DROUGHT RESISTANCE

Conventional breeding approaches, involving inter-specific and inter-generic hybridizations and mutagenesis, have been used but with limited success. Major problems have been the complexity of drought tolerance, low genetic variance for yield components under drought conditions, and the lack of efficient selection procedures. With the advent of innovative approaches of biotechnology, our understanding of the processes underlying plant responses to drought at the molecular and whole plant levels has rapidly progressed. Hundreds of genes that are induced under drought have been identified and some of these have been cloned. A range of tools from gene expression patterns to transgenic plants has now become available to better understand drought tolerance mechanisms. New techniques, such as genome-wide tools, proteomics, stable isotopes, and thermal or fluorescence imaging, may help bridge the genotype–phenotype gap. There are two main biotechnological approaches, i.e., plant genetic engineering and molecular-marker technology, which are being followed to develop drought-tolerant germplasm.

14.2.1 Plant genetic engineering and development of transgenics

Following the availability of genetic-engineering techniques, useful gene(s) cloned from viruses, bacteria, fungi, insects, animals, and human beings, as well as genes synthesized in the laboratory, can be introduced into plants. Unlike conventional plant breeding, only the specific cloned gene(s) are being introduced without the co-transfer of undesirable genes from donors and there is no need for repeated backcrossing. Gene pyramiding or gene stacking through co-transformation of different genes with similar effects can also be achieved. Transgenic plants carrying genes for abiotic stress tolerance are being developed for water management. During the past 15 years, combined use of recombinant-DNA technology, gene-transfer methods, and tissue-culture techniques has led to efficient transformation and production of transgenics (genetically modified organisms or GMOs) in a wide variety of crop plants (James, 2007). In fact, transgenesis has emerged as an additional tool to carry out single-gene breeding or transgenic breeding of crops.

14.2.1.1 Types of genes used for developing abiotic stress resistance through genetic engineering

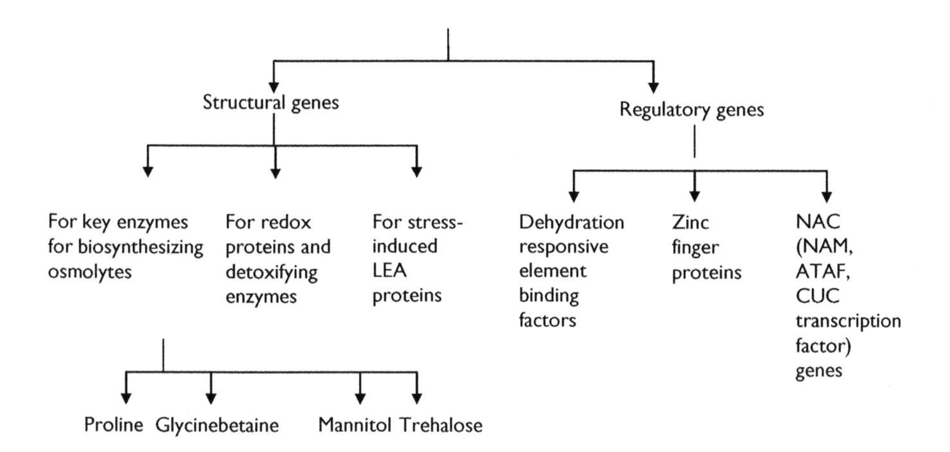

14.3 STRUCTURAL GENES

14.3.1 Key enzymes for biosynthesizing osmolytes

14.3.1.1 Proline

Amino acid proline is the most commonly distributed compatible osmolyte in plants. Proline synthesis pathway in plants, which takes place in cytoplasm, is from glutamate through γ-glutamyl phosphate and glutamyl-γ-semialdehyde. Enzyme pyrroline-5-carboxylate synthetase (P5CS) in plants and animals catalyzes this reaction in two steps. The glutamyl-γ-semialdehyde is naturally cyclised to pyrroline-5-carboxylate, which is converted to proline by pyrroline-5-carboxylate reductase (Taylor, 1996). The role of proline in stress tolerance has been proved through the production of transgenic plants overexpressing proline in many crop plants and forest trees (Kishore et al., 1995; Zhu et al., 1998; Gleeson, Walter & Parkinson, 2005; Molinari et al., 2007). Overexpression of *Vigna aconitifolia* gene P5CS in tobacco resulted in a 10- to 18-fold increase in proline synthesis. This proline acted as an osmoprotectant and its overproduction resulted in increased tolerance to osmotic stress in plants (Kishore et al., 1995). The P5CS gene from *V. aconitifolia* has also been transferred into rice under the control of ABA-responsive element from barley *HVA22* gene (promoter elements). Transgenic plants showed stress-inducible proline accumulation under water stress (Zhu et al., 1998). The root biomass of transgenic plants was significantly higher than that of wild-types. After a limited period of salt stress, the transgenic plants showed quick recovery (Zhu et al., 1998). Likewise, *Agrobacterium*-mediated transfer of *V. aconitifolia* P5CS gene into wheat resulted in improved salt tolerance, thus proving the osmoprotectant nature of proline in wheat (Sawahel and Hassan, 2002). Further, transformation experiments with some other plant species, such as *Arabidopsis* (Nanjo et al., 1999), soybean (De Ronde et al., 2004), and tobacco (Yonamine et al., 2004), have also indicated the role of proline in preventing osmotic stress. A forest tree, *Larix leptoeuropaea*, has also been transformed with P5CS gene from *V. aconitifolia* through *Agrobacterium*-mediated gene transfer. Transgenic plants showed a 30-fold increase in proline level against non-transformed control. Transgenic tissue lines were significantly more tolerant to cold, salt, and freezing stresses, and grew under 200 mM NaCl or 4°C, which completely inhibited the growth of control cell lines (Gleeson, Walter & Parkinson, 2005). Further, P5CS from *V. aconitifolia* was introduced through the 'particle gun' method of gene transfer into *Saccharum officinarum* under the action of AIPC (ABA-inducible promoter complex) promoter. Stress-inducible proline accumulation in transgenic sugarcane plants under water-deficit stress acted as a component of an antioxidative defense system rather than as an osmotic-adjustment mediator (Molinari et al., 2007; Table 14.1).

14.3.1.2 Glycinebetaine

Glycinebetaine is another common compatible solute in various organisms, including higher plants (Csonka and Hanson, 1991; Le Rudulier, 1993; Rhodes and Hanson, 1993). Plant species accumulate betaine in response to drought and salinity stress (Rhodes and Hanson, 1993). Betaine protects plants through its action as an osmolyte, which helps maintain water balance between the plant cell and the environment (Robinson and Jones, 1986) and through stabilization of macromolecules during cell dehydration (Incharoensakdi, Takabe & Akazawa, 1986). In plants, a two-step oxidation of choline via betainealdehyde leads to the synthesis of betaine. In spinach, choline monooxygenase (CMO), a ferredoxin-dependent enzyme, catalyzes the first step i.e., conversion of choline into betainealdehyde (Figure 14.1; Rhodes and Hanson 1993). While the second step i.e., conversion of betainealdehyde into betaine, is mediated by betainealdehyde dehydrogenase (BADH; Rhodes and Hanson, 1993).

Although many plant species, such as maize, sorghum, sugar beet, barley, and *Artiplex* synthesize glycinebetaine (Rhodes and Hanson, 1993), some plants, e.g., *Brassica juncea, Arabidopsis*, tobacco, and rice, do not accumulate glycinebetaine; rice

Table 14.1 Genetic engineering of crop plants for abiotic stress tolerance.

Mechanism	*Transgene (s)*	*Plant species*	*Transformation method*	*Promoter*	*Remarks*	*Reference*
Proline	*P5CS* (Pyrroline - 5-carboxylate synthetase) from *Vigna aconitifolia*	*N. tabacum*	*Agrobacterium*	CaMV 35S	Transgenic plants produced 10–18 fold more proline than control plants. Over production of proline also enhanced root biomass and flower development in transgenic plants.	Kishore et al., 1995
	P5CS from *Vigna aconitifolia*	*Larix leptoeuropaea*	*Agrobacterium*	–	The integration of the gene into the plant genome was confirmed by Southern blot and by proline content analysis. There was an approximately 30-fold increase in proline level in transgenic tissue compared to non-transformed controls. The transgenic tissue lines were significantly more resistant to cold, salt, and freezing stresses and grew under conditions (200 mM NaCl or 4°C) that completely inhibited the growth of control cell lines.	Gleeson, Walter & Parkinson, 2005
	P5CS from *Vigna aconitifolia*	*Saccharum officinarum*	Particle gun	*AIPC*	Stress-inducible proline accumulation in transgenic sugarcane plants under water-deficit stress acts as a component of antiox idative defense system rather than as an osmotic adjustment mediator.	Molinari et al., 2007
Glycinebetaine	*codA from A. globiformis*	*Oryza sativa*	–	CaMV 35S	Transgenic plants had high levels of glycinebetaine and grew faster as compared to wild-types on removal of stress.	Sakamoto et al., 1998

(*Continued*)

Table 14.1 (Continued)

Mechanism	*Transgene (s)*	*Plant species*	*Transformation method*	*Promoter*	*Remarks*	*Reference*
	bet *A and bet B* from *E. coli*	*N. tabacum*	*Agrobacterium*	*rbc SIA*	Transgenic plants showed increased tolerance to salt stress as measured by biomass production of green house grown plants	Holmstrom et al., 2000
	CMO(choline monooxygenase) from *Spinacia olearceae*	*Oryza sativa*	*Agrobacterium*	Maize *ubi*	Transgenic plants were tolerant to salt and temperature stress at seedling stage. CMO expressing rice plants were not effective for accumulation of glycinebetaine and improvement of productivity	Shirasawa et al., 2006
Mannitol	*mtlD* from *E. coli*	*Triticum aestivum*	Particle gun	*Maize ubi-1*	Ectopic expression of the *mtlD* gene for the biosynthesis of mannitol in wheat improves tolerance to water stress and salinity	Abebe et al., 2003
Trehalose	*TPSP* from *E. coli*	*Oryza sativa*	*Agrobacterium*	rbcS, and ABA	Transgenic rice plants accumulate trehalose at levels 3–10 times that of the non transgenic (wild-type) controls. Compared with non transgenic (wild-type) rice, several independent transgenic lines exhibited sustained plant growth, less photo-oxidative damage, and more favorable mineral balance under salt, drought, and low-temperature stress conditions.	Garg et al., 2002
	TPS1-TPS-2 from *E. coli*	*A. thaliana*	*Agrobacterium*	CaMV 35S	No morphological growth alterations were observed in lines over-expressing the *TPS1-TPS2* construct, while the plants over expressing the *TPS1* alone under the control of 35S promoter had abnormal growth, color and shape	Miranda et al., 2007

Redox Proteins	*Mn-SOD* Superoxide dismutase from *N. plumbaginifolia*	*Medicago sativa.*	*Agrobacterium*	CaMV 35S	Transgenic plants had reduced injury from water-deficit stress. A three-year field trial indicated that yield and survival of transgenic plants were significantly improved.	Mckersie et al., 1996
	GmTP55 from *Glycine max*	*A. thaliana and N. tabacum*	*Agrobacterium*	CaMV 35S	Ectopic expression of GmTP55 in both *Arabidopsis* and tobacco conferred tolerance to salinity during germination and to water deficit during plant growth. Antiquitin may be involved in adaptive responses mediated by a physiologically relevant detoxification pathway in plants.	Rodrigues et al., 2006
	GST (Glutathione S-t ransferase) and *CAT I* (Catalase) from *Suaeda salsa*	*Oryza sativa*	*Agrobacterium*	–	Transgenic rice over-expressing the *GST* (Glutathione S-transferase) and *CAT I* (Catalase) from *Suaeda* salsa showed increased tolerance to oxidative stress caused by salt and paraquat. Double transgenic (*GST* + *CAT I*) showed higher abiotic stress tolerance as compared with a single GST.	Zhao and Zhang, 2006a, 2006b
	MDAR from A. thaliana	*N. tabacum*	*Agrobacterium*	–	Transgenic plants exhibited 2.1 fold higher MDAR activity and 2.2 fold higher level of reduced AsA compared to wild-type control plants, which increased photosynthetic rates under ozone, salt and PEG stress. In addition, these transgenic plants showed significantly lower hydrogen peroxide level when tested under salt stress.	Eltayeb et al., 2007

(Continued)

Table 14.1 (Continued)

Mechanism	*Transgene (s)*	*Plant species*	*Transformation method*	*Promoter*	*Remarks*	*Reference*
	VTE1 from *Arabidopsis*	*N. tabacum*	*Agrobacterium*	–	Transgenic lines showed enhanced tolerance to drought stress.	Liu et al., 2008
Lea Proteins	*HVA1* from *Hordeum vulgare*	*Oryza sativa*	Particle gun	Rice *actin 1*	Second-generation transgenic plants showed significantly increased tolerance to water deficit and salinity stress.	Xu et al., 1996
	HVA1 from *Hordeum vulgare*	*Triticum aestivum*	Particle gun	Maize *ubi 1*	Transgenic plants showed improved growth characteristics in response to soil-water deficits. Field trials showed that *HVA1* gene had the potential to confer drought-stress protection on transgenic spring wheat.	Sivamani et al., 2000 Bahieldin et al., 2005
	ME.LEA N4 Brassica napus	*Lactuca sativa*	*Agrobacterium*	CaMV 35S	Transgenic plants showed improved growth characteristics under salt and water-deficit stress.	Park et al., 2005a
	ME.LEA N4 Brassica napus	*Brassica campestris*	*Agrobacterium*	CaMV 35S	Transgenic plants showed increased growth ability under salt and water deficit.	Park et al., 2005b
	Os LEA-3–1 from *Oryza sativa*	*Oryza sativa*	*Agrobacterium*	CaMV 35S, HVA1-like	Transgenic rice plants showed increased growth ability under salt and water-deficit stress.	Xiao et al., 2007
Regulatory genes	*CBF1* from *A. thaliana*	*Lycopersicon esculentum*	*Agrobacterium*	CaMV 35S	Transgenic tomato plants were more resistant to water-deficit stress than the wild-type plants.	Hsieh et al., 2002
	DREB 1A from *A. thaliana*	*Triticum aestivum*	Particle gun	*rd 29 A*	Transgenic plants expressing *DREB1A* gene demonstrated substantial resistance to water stress under greenhouse conditions.	Pellegrineschi et al., 2004

NPK1 from *N. tabacum*	*Zea mays*	*Agrobacterium, Particle gun*	35SC4PPDK	Transgenic maize plants showed enhanced drought tolerance. The *Agrobacterium*-derived events contained fewer than five copies of the NPK1 transgene, whereas the bombardment-derived events carried more than 20 copies of the transgene.	Shou et al., 2004
CBF3/ DREB 1A from *A. thaliana*	*Oryza sativa*	–	Rice *Ubi 1*	Tolerance to drought and high salinity without growth retardation or any phenotypic alteration.	Oh et al., 2005
MBF1c from *A. thaliana*	*A. thaliana*	*Agrobacterium*	CaMV 35S	Constitutive expression of the stress-response transcriptionalcoactivator multiprotein bridging factor 1c (MBF1c) in *Arabidopsis* enhances the tolerance of transgenic plants to bacterial infection, heat, and osmotic stress. Moreover, the enhanced tolerance of transgenic plants to osmotic and heat stress was maintained even when these two stresses were combined.	Suzuki et al., 2005
DREB1A from *A. thaliana*	*Solanum tuberosum*	*Agrobacterium*	rd29A	Transgenic plants showed significant tolerance against salinity stress (1M NaCl)	Benham et al., 2006
SNAC 1 from *Oryza sativa*	*Oryza sativa*	*Agrobacterium*	CaMV 35S	Transgenic plants showed improved drought resistance under favorable conditions and strong tolerance to salt stress.	Hu et al., 2006
Os DREB 1A / Os DREB 1B from *Oryza sativa*	*Oryza sativa*	*Agrobacterium*	Ca MV 35S	DREB 1 type genes are quite useful for improving stress tolerance in crop plants, including rice.	Ito et al., 2006

(*Continued*)

Table 14.1 (Continued)

Mechanism	Transgene (s)	Plant species	Transformation method	Promoter	Remarks	Reference
	DREB2A from *A. thaliana*	*A. thaliana*	*Agrobacterium*	CaMV 35S	Significant drought-stress tolerance but slight freezing tolerance in transgenic *Arabidopsis* plants.	Sakuma et al., 2006
	DREB 1A from *A. thaliana*	*Arachis hypogaea*	*Agrobacterium*	CaMV 35S, rd 29 A	Transgenic plants showed increased transpiration efficiency, an important feature of drought tolerance.	Bhatnagar-Mathur et al., 2007
	Os NAC 6 from *Oryza sativa*	*Oryza sativa*	*Agrobacterium*	–	Transgenic plants showed tolerance to dehydration and high salt stress.	Nakashima et al., 2007
	HvCBF4 from *Hordeum vulgare*	*Oryza sativa*	*Agrobacterium*	*Ubi1*	Transgenic rice resulted in an increase in tolerance to drought, high salinity and low-temperature stresses without stunting growth.	Oh et al., 2007
	Zm DREB 2A from *Zea mays*	*A. thaliana*	*Agrobacterium*	CaMV 35S	Drought stress tolerance in plants was improved.	Qin et al., 2007
	SDIR1 from *Arabidopsis*	*Arabidopsis*	–	–	Transgenic plants showed ABA hypersensitivity and ABA-associated phenotypes, such as salt hypersensitivity in germination, enhanced ABA-induced stomatal closure, and enhanced drought tolerance.	Zhang et al., 2007
	AtMYB44 from *Arabidopsis*	*Arabidopsis*	*Agrobacterium*	CaMV 35S	Transgenic plants exhibited a reduced rate of water loss, asmeasured by the fresh-weight loss of detached shoots, and remarkably enhanced tolerance to drought and salt stress comparedto wild-type plants.	Jung et al., 2008

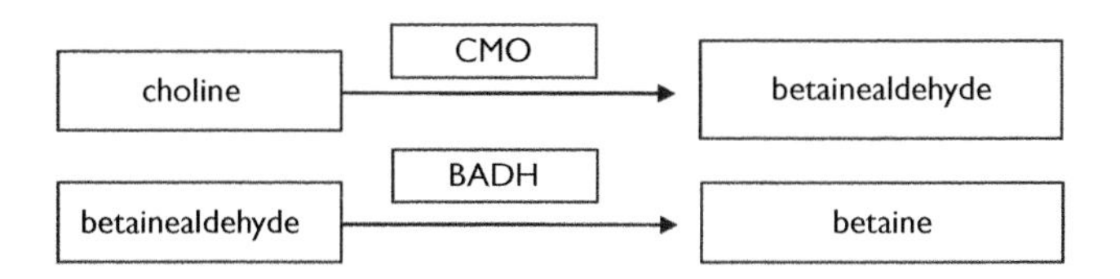

Figure 14.1 A two-step oxidation of choline via betainealdehyde to form betaine. CMO: choline monooxygenase, BADH: Betainealdehyde dehydrogenase.

is the only important cereal crop that does not accumulate glycinebetaine (Shirasawa et al., 2006).

Efforts have been made to produce transgenic plants that accumulate glycinebetaine by introducing genes cloned from different plant species and microorganisms (Sakamoto & Murata, 2002). Initial attempts at producing transgenic plants through the introduction of Choline monooxygenase (CMO) and BADH pathway were made in tobacco. Tobacco plants were transformed with cDNA for BADH from spinach (*Spinacia oleracea*) and sugar beet (*Beta vulgaris*) under the control of CaMV 35 S promoter (Rathinasabapathi et al., 1994). The BADH was produced in chloroplasts of tobacco. Betaine aldehyde was converted to betaine by BADH, thus conferring resistance to betaine aldehyde. However, transgenic plants were not able to accumulate betaine in the absence of exogenously supplied betaine aldehyde. This was because of the absence of the enzyme required for oxidation of choline. This showed that expression of *BADH* alone was not sufficient for synthesis of betaine in transgenic plants. In another attempt, cDNA for CMO from *Spinacia oleracea* was introduced into tobacco and the enzyme thus synthesized was transported to its functional site, i.e., chloroplasts, but the leaves of tobacco accumulated betaine at a very low concentration, i.e., 10- to 100-fold lower than expected (Nuccio et al., 1998). The most probable reason for insufficient synthesis of betaine was the absence of engineered BADH activity in chloroplasts. Therefore, both CMO and BADH must be present in the chloroplasts for efficient synthesis of betaine in transgenic plants that do not accumulate glycinebetaine.

Genetic transformation of tobacco with *E. coli* genes resulted in transgenic plants accumulating glycinebetaine. The plants that produced both CDH and BADH generally accumulated higher amounts of glycinebetaine than plants producing CDH alone. Increased tolerance to salt stress was visible from increased biomass production of transgenic tobacco lines under greenhouse conditions. In addition, the transgenic plants showed enhanced recovery from photo-inhibition caused by light, salt stress, and low-temperature conditions (Holmstrom et al., 2000). Thus, the theory that both CMO and BADH should be present in chloroplasts was again confirmed. Further, a gene for CMO cloned from spinach was introduced into rice through *Agrobacterium*-mediated transformation. The level of glycinebetaine in rice was lower than expected. Two of the reasons (Shirasawa et al., 2006) given for the low productivity of rice and low glycinebetaine accumulation were: First, the position of spinach CMO and endogenous BADH might be different; second, the catalytic activity of spinach CMO in rice plants might be lower than it was in spinach (Table 14.1). Thus, it was concluded that glycinebetaine had a role as a compatible solute and its engineering into non-accumulations would be a success only if both CMO and BADH pathways were introduced and if both CMO and BADH were localized in chloroplasts.

14.3.1.3 Mannitol

Mannitol is an osmolyte that is normally synthesized in numerous plant species. On exposure to low water potential, plants accumulate mannitol at an increased rate (Patonnier, Peltier & Marigo, 1999). Mannitol is synthesized from fructose-6-P. The latter is converted to mannitol-1-P by mannose-6-P isomerase. Mannitol-1-P is then reduced to mannitol by mannitol-1-phosphate dehydrogenase (Chinnusamy, Xiong & Zhu, 2006). Attempts have been made to produce transgenic plants that otherwise do not accumulate mannitol. (Tarczynski Jensen & Bohnert, 1992, 1993; Thomas et al., 1995; Karakas et al., 1997; Shen, Jensen & Bohnert, 1997; Abebe et al., 2003). Mannitol accumulated to more than 6 μ mol g^{-1} fw in the leaves of transgenic tobacco plants (Tarczynski, Jensen & Bohnert, 1992) transformed with *mtlD* – a gene for mannitol1-phosphate dehydrogenase from *E. coli.* These transgenic plants showed increased tolerance to high salinity (Tarczynski, Jensen & Bohnert, 1993). Mannitol-accumulating transgenic plants were 20% to 25% shorter than wild plants under non-stressed conditions. The dry weight of the wild-type plants was reduced by 44% under 150 mM NaCl stress, but the dry weight of transgenic plants remained unchanged. In contrast to wild-type plants, transgenic plants adjusted their growth in response to osmotic stress (Karakas et al., 1997). In another experiment, an *mtlD* gene construct was transferred into tobacco chloroplasts. Mannitol accumulation ranged from 2.5–7.0 μ mol g^{-1} fw This resulted in increased resistance to methyl viologen-induced oxidative stress. Such resistance was due to increased capacity to scavenge hydroxyl radicals (Shen, Jensen & Bohnert, 1997). The concentration of mannitol reached 10 μ mol g^{-1} dw in the seeds of *Arabidopsis* plants accumulating mannitol (Thomas et al., 1995). At a salinity stress of 400 mM NaCl, the mannitol accumulating seeds were able to germinate, whereas the control seeds were unable to germinate at even 100 mM NaCl.

Genetic transformation of *Arabidopsis* plants with *M6PR* gene from celery under the control of CaMV35S promoter resulted in transgenic plants that accumulated 0.5–6 mM g^{-1} fw mannitol. Transgenic *Arabidopsis* plants were tolerant to salt stress as tested via an application of irrigation water containing 300 mM NaCl in nutrient solution (Zhifang and Loescher, 2003). The *mtlD* gene from *E. coli* has also been transferred into wheat through particle gun under the control of maize ubiquitin 1 promoter. Transgenic plants lacked the osmotically significant quantities of mannitol. However, transgenics showed tolerance to water stress and salinity (Abebe et al., 2003; Table 14.1). These studies clearly indicate that mannitol has a role as a compatible osmolyte. Further osmoprotection through mannitol is one of the ways to produce salt- and water-stress-tolerant transgenic plants.

14.3.1.4 Trehalose

Trehalose (α-D- glucopyranosyl -1, 1-α-D glucopyranoside) is a non-reducing disaccharide commonly found in bacteria, fungi, and some vertebrates (Elbein, 1974). Biological synthesis of trehalose is a two-step pathway. It is formed from glucose-6-phosphate and uridine diphosphoglucose via trehalose-6-phosphate. The first and the second steps are mediated by trehalose phosphate synthase (*TPS*) and trehalose-6-phosphate phosphatase (*TPP*). In plants, *TPS* level is highly regulated by enzymes that directly metabolize it or by trehalase that breaks down trehalose (Rathinasabapathi and Kaur, 2006).

The introduction of trehalose genes into plants has led to improved stress tolerance. When the TPS-coding gene was expressed alone, it resulted in striking morphological changes (Pilon-Smits et al., 1998; Yeo et al., 2000). To overcome this problem (morphological changes) both *TPS1* and *TPS2* homologues were used in a stress-inducible expression in rice. Bacterial genes *otsA* and *otsB* were transferred into rice. Transgenic rice accumulated trehalose three to ten times more than non-transgenic controls and exhibited tolerance to salt, drought, and low-temperature without stunting growth. Thus, stress-inducible transgene expression is important in recovering trehalose-accumulating transgenics without deleterious effects (Garg et al., 2002). Further, *TPS1-TPS2* fusion gene construct was introduced into *Arabidopsis* through *Agrobacterium*-mediated gene transfer under the control of CaMV 35S or stress regulated *rd29A* promoter. No morphological growth alterations were observed in lines overexpressing the *TPS1-TPS2* construct, whereas the plants overexpressing the *TPS1* alone under the control of 35S promoter showed abnormal growth, color, and shape (Miranda et al., 2007; Table 14.1). Thus, it can be concluded that engineering trehalose metabolism in plants can substantially increase their capacity to tolerate abiotic stresses.

14.3.2 Redox proteins and detoxifying enzymes

Although oxygen is important for living organisms to survive, it makes the survival of aerobic living beings difficult through the production of reactive oxygen species (ROS), such as superoxide radicals (O_2^-), hydrogen peroxide (H_2O_2), and hydroxyl radicals (OH^-). The imbalance between the production and removal of ROS leads to oxidative stress (Halliwell, 1997). The increased production of reactive oxygen leads to drought and salinity stresses. The quick buildup of toxic products results in oxidative stress (Rathinasabapathi and Kaur, 2006). The increasing knowledge on the genetic, molecular, and sub-molecular basis of plant responses to stress factors may be exploited for the development of oxidative stress-tolerant crops (Edreva, 2005). In this context, many attempts have been made to produce transgenic plants overexpressing ROS-scavenging enzymes – catalase (CAT), ascorbate peroxidase (APX), superoxide dismutase (SOD), glutathione reductase (GR) – and improvement in stress tolerance has been achieved. *SOD*-overexpressing alfalfa plants were tolerant to water deficit (McKersie et al., 1996), and tobacco plants were tolerant to high salt (Van Camp et al., 1996). Field trials have confirmed tolerance of transgenic plants to oxidative stress in a drought environment. Results similar to those of McKersie et al. (1996) and Van Camp et al. (1996) were obtained with overproduction of *SOD* in lucerne mitochondria (McKersie, Murnaghan & Bowley, 1997) and in cytosol of potato (Perl et al., 1993). In another study, *Mn SOD* overexpression in chloroplasts of tobacco resulted in transgenic plants that were tolerant to Mn deficiency-mediated oxidative stress (Yu and Rengel, 1999). The overexpression of cell wall peroxidase (*TPX2*) in tobacco plants improved germination under oxidative stress (Amaya et al., 1999).

Ascorbate peroxidase gene from *A. thaliana* was transferred into tobacco chloroplasts following *Agrobacterium*-mediated gene transfer. The first generation of transgenic lines showed enhanced tolerance to polyethylene glycol (PEG) and water stress, as determined by net photosynthesis. This demonstrated that overexpression of cytosolic APX in tobacco chloroplasts reduced the toxicity of hydrogen peroxide (H_2O_2) (Badawi et al., 2004).

Ascorbate (AsA) is a major antioxidant and free-radical scavenger in plants. Mono-dehydro-ascorbate reductase (MDAR) is crucial for AsA regeneration and essential for maintaining a reduced pool of AsA. Further, MDAR gene from *A. thaliana* was overexpressed in tobacco cytosol. Transgenic plants exhibited a 2.1-fold higher MDAR activity and 2.2-fold higher level of reduced AsA as compared to wild-type control plants. Transgenic plants also showed enhanced stress tolerance as shown by significantly higher net photosynthesis rates under ozone, salt, and PEG stresses. In addition, these transgenic plants showed significantly lower hydrogen peroxide levels when tested under salt stress (Eltayeb et al., 2007). Transgenic plants showing tolerance to H_2O_2 and paraquat-induced oxidative stress have been produced. *Gm TP55* antiquitin homologue gene from soybean has been overexpressed in *A. thaliana* and *N. tabacum*. Both the transgenic plants possessed salinity tolerance during germination and water-deficit tolerance during plant growth. These transgenic lines also exhibited a lower concentration of lipid peroxidation-derived reactive aldehydes under oxidative stress than control leaves (Rodrigues et al., 2006). The transgenic plants exhibited an enhanced tolerance to paraquat-induced oxidative stress (Rodrigues et al., 2006). Antioxidant genes are significant in promoting an understanding of the roles of specific antioxidant defenses under stress conditions (Zhao and Zhang, 2007). Transgenic rice overexpressing the *GST* (glutathione S-transferase) and *CAT 1* (catalase) from *Suaeda salsa* showed increased tolerance to oxidative stress caused by salt and paraquat (Zhao and Zhang, 2006a,b). Double transgenic (*GST* + *CAT 1*) showed better abiotic stress tolerance as compared with *GST* alone transgenic. This indicates the need for gene pyramiding or multigene transfer for the complex trait of abiotic stress.

14.3.3 Stress-Induced LEA Proteins

Late embryogenesis-abundant (LEA) proteins are mainly low molecular weight (10–30 kDa) proteins, which are involved in protecting higher plants from damage caused by environmental stresses, especially drought (dehydration) (Hong, Zong-Suo & Ming-An, 2005). Over-expression of barley (*Hordeum vulgare* L.) group 3 LEA protein HVA1 in rice has resulted in transgenic plants constitutively accumulating HVA1 protein both in leaves and roots. Second generation transgenic plants have shown increased tolerance to water deficit and salinity stress (Xu et al., 1996). Transgenic plants also maintained higher growth rates than the non-transformed, wild-type control plants. A similar attempt has improved biomass productivity and water-use efficiency under water deficit conditions in transgenic wheat constitutively expressing the barley *HVA1* gene (Sivamani et al., 2000). In this case, the gene under the control of maize *ubi1* promoter was transferred into immature embryos of greenhouse-grown wheat by particle gun transformation method After five years, field evaluation showed a yield increase in the transgenic lines under drought conditions (Bahieldin et al., 2005; Table 14.1).

Overexpression of *COR15a*, a group-II LEA protein that was targeted to the chloroplasts, increased freezing tolerance of chloroplasts *in vivo*, and of protoplasts *in vitro* (Artus et al., 1996). This increase most likely resulted from the membrane-stabilizing effect of *COR15a* (Artus et al., 1996, Steponkus et al., 1998). However, the protective effect of *COR15a* was insignificant for the survival of whole plants during freezing (Jaglo-Ottosen et al., 1998).

An investigation has been done on the third generation of transgenic oat (*Avena sativa* L.) expressing barley *HVA1* stress tolerance (*uidA*; *GUS*) and *bar* (herbicide resistance) genes. Accordingly, transgenic plants showed normal 9:7 ratio, third-generation inheritance for glufosinate ammonium herbicide resistance. Molecular and histochemical studies confirmed the presence and stable expression of all three genes. Compared with the non-transgenic control plants, transgenic R3 plants exhibited greater growth and showed a significant increase in tolerance to salt stress (200 mM NaCl) for various traits, including number of days to heading, plant height, flag leaf area, root length, panicle length, number of spikelets/panicle, number of tillers/plant, number of kernels/panicle, 1000-kernel weight, and kernel yield/plant (Oraby et al., 2005).

In separate experiments, *ME-lea N4* gene from *Brassica napus* was introduced into lettuce (*Lactuca sativa* L.; Park et al., 2005a) and Chinese cabbage (*Brassica campestris. Pekinensis*; Park et al., 2005b) through *Agrobacterium*-mediated gene transfer under the control of CaMV 35S promoter. In both the cases, transgenic plants showed enhanced growth ability as compared to non-transgenic controls under salt and water-deficit stress.

Further, rice has been transformed with *OsLEA 3–1* gene through *Agrobacterium*-mediated gene transfer method under the control of different promoters. Constitutive and stress-inducible expression of *Os LEA 3–1* under the control of CaMV 35S and HVA1-like promoter, respectively, resulted in transgenic rice plants showing increased tolerance to drought under field conditions (Xiao et al., 2007). Thus, the above findings show that LEA genes have the potential to confer environmental stress protection on various crop plants.

14.4 REGULATORY GENES

The gene-regulating protein factors that regulate gene expression and signal transduction and function under stress responses may be useful for improving the abiotic stress tolerance in plants. These genes comprise regulatory proteins, i.e., transcription factors (*bZip, MYC, MYB, DREB, NACs NAM, ATAF*, and *CUC*); protein kinases (*MAP* kinase, CDP kinase, receptor protein kinase, ribosomal protein kinase, and transcription regulation protein kinase, etc.); and proteinases (phosphoesterases and phospholipase) (Katiyar-Agarwal et al., 2006). Using regulatory genes, transgenic plants have been developed and tested for abiotic stress tolerance.

14.4.1 Dehydration-Responsive Element-Binding Factors (DREB)

The transcription factors activate cascades of genes that act together in enhancing tolerance towards multiple stresses. Dozens of transcription factors are involved in the plant response to drought stress (Vinocur and Altman, 2005; Bartels and Sunkar, 2005). Stress tolerance is a complex trait, and as such it is unlikely to be under a single-gene control. A wise strategy may be the use of transcription factors regulating the expression of several genes related to abiotic stress. Initial attempts at genetic transformation using these *DREB* genes started with *Arabidopsis*. Over-expression of *DREB1/CBF* in *Arabidopsis* resulted in the activation of expression of many stress-tolerance genes and tolerance of the plant to drought, high salinity, and/or freezing was improved (Jaglo-Ottosen et al., 1998, Liu et al., 1998). Transgenic plants overexpressing *DREB1/CBF3* under the control of CaMV 35S promoter also showed increased tolerance to drought, high salinity and freezing stress. (Kasuga et al., 1999; Gilmour et al., 2000). Constitutive expression of *DREB1A* protein led to growth retardation under normal growth conditions. Thus, the use of stress-inducible *rd 29A* promoter for the over-expression of *DREB1A* minimizes negative effects on plant growth (Kasuga et al., 1999).

In their first attempt to transform wheat with *DREB1A* transcription factor, Pellegrineschi et al. (2004) used the particle-gun method as the means for gene transfer. *DREB1A* was introduced into wheat under the control of stress inducible *rd 29A* promoter. Under greenhouse conditions, substantial resistance to water deficit was demonstrated. Oh et al. (2005) ectopically expressed *Arabidopsis DREB1A /(CBF3)* in transgenic rice plants under the control of CaMV 35S promoter. *DREB1A* transgenic rice plants showed enhanced tolerance to drought and salinity but only to a little extent to low-temperature stress without any stunted phenotype despite its constitutive expression.

Four rice *CBF/DREB1A* orthologs, *OsDREB1A*, *OsDREB1B*, *OsDREB1C* and *OsDREB1D*, have been isolated (Dubouzet et al., 2003). *OsDREB1* transgenic rice plants had improved tolerance to drought, salt, and low temperatures. On further analysis, a large portion of stress-inducible genes were identified, which provided the confirmation that *DREB1/CBF* cold-responsive pathways are conserved in *Arabidopsis* and rice (Ito et al., 2006). In another attempt, constitutive or stress-inducible expression of *ZmDREB2A* resulted in an improved drought-stress tolerance in plants. In addition, transgenic plants showed enhanced thermo tolerance, which indicated that *ZmDREB2A* had a dual function of mediating expression of genes responsive to both water and heat stress (Qin et al., 2007). Similar results were earlier found in transgenic *Arabidopsis* plants overexpressing constitutively active *DREB2A*, in which transgenic plants showed increased thermo tolerance in addition to tolerance against water stress (Sakuma et al., 2006).

Transpiration efficiency (TE) is the most important component trait of drought. About 90% of the rooted peanut shoots transformed with rd *29A:DREB1A* construct survived and appeared to be phenotypically normal. Fifty-three *rd 29A: DREB1A* and 18 *35S:DREB1A* plants were successfully transferred to greenhouse and their seeds were collected. Transgenic peanut had higher TE than the wild type under well-watered conditions. Some transgenic events showed a 70% increase over wild type; JL 24; RD 2 had 40% more TE than WT JL 24 under water-limited conditions (Bhatnagar-Mathur et al., 2007). Transgenic overexpression of *HvCBF4* from barley in rice increased tolerance to drought, high-salinity, and low-temperature stresses without stunting growth. Using the 60 K rice whole-genome microarrays, fifteen rice genes were identified that were activated by *HvCBF4*. When compared with twelve target rice genes of *CBF3/DREB1A*, five genes were common to both

HvCBF4 and *CBF3/DREB1A*, and ten and seven genes were specific to *HvCBF4* and CBF3/*DREB1A*, respectively (Oh et al., 2007).

14.4.2 Zinc finger proteins

Zinc finger proteins (ZFPs) play an important role in growth and development in both animals and plants. *Arabidopsis* genes encoding distinct ZFPs have been identified. However, the physiological role of their homologues with putative zinc finger motif remains unclear. *ThZF1*, a novel gene, was characterized from salt-stressed cress (*Thellungiella halophila*, Shan Dong), which encoded a functional transcription factor. *ThZF1* contains two conserved C2H2 regions and shares conserved domains, including DNA-binding motif, with *Arabidopsis thaliana ZFP* family members. The transcript of the *ThZF1* gene was induced by salinity and drought. Transient expression analysis of *ThZF1–GFP* fusion protein revealed that *ThZF1* was localized preferentially in the nucleus. A gel-shift assay showed that *ThZF1* specially binds to the wild-type (WT) *EP2* element, a *cis*-element present in the promoter regions of several target genes regulated by ZFPs. Furthermore, a functional analysis demonstrated that *ThZF1* was able to activate HIS marker gene in yeast. Finally, ectopic expression of *ThZF1* in *Arabidopsis* mutant azf2 suggested that *ThZF1* may have similar roles as *Arabidopsis AZF2* in plant development as well as regulation of downstream gene (Xu et al., 2007).

14.5 *NAM, ATAF* AND *CUC* TRANSCRIPTION FACTOR (NAC) GENES

The NAC gene family encodes one of the largest families of plant specific transcription factors and is absent in other eukaryotes (Gao, Chao & Lin, 2007). Rice and *Arabidopsis* genomes contain 75 and 105 putative NAC genes, respectively (Ooka et al., 2003). The role of the NAC gene family in abiotic stresses was discovered in *Arabidopsis*. Three NAC genes were induced under salt and/or drought stress. Over-expression of three genes, *ANACO19*, *ANACO55*, and *ANACO72*, greatly enhanced drought tolerance in *Arabidopsis* (Tran et al., 2004). Multiple rice transcription factors, including a NAC gene, were induced in the early stages of salt stress (Chao et al., 2005). *OsNAC6*, a member of ATAF family, was also induced by cold, salt, drought, and abscisic acid (ABA; Ohnishi et al., 2005). Over-expression of stress-responsive gene *SNAC1* (stress-responsive NAC 1) significantly enhanced drought resistance in transgenic rice (22% to 34% higher seed setting than control) in the field under severe drought-stress conditions at the reproductive stage while showing no phenotypic changes or yield penalty. The transgenic rice exhibited significant improvement in drought resistance and salt tolerance at the vegetative stage. Compared with wild type, the transgenic rice was more sensitive to abscisic acid and lost water more slowly by closing more stomatal pores, yet displayed no significant difference in the rate of photosynthesis. DNA-chip analysis revealed that a large number of stress-related genes were upregulated in the *SNAC1*-overexpressing rice plants. *SNAC1* holds promise in improving drought and salinity tolerance in rice (Hu et al., 2006).

We have standardized transformation procedures for sugarcane (Kaur et al., 2007) and for *indica* rice (Gosal et al., 2001; Grewal, Gill & Gosal, 2006). Attempts are now being made to introduce transgene(s) such as *Gly I, Gly II, DREB1A, and ZAT 12* in different combinations into rice, sugarcane, and maize for developing resistance to abiotic stresses.

14.6 MOLECULAR-MARKER TECHNOLOGY

Molecular techniques for detecting differences in the DNA of individual plants have many applications of value to crop improvement. These differences are known as molecular markers because they are often associated with specific genes and act as 'signposts' to those genes. Several types of molecular markers that have been developed and are being used in plants include restriction fragment-length polymorphisms (RFLPs), amplified fragment-length polymorphism (AFLP), random amplification of polymorphic DNA (RAPD), cleavable amplified polymorphic sequences (CAPS), single strand conformation polymorphisms (SSCP), sequence-tagged sites (STS), simple sequence repeats (SSRs) or microsatellites, and single-nucleotide polymorphisms (SNPs) (Mohan et al., 1997; Rafalski, 2002). Such markers, closely linked to genes of interest, can be used to select indirectly for the desirable allele, which represents the simplest form of marker-assisted selection (MAS), now being exploited to accelerate the backcross breeding and to pyramid several desirable alleles (Singh et al., 2001). Selection of a marker flanking a gene of interest allows selection for the presence (or absence) of a gene in progeny; thus, molecular markers can be used to follow any number of genes during the breeding program (Paran and Michelmore, 1993). The discovery of molecular markers has enabled dissection of quantitative traits into their single genetic components (Tanksley, 1993; Pelleschi et al., 2006; Bernier et al., 2007; Kato et al., 2008) and helped in the selection and pyramiding of QTL alleles through MAS (Ribaut et al., 2004; Neeraja et al., 2007; Ribaut & Ragot, 2007).

14.6.1 Rice

A doubled haploid (DH) population of 154 lines was obtained from a cross (CT9993–5–10–1–M/IR62266–42–6–2). Rice QTL linked to plant water-stress indicators, phenology, and production traits under irrigated and drought-stress conditions were mapped by using the above population. Water stress was applied to these DH lines before anthesis in three field experiments at two locations. Under irrigated and water-stress conditions, there was significant variation between the DH lines for plant water-stress indicators, phenology, and production traits. Forty-seven (47) QTL were identified for various plant water-stress indicators, phenology, and production traits under control and water-stress conditions in the field, which individually explained 5% to 59% of the phenotypic variation. There was a positive correlation between root traits and yield under drought stress (Babu et al., 2003) (Table 14.2).

A high-density map for a cross between upland (CT9993) and a lowland variety (IR62266) revealed QTL across the genome for osmotic adjustment (OA) and root physiological and morphological traits. This map has been used to locate expressed genes and identify putative candidates for these traits. Roots have been the focus of many physiological and QTL-mapping studies aimed at improving drought tolerance. In rice, some of the many QTL for roots are common across different genetic back-

Table 14.2 QTL in Relation to drought tolerance in different plant species.

Species	*Cross*	*Environment*	*Main traits*	*Reference*
Maize	Lo964 × Lo1016	Field	Root traits and yield	Tuberosa et al., 2002
Rice	CT9993–5–10–1–M × IR62266–42–6–2	Field	Root morphology, plant height, grain yield	Babu et al., 2003
Rice	IR62266 × IR60080	Greenhouse	Osmotic adjustment	Robin et al., 2003
Maize	F2 × F252	Field	Silking date, grain yield, yield stability	Moreau et al., 2004
Rice	Zhenshan 97 × Wuyujing 2	Field	Stay-green	Jiang et al., 2004
Rice	Bala × Azucena	Field	Grain yield, yield components	Laffitte et al., 2004
Rice	CT933 × IR62266	Field	Grain yield	Lanceras et al., 2004
Rice	Teqing × Lemont	Field	Yield components, phenology	Xu et al., 2005
Cotton	*G. hirsutum* × *G. barbadense*	Field	$^{12}C/^{13}C$, osmotic potential, canopy temperature, dry matter, seed yield	Saranga et al., 2004
Wheat	Beaver × Soissons	Field	Flag leaf senescence	Verma et al., 2004
Arabidopsis	Landsberg × Cape Verde	Greenhouse	$^{12}C/^{13}C$, flowering time, stomatal conductance, transpiration efficiency	Juenger et al., 2005
Barley	Tadmor × Er/Apm	Field	$^{12}C/^{13}C$, osmotic adjustment, leaf relatives water content, grain yield	Diab et al., 2004
Wheat	SQ1 × Chinese Spring	Field	Water-use efficiency, grain yield	Quarrie et al., 2005
Rice	Kalinga III × Azucena	Field	Root length	Steele et al., 2006
Rice	Vandana × Way Rarem	Field	Reproductive-stage drought, grain yield	Bernier et al., 2007
Pearl millet	ICMB 841 × 863 B	Field	Postflowering moisture, grain yield	Bidinger et al., 2007
Rice	Akihikari × IRAT 109	Screening facility	Specific water use, (SWU), Water-use efficiency (WUE)	Kato et al., 2008

grounds (Li et al., 2005). A QTL for root length and thickness on chromosome 9 has been mapped in several populations and is expressed across a range of environments. It was only one of the four target root QTL that significantly increased root length when introgressed into a novel genetic background (Steele et al., 2006). Further, four near-isogenic lines (NILs) were selected and characterized in replicated field experiments in eastern and western India across three years. They were tested by upland farmers in a target population of environments (TPE) in three states of eastern India across two years. The NILs outperformed Kalinga III for grain and straw yield, and there was interaction between the genotypes and the environments (G × E). No effect was found for the root QTL9 on grain or straw yield; however, several introgressions significantly improved both traits. Some of this effect was due to introgression of Azucena alleles at non-target regions. (Steele et al., 2007). Co-segregation with improved yield and stability across environments is a hot area of rice QTL research. In a population obtained from Vandana/Way Rarem, there was a co-location for QTL for grain yield under drought stress with QTL for maturity, panicle number, and plant height (Bernier et al., 2007). There was stronger association with maturity than with other traits for plant water relations. Further, genetic control of drought tolerance was observed without the effect of drought avoidance using hydroponic culture. A backcross inbred population of 'Akihikari' (lowland cultivar) × 'IRAT109' (upland cultivar) was cultured with (stress) and without (non- stress) polyethylene glycol (PEG) at seedling stage. There was a significant G × E interaction for relative growth rate (RGR), specific water use (SWU), and water-use efficiency (WUE) with or without PEG, which showed that each line responded differently to water stress. These interactions were QTL-specific, as shown by the QTL analysis. A total of three QTL on chromosomes 2, 4, and 7 were detected for RGR. The QTL on chromosome 7 had a constant effect across environments, whereas the QTL on chromosome 4 had an effect only under non-stressed condition and that on chromosome 2 only under stressed condition. Thus, it was concluded that PEG-treated hydroponic culture can be very effectively used in genetic analyses of drought tolerance at seedling stage (Kato et al., 2008).

14.6.2 Wheat

A cross between photoperiod-sensitive variety Beaver and photoperiod-insensitive variety Soissons of wheat was made, and a doubled haploid (DH) population was derived from this cross. This DH population varied significantly for the trait, measured as the percent green flag-leaf area remaining at 14 days and 35 days after anthesis. This trait also showed a significantly positive correlation with yield under variable environmental regimes. The genetic control of this trait was revealed by QTL analysis based on a genetic map derived from 48 doubled haploid lines using AFLP and SSR markers. Complex genetic mechanism of this trait was clear due to coincidence of QTL for senescence on chromosomes 2B and 2D under drought-stressed and optimal environments, respectively (Verma et al., 2004). Further, a population of 96 DHLs was developed from F_1 plants of the hexaploid wheat cross Chinese Spring × SQ1 (a high abscisic acid-expressing breeding line) and was mapped with 567 RFLP, AFLP, SSR, morphological, and biochemical markers covering all 21 chromosomes, with a total map length of 3,522 cM. The map was used to identify QTL for yield and yield components from a combination of 24 site × treatment × year combinations, including nutrient stress, drought stress, and salt stress treatments. Due to the variation in grain number per ear, the strongest yield QTL effects were on chromosomes 7AL and 7BL. Three of the yield QTL clusters were coincident with the dwarfing gene *Rht-B1* on 4BS and with the vernalization genes *Vrn-A1* on 5AL and *Vrn-D1* on 5DL. Yields of each DHL were calculated for trial mean yields of 6 g per plant and 2 g per plant (equivalent to about 8 t per ha and 2.5 t per ha, respectively), representing optimum and moderately stressed conditions (Quarrie et al., 2005) (Table 14.2).

14.6.3 Pearl Millet

Pearl millet (*Pennisetum glaucum* L.) is the staple cereal of the hottest, driest areas of the tropics and subtropics. Drought stress is a regular occurrence in these regions, making stress tolerance an essential attribute of new pearl millet cultivars. Pearl millet has a broad range of adaptive traits to intermittent drought stress because of its evolution from a desert grass species to a crop species and because of its long history of cultivation at the margins of arable agriculture (Bidinger & Hash, 2004). Analysis of co-mapping of QTL for individual traits and grain yield under stress suggested a linkage between the ability to maintain grain yield under stress and the ability to maintain both panicle harvest index (primarily grain filling) and harvest index under terminal stress, and confirmed the benefits of drought escape achieved through early flowering (Yadav et al., 2002, 2004). An initial evaluation of the putative drought tolerance QTL on LG 2 as a selection criterion was made by comparing hybrids made with topcross pollinators bred from progenies selected from the original mapping population for presence of the tolerance allele at the target QTL versus for field performance in the phenotyping environments (Bidinger et al., 2005). A more rigorous evaluation of the putative drought tolerance QTL is being done using near-isogenic versions of the more drought-sensitive parent H 77/833–2, bred by marker-assisted introgression of various segments of LG 2 from donor parent PRLT 2/89–33 in the region of the putative drought tolerance QTL. $BC4F_3$ progenies from selected $BC4F_2$ plants homozygous for various portions of the LG 2 target region were crossed to each of five related, early-maturing seed parents, and the resulting hybrids were evaluated under a range of terminal stress environments (Serraj et al., 2005). Further three major QTL for grain yield (on linkage group LG 2, LG 3, and LG 4) with low QTL × environment interactions were identified from an extensive data set including both stressed and unstressed post-flowering environments in pearl millet. Selection of these positive alleles by MAS could be useful in breeding programs (Bidinger et al., 2007) (Table 14.2).

14.6.4 Barley

To identify allelic variation in wild barley (*Hordeum vulgare ssp. spontaneum*), advanced backcross QTL (AB-QTL) analysis was deployed and found to be of value in the improvement of grain yield and other agronomically important traits in barley (*Hordeum vulgare ssp. vulgare*) grown under conditions of water deficit in

Mediterranean countries. From a cross between Barke (European two-row cultivar) and HOR11508 (a wild barley accession) $BC1F_2$ plants were derived. A population of 123 doubled-haploid (DH) lines was further obtained from the above $BC1F_2$ plants and was tested in replicated field trials under varying conditions of water availability in Italy, Morocco, and Tunisia for seven quantitative traits (Talamè et al., 2004). For all the seven traits, significant QTL effects at one ($P \leq 0.001$) or more trial sites ($P \leq 0.01$) were identified. Although most of the QTL alleles (67%) increasing grain yield were contributed by *H. vulgare, H. spontaneum* contributed the alleles increasing grain yield located in six regions on chromosomes 2H, 3H, 5H, and 7H. Among them, two QTL (associated to Bmac 0093 on chromosome 2H and to Bmac 0684 on chromosome 5H) were identified in all three locations and had the highest additive effects, thus depicting the validity of deploying advanced backcross QTL (AB-QTL) analysis for identifying favorable QTL alleles and its potential as a strategy to improve the adaptation of cultivars to drought-prone environments.

14.6.5 Maize

Initial attempts in MAS for drought tolerance in maize were carried out at CIMMYT, Mexico, in 1994. In this case, MAS was utilized to introgress QTL alleles for reducing the interval between the extrusion of the anthers and silks (ASI). ASI is negatively associated with grain yield (Bolaños and Edmeades, 1996; Ribaut et al., 1997, 2002a) under water-deficit conditions. The availability of molecular markers linked to the QTL for ASI allows for a more effective selection under drought as well as when drought fails to occur at flowering (Ribaut et al., 2002a, 2004). A backcross-marker-assisted selection (BC MAS) was started by Ribaut and coworkers based on the manipulation of five QTL affecting ASI. In this program, line CML247 was used as the recurrent parent and Ac7643 as the drought-tolerant donor. CML247, an elite line with high yield per se under well-watered conditions, is drought susceptible and shows long ASI under drought. MAS was used to introgress the QTL regions carrying alleles for short ASI from Ac7643 into CML247. A number of lines (ca. 70) derived through BC MAS were crossed with two testers and were evaluated for three consecutive years under several water regimes. Under severe stress conditions that reduced yield by at least 80%, the selected lines were superior to the control.

Marker-assisted recurrent selection (MARS) offers a refinement of MAS. This scheme is based on successive generations of crossing individuals based on their molecular profile, the objective being to attain an ideal genotype at different target QTL regions (Peleman and van der Voort, 2003). MARS allows for the selection of additional favorable alleles besides those targeted by BC-MAS. As to the applications of MARS, while Moreau et al. (2004) and Openshaw and Frascaroli (1997) showed limited advantage (Table 14.2) for MARS when compared to conventional selection, others (Ragot et al., 2000; Johnson, 2004; Eathington, 2005; Crosbie et al., 2006) have reported more successful applications of MARS in maize breeding programs.

Marker-assisted backcross (MABC) provides another opportunity to refine our research on complex traits. The basis of a marker-assisted backcrossing (MAB) strategy is to transfer a specific allele at the target locus from a donor line to a recipient line while selecting against donor introgressions across the rest of the genome. The use of molecular markers, which permit the genetic dissection of the progeny at each generation, increases the speed of the selection process, thus increasing genetic gain per unit time. The main advantages of MAB are: 1) efficient foreground selection for the target locus; 2) efficient background selection for the recurrent parent genome; 3) minimization of linkage drag surrounding the locus being introgressed; and 4) rapid breeding of new genotypes with favorable traits. The effectiveness of MAB depends on the availability of closely linked markers and/or flanking markers for the target locus, the size of the population, the number of backcrosses, and the position and number of markers for background selection (Neeraja et al., 2007). Selected MABC-derived BC2F_3 families were crossed with two testers and evaluated under different water regimes. Mean grain yield of MABC-derived hybrids was consistently higher than that of control hybrids under severe water-stress conditions. Under those conditions, the best five MABC-derived hybrids yielded, on average, at least 50% more than control hybrids. Under mild water stress, defined as resulting in <50% yield reduction, no difference was observed between MABC-derived hybrids and the control plants, thus confirming that the genetic regulation for drought tolerance is dependent on stress intensity. MABC conversions involving several target regions are likely to result in partial rather than complete line conversion. Simulations were conducted to assess the utility of such partial conversions, i.e., containing favorable donor alleles at non-target regions, for subsequent phenotypic selection. The results clearly showed that selecting several genotypes (10–20) at each MABC cycle was most efficient. Given the current approaches for MAS and the choices of marker technologies available now and potential for future developments, the use of MAS techniques in further improving grain yield under drought stress is very promising. (Ribaut and Ragot, 2007).

14.7 CONCLUSIONS

By use of biotechnology tools, our understanding of the processes underlying plant responses to drought at molecular and whole plant levels has rapidly progressed. Recent success on laboratory-production of drought stress-tolerant transgenic plants has been achieved, which must be exploited in the future. While insect-, viral- and herbicide-resistant transgenic plants are being commercially grown, drought stress-tolerant transgenic plants are still under pot experiments or under field evaluation. Molecular markers are being used to identify drought-related QTL and efficiently transfer them into commercially grown crop varieties of rice, wheat, maize, pearl millet, and barley.

REFERENCES

Abebe, T., A.C. Guenzi, B. Martin, and J.C. Cushman. 2003. Tolerance of mannitol-accumulating transgenic wheat to water stress and salinity. *Plant Physiol.* 131: 1748–1755.

Amaya, I., M.A. Botella, M.D.L. Calle, M.I. Medina, A. Heredia, R.A. Bressan, P.M. Hasegawa, M.A. Quesada, and V. Valpuesta. 1999. Improved germination under osmotic stress of tobacco plants overexpressing a cell wall peroxidase. *FEBS Lett.* 457: 80–84.

Artus, N.N., M. Uemura, P.L. Steponkus, S.J. Gilmour, C. Lin, and M.F. Thomashow. 1996. Constitutive expression of the cold-regulated *Arabidopsis thaliana* COR15a gene affects both chloroplast and protoplast freezing tolerance. *Proc. Natl. Acad. Sci.* 93: 13404–13409.

Babu, C.R., B.D. Nguyen, V. Chamarerk, P. Shanmugasundaram, P. Chezhian, P. Juyaprakash, S.K. Ganesh, A. Palchamy, S. Sadasivam, and S. Sarkarung. 2003. Genetic analysis of drought resistance in rice by molecular markers: association between secondary traits and field performance. *Crop Sci.* 43: 1457–1469.

Badawi, G.H., N. Kawano, Y. Yamauchi, E. Shimada, R. Sasaki, A. Kubo, and K. Tanaka. 2004. Over-expression of ascorbate peroxidase in tobacco chloroplasts enhances the tolerance to salt stress and water deficit. *Physiol. Plant.* 121(2): 231–238.

Bahieldin, A., H.T. Mahfouz, H.F. Eissa, O.M. Saleh, A.M. Ramadan, I.A. Ahmed, W.E. Dyer, H.A. El-Itriby, and M.A. Madkour. 2005. Field evaluation of transgenic wheat plants stably expressing the *HVA1* gene for drought tolerance. *Physiol. Planta.* 123: 421–427.

Bartels, D., and R. Sunkar. 2005. Drought and salt tolerance in plants. *Crit. Rev. Plant Sci.* 21: 1–36.

Behnam, B., A. Kikuchi, F. Celebi-Toprak, S. Yamanaka, M. Kasuga, K. Yamaguchi-Shinozaki, and K.N. Watanabe. 2006. The *Arabidopsis* DREB1A gene driven by the stress-inducible rd 29A promoter increases salt-stress tolerance in proportion to its copy number in tetrasomic tetraploid potato (*Solanum tuberosum*). *Plant Biotech.* 23: 169–177.

Bernier, J., A. Kumar, V. Ramaiah, D. Spaner, and G. Atlin. 2007. A large-effect QTL for grain yield under reproductive-stage drought stress in upland rice. *Crop Sci.* 47: 505–516.

Bhatnagar-Mathur, P., M.J. Devi, D.S. Reddy, M. Lavanya, V. Vadez, R. Serraj, K. Yamaguchi-Shinozaki, and K.K. Sharma. 2007. Stress inducible expression of *At DREB1A* in transgenic peanut (*Arachis hypogaea*) increases transpiration efficiency. *Pl. Cell Rep.* 26: 2071–2082.

Bidinger, F.R., R.S. Serraj, S.M.H. Rizvi, C. Howarth, R.S. Yadav, and C.T. Hash. 2005. Field evaluation of drought tolerance QTL effects on phenotype and adaptation in pearl millet [*Pennisetum glaucum* (L.) R. Br.] topcross hybrids. *Field Crops Res.* 94: 14–32.

Bidinger, F.R., and Hash, C.T. 2004. Pearl millet. In *Physiology and biotechnology integration in plant breeding*, H.T. Nguyen and A. Blum (eds), pp. 225–270. New York: Marcel Dekker.

Bidinger F.R., T. Nepolean, C.T. Hash, R.S. Yadav, and C.J. Howarth. 2007. Quantitative trait loci for grain yield in pearl millet under variable postflowering moisture conditions. *Crop Sci.* 47: 969–980.

Bolaños, J., and G.O. Edmeades. 1996. The importance of the anthesis-silking interval in breeding for drought tolerance in tropical maize. *Field Crops Res.* 48: 65–80.

Borlaug, N.E., and C.R. Dowswell. 2005. Feeding a world of ten billion people: A 21st century challenge. In proc. of "In the wake of double helix: From the green revolution to the gene revolution" 27–31st May 2003 Bologna, Italy.

Chao, D.Y., Y.H. Luo, M. Shi, D. Luo, and H.X. Lin. 2005. Salt-responsive genes in rice revealed by cDNA microarray analysis. *Cell Res.* 15: 796–810.

Chinnusamy, V., L. Xiong, and J.K. Zhu. 2006. Use of genetic engineering and molecular biology approaches for crop improvement for stress environments. In *Abiotic stresses, plant resistance through breeding and molecular approach*, eds. M. Ashraf and P.J.C. Harris, 47–89. U.P. India: IBDC.

Crosbie, T.M., S.R. Eathington, G.R. Johnson, M. Edwards, R. Reiter, S. Stark, R.G. Mohanty et al. 2006. Plant breeding: Past, present, and future. In *Plant breeding: The Arnel R. Hallauer Int. Symp., Mexico City*. 17–22 Aug. 2003, eds. K.R. Lamkey and M. Lee. Ames, IA: Blackwell.

Csonka, L.N., and A.D. Hanson. 1991. Prokaryotic osmoregulation: genetics and physiology. *Ann. Rev. Microbiology* 45: 569–606.

De Ronde, J.A., R.N. Laurie, T. Caetano, M.M. Greyling, and I. Kerepesi. 2004. Comparative study between transgenic and non-transgenic soybean lines proved transgenic lines to be more drought tolerant. *Euphytica* 138: 123–132.

Diab, A.A., B. Teulat-Merah, D. This, N.Z. Ozturk, D. Benscher, and M.E. Sorrells. 2004. Identification of drought-inducible genes and differentially expressed sequence tags in barley. *Theor. Appl. Genet.* 109(7): 1417–1425.

Dubouzet, J.G., Y. Sakuma, Y. Ito, M. Kasuga, E.G. Dubouzet, and S. Miura. 2003. OsDREB genes in rice, *Oryza sativa* L. encode transcription activators that function in drought, high salt and cold responsive gene expression. *Plant J.* 33: 751–763.

Eathington, S. 2005. Practical applications of molecular technology in the development of commercial maize hybrids. In Proc. of the 60th Annual Corn and Sorghum Seed Res. Conf., Chicago. 7–9 Dec. 2005.Washington, D.C.: Am. Seed Trade Assoc.

Edreva, A. 2005. Generation and scavenging of reactive oxygen species in chloroplasts: a submolecular approach. *Agric. Ecosyst. Environ.* 106: 119–133.

Elbein, A.D. 1974. The metabolism of ά - ά-trehalose. *Adv. Carbohydr. Chem. Biochem.* 30: 227–257.

Eltayeb, A.E., N. Kawano, G.H. Badawi, H. Kaminaka, T. Sanekata, T. Shibahara, S. Inanaga, and K. Tanaka. 2007. Overexpression of monodehydroascorbate reductase in transgenic tobacco confers enhanced tolerance to ozone, salt and polyethyleneglycol stresses. *Planta* 225(5): 1255–1264.

Flowers, T.J., and Yeo A.R. 1995. Breeding for salinity resistance in crop plants – where next? *Aust. J. Plant Physiol.* 22: 875–884.

Gao, J.P., D.Y. Chao, and H.X. Lin. 2007. Understanding abiotic stress tolerance mechanisms: recent studies on stress response in rice. *J. Integrative Plant Bio.* 49(6): 742–750.

Garg, A.K., J.K. Kim, T.G. Owens, A.P. Ranwala, Y.D. Choi, L.V. Kochian, and R.J. Wu. 2002. Trehalose accumulation in rice plants confers high tolerance levels to different abiotic stresses. *Proc. Natl. Acad. Sci.* 99(25): 15898–15903.

Gilmour, S.J., A.M. Sebolt, M.P. Salazar, J.D. Everard, and M.F. Thomashow. 2000. Overexpression of the *Arabidopsis* CBF3 transcriptional activator mimics multiple biochemical changes associated with cold acclimation. *Plant Physiol.* 124: 1854–1865.

Gleeson, D.M., A.L. Walter, and M. Parkinson. 2005. Overproduction of proline in transgenic hybrid larch (*Larix leptoeuropaea* (Dengler)) cultures renders them tolerant to cold, salt and frost. *Mol. Breed.* 15: 21–29.

Gosal, S.S., R. Gill, A.S. Sindhu, D. Kaur, N. Kaur, and H.S. Dhaliwal. 2001. Transgenic basmati rice carrying genes for stem borer resistance and bacterial leaf blight resistance. In *Rice Research for Food Security and Poverty Alleviation*, eds. S. Peng and B. Hardy, pp. 353–360. Philippines: IRRI.

Grewal, D.K., R. Gill, and S.S. Gosal. 2006. Genetic engineering of *Oryza sativa* by particle bombardment. *Biol. Plant.* 50(2): 311–314.

Halliwell, B. 1997. Introduction: free radicals and human disease – trick or treat? In *Oxygen Radicals and the Disease Process*, eds. C.E. Thomas and B. Kalyanaraman, 1–14. Newark, NJ: Harwood Academic.

Holmstrom, K.O., S. Somersalo, A. Mandal, T.E. Palva, and W. Bjorn. 2000. Improved tolerance to salinity and low temperature in transgenic tobacco producing glycinebetaine. *J. Exp. Bot.* 51(343): 177–85.

Hong, B.S., L. Zong-Suo, and S. Ming-An. 2005. LEA proteins in higher plants: Structure, function, gene expression and regulation. *Colloids Surf. B: Biointerf.* 45: 131–135.

Hsieh, T.H., J.T. Lee, Y.Y. Charng, and M.T. Chan. 2002. Tomato plants ectopically expressing *Arabidopsis CBF1* show enhanced resistance to water deficit stress. *Plant Physiol.* 130: 618–626.

Hu, H., M. Dai, J. Yao, B. Xiao, X. Li, Q. Zhang, and L. Xiong. 2006. Overexpressing a NAM, ATAF, and CUC (NAC) transcription factor enhances drought resistance and salt tolerance in rice. *Proc. Natl. Acad. Sci.* 103: 12987–12992.

Ikuta, S., S. Imamura, H. Misaki, and Y. Horiuti. 1977. Purification and characterization of choline oxidase from *Arthrobacter globiformis*. *J. Biochem.* 82: 1741–1749.

Incharoensakdi, A., T. Takabe, and T. Akazawa. 1986. Effect of betaine on enzyme activity and subunit interaction of ribulose 1,5-bisphosphate carboxylase/oxygenase from *Aphonothece halophytica*. *Plant Physiol.* 81: 1044–1049.

Ito, Y., K. Katsura, K. Maruyama, T. Taji, M. Kobayashi, and M. Seki. 2006. Functional analysis of rice DREB1/CBF-type transcription factors involved in cold-responsive gene expression in transgenic rice. *Plant Cell Physiol.* 47: 141–153.

Jaglo-Ottosen, K.R., S.J. Gilmour, D.G. Zarka, O. Schabenberger, and M.F. Thomashow. 1998. *Arabidopsis* CBF1 overexpression induces COR genes and enhances freezing tolerance. *Science* 280:104–106.

James, C. 2007. *Global status of commercialized biotech / GM crops*, 1–115. ISAAA Brief No. 37. Ithaca, New York.

Jiang, G.H., Y.Q. He, C.G. Xu, X.H. Li, and Q. Zhang. 2004. The genetic basis of stay-green in rice analyzed in a population of doubled haploid lines derived from an *indica* by *japonica* cross. *Theor. Appl. Genet.* 108(4): 688–699.

Johnson, R. 2004. Marker-assisted selection. *Plant Breed. Rev.* 24:293–309.

Juenger, T.E., J.K. Mckay, N. Hausmann, J.J.B. Keurentjes, S. Sen, K.A. Stowe, T.E. Dawson, E.L. Simms, and J.H. Richards. 2005. Identification and characterization of QTL underlying whole plant physiology in *Arabidopsis thaliana*: d13C, stomatal conductance and transpiration efficiency. *Plant Cell Environ.* 28: 697–708.

Jung, C., J.S. Seo, S.W. Han, Y.J. Koo, C.H. Kim, S.I. Song, B.H. Nahm, Y.D. Choi, and J.J. Cheong. 2008. Overexpression of AtMYB44 enhances stomatal closure to confer abiotic stress tolerance in transgenic *Arabidopsis*. *Plant Physiol.* 146: 623–635.

Karakas, B., P. Ozias-Akins, C. Stushnoff, M. Suefferheld, and M. Rieger. 1997. Salinity and drought tolerance of mannitol-accumulating transgenic tobacco. *Plant Cell Environ.* 20: 609–616.

Kasuga, M., Q. Liu, S. Miura, K. Yamaguchi-Shinozaki, and K. Shinozaki. 1999. Improving plant drought, salt, and freezing tolerance by gene transfer of a single stress inducible transcription factor. *Nat. Biotechnol.* 17: 287–291.

Katiyar-Agarwal, P., P. Agarwal, M.K. Reddy, and S.K. Sopory. 2006. Role of DREB transcription factors in abiotic and biotic stress tolerance in plants. *Plant Cell Rep.* 25: 1263–1274.

Kato, Y., S. Hirotsu, K. Nemoto, and J. Yamagishi. 2008. Identification of QTL controlling rice drought tolerance at seedling stage in hydroponic culture. *Euphytica* 160(3): 423–430.

Kaur, A., M.S. Gill, R. Gill, and S.S. Gosal. 2007. Standardization of different parameters for 'particle gun' mediated genetic transformation of sugarcane (*Saccharum officinarum*). *Indian J. Biotech.* 6(1): 31–34.

Kishore, P.B.K., Z. Hong, G.H. Miao, C.A.A. Hu, and D.P.S. Verma. 1995. Overexpression of Δ-1-pyrroline-5-carboxylate synthetase increases proline production and confers osmotolerance in transgenic plants. *Plant Physiol.* 108: 1387–1394.

Lafitte, H.R., A.H. Price, and B. Courtois. 2004. Yield response to water deficit in an upland rice mapping population: associations among traits and genetic markers. *Theor. Appl. Genet.* 109: 1237–1246.

Lanceras, J.C., G.P. Pantuwan, B. Jongdee, and T. Toojinda. 2004. Quantitative trait loci associated with drought tolerance at reproductive stage in rice. *Plant Physiol.* 135: 1–16.

Le Rudulier, D. 1993. Elucidation of the role of osmoprotective compounds and osmoregulatory genes: the key role of bacteria. In *Towards the rational use of high salinity-tolerant plants, Proc. of the first ASWAS conference*, eds. H. Leith and A. Al Masoom, pp. 313–322. Boston: Kluwer Academics.

Li. Z., P. Mu, C. Li, H. Zhang, Z.K. Li, Y. Gao, and X. Wang. 2005. QTL mapping of root traits in a doubled haploid population from a cross between upland and lowland *japonica* rice in three environments. *Theor. Appl. Genet.* 110(7): 1244–1252.

Liu, Q., M. Kasuga, Y. Sakuma, H. Abe, S. Miura, K. Yamaguchi-Shinozaki, and K. Shinozaki. 1998. Two transcription factors, DREB1 and DREB2, with an EREBP/AP2

DNA binding domain separate two cellular signal transduction pathways in drought and low temperature-responsive gene expression, respectively, in *Arabidopsis. Plant Cell* 10: 1391–1406.

Liu, X., X. Hua, J. Guo, D. Qi, L. Wang, Z. Liu, Z. Jin, S. Chen, and G. Liu. 2008. Enhanced tolerance to drought stress in transgenic tobacco plants overexpressing VTE1 for increased tocopherol production from *Arabidopsis thaliana. Biotech. Lett.* (in press).

McKersie, B.D., J. Murnaghan, and S.R. Bowley. 1997. Manipulating freezing tolerance in transgenic plants. *Acta Physiol. Plant.* 19: 485–495.

McKersie, B.D., S.R. Bowley, E. Harjanto, and O. Leprince. 1996. Water deficit tolerance and field performance of transgenic alfalfa overexpressing superoxide dismutase. *Plant Physiol.* 111: 1177–1181.

Miranda, J.A., N. Avonce, R. Suárez, J.M. Thevelein, P.V. Dijck, and G. Iturriaga. 2007. A bifunctional TPS-TPP enzyme from yeast confers tolerance to multiple and extreme abiotic-stress conditions in transgenic *Arabidopsis. Planta* 226(6): 1411–1421.

Mohan, M., S. Nair, A. Bhagwat, T.G. Krishna, M. Yano, C.R. Bhatia, and T. Sasaki. 1997. Genome mapping, molecular markers and marker-assisted selection in crop plants. *Mol. Breed.* 3: 87–103.

Molinari, H.B.C., C.J. Marura, E. Darosb, M.K. Freitas de Camposa, J.F.R. Portela de Carvalhoa, J.C.B. Filhob, L.F.P. Pereirac, and L.G.E. Vieira. 2007. Evaluation of the stress-inducible production of proline in transgenic sugarcane (*Saccharum* spp.) osmotic adjustment, chlorophyll fluorescence and oxidative stress. *Physiol. Planta.* 130: 218–229.

Moreau, L., A. Charcosset, and A. Gallais. 2004. Use of trial clustering to study QTL3 environment effects for grain yield and related traits in maize. *Theor. Appl. Genet.* 110: 92–105.

Nakashima, K., L.S.P. Tran, D.V. Nguyen, M. Fujita, K. Maruyama, D. Todaka, Y. Ito, N. Hayashi, K. Shinozaki, and K. Yamaguchi-Shinozaki. 2007. Functional analysis of a NAC-type transcription factor OsNAC6 involved in abiotic and biotic stress-responsive gene expression in rice. *Plant J.* 51(4): 617–630.

Nanjo, T., K. Masatomo, Y. Yoshiba, Y. Sanada, K. Wada, H. Tsukaya, Y. Kakubari, K. Yamaguchi-Shinozaki, and K. Shinozaki. 1999. Biological functions of proline in morphogenesis and osmotolerance revealed in antisense transgenic *Arabidopsis thaliana. Plant J.* 18: 185–193.

Neeraja, C.N., R. Maghirang-Rodriguez, A. Pamplona, S. Heuer, B.C.Y. Collard, E.M. Septiningsih, and G. Vergara et al. 2007. A marker-assisted backcross approach for developing submergence-tolerant rice cultivars. *Theor. Appl. Genet.* 115: 767–776.

Nuccio, M.L., B.L. Russell, K.D. Nolte, B. Rathinasabapathi, D.A. Gage, and A.D. Hanson. 1998. The endogenous choline supply limits glycinebetaine synthesis in transgenic tobacco expressing choline monooxygenase. *Plant J.* 16: 487–496.

Oh, S.J., C.W. Kwon, D.W. Choi, S.I. Song, and J.K. Kim. 2007. Expression of barley HvCBF4 enhances tolerance to abiotic stress in transgenic rice. *Plant Biotech. J.* 5: 646–656.

Oh, S.J., S.I. Song, Y.S. Kim, H.J. Jang, S.Y. Kim, and M. Kim. 2005. *Arabidopsis* CBF3/DREB1A and ABF3 in transgenic rice increased tolerance to abiotic stress without stunting growth. *Plant Physiol.* 138: 341–351.

Ohnishi, T., S. Sugahara, T. Yamada, K. Kikuchi, Y. Yoshiba, and H.Y. Hirano. 2005. OsNAC6, a member of the NAC gene family, is induced by various stresses in rice. *Genes Genet. Syst.* 80: 135–139.

Ooka, H., K. Satoh, K. Doi, T. Nagata, Y. Otomo, and K. Murakami. 2003. Comprehensive analysis of NAC family genes in *Oryza sativa* and *Arabidopsis thaliana. DNA Res.* 10: 239–247.

Openshaw, S., and E. Frascaroli. 1997. QTL detection and marker assisted selection for complex traits in maize. In Proc. of the 52nd Annu. Corn and Sorghum Res. Conf., Chicago. 9–12 Dec. 1997. Washington, D.C.: Am. Seed Trade Assoc. pp. 44–53.

Oraby, H.F., C.B. Ransom, A.N. Kravchenko, and M.B. Sticklen. 2005. Barley Hva1 gene confers salt tolerance in R_3 transgenic oat. *Crop Sci.* 45: 2218–2227.
Paran, I., and R.W. Michelmore. 1993. Identification of reliable PCR-based markers linked to disease resistance genes in lettuce. *Theor. Appl. Genet.* 85: 985–993.
Park, B.J., Z. Liu, A. Kanno, and T. Kameya. 2005a. Increased tolerance to salt and water deficit stress in transgenic lettuce (*Lactuca sativa* L.) by constitutive expression of LEA. *Plant Growth Regul.* 45: 165–171.
Park, B.J., Z. Liu, A. Kanno, and T. Kameya. 2005b. Genetic improvement of Chinese cabbage for salt and drought tolerance by constitutive expression of a *B. napus* LEA gene. *Plant Sci.* 169: 553–558.
Patonnier, M.P., J.P. Peltier, and G. Marigo. 1999. Drought-induced increase in xylem malate and mannitol concentrations and closure of *Fraxinus excelsior* L. stomata. *J. Exp. Bot.* 50: 1223–1229.
Peleman, J.D. and J.R. van der Voort. 2003. Breeding by design. *Trends Plant Sci.* 8: 330–334.
Pellegrineschi, A., M. Reynolds, M. Pacheco, R.M. Brito, R. Almeraya, K. Yamaguchi-Shinozaki, and D. Hoisington. 2004. Stress-induced expression in wheat of the *Arabidopsis* thaliana DREB1A gene delays water stress symptoms under greenhouse conditions. *Genome* 47: 493–500.
Pelleschi, S., A. Leonardi, J.P. Rocher, G. Cornic, D. de Vienne, C. Thévenot, and J.L. Prioul. 2006. Analysis of the relationships between growth, photosynthesis and carbohydrate metabolism using quantitative trait loci (QTLs) in young maize plants submitted to water deprivation. *Mol. Breed.* 17: 21–39.
Perl, A., R. Perl-Treves, S. Galili, D. Aviv, E. Shalgi, S. Malkin, and E. Galun. 1993. Enhanced oxidative-stress defense in transgenic potato overexpressing tomato Cu, Zn superoxide dismutase. *Theor. Appl. Genet.* 85: 568–576.
Pilon-Smits, E.A.H., N. Terry, T. Sears, H. Kim, A. Zayed, S. Hwang, K. van Dun et al. 1998. Trehalose-producing transgenic tobacco plants show improved growth performance under drought stress. *J. Plant Physiol.* 152: 525–532.
Postel, S.L., G.C. Daily, and P.R. Ehrlich. 1996. Human appropriation of renewable fresh water. *Science* 271: 785–788.
Qin, F., M. Kakimoto, Y. Sakuma, K. Maruyama, Y. Osakabe, L.S.P. Tran, K. Shinozaki, and K. Yamaguchi-Shinozaki. 2007. Regulation and functional analysis of *ZmDREB2A* in response to drought and heat stress in *Zea mays* L. *Plant J.* 50: 54–69.
Quarrie, S.A., A. Steed, C. Calestani, A. Semikhodskii, C. Lebreton, C. Chinoy, N. Steele, et al. 2005. A high-density genetic map of hexaploid wheat (*Triticum aestivum* L.) from the cross Chinese Spring × SQ1 and its use to compare QTLs for grain yield across a range of environments. *Theor. Appl. Genet.* 110(5): 865–880.
Rafalski, J.A. 2002. Applications of single nucleotide polymorphisms in crop genetics. *Curr. Opin. Plant Biol.* 5: 94–100.
Ragot, M., G. Gay, J.P. Muller, and J. Durovray. 2000. Efficient selection for the adaptation to the environment through QTL mapping and manipulation in maize. In *Molecular Approaches for Genetic Improvement of Cereals for Stable Production in Water-Limited Environments.* A Strategic Planning Workshop, El-Batan, Mexico. 21–25 June 1999, eds. M. Ribaut. and D. Poland, 128–130. El-Batan, Mexico: CIMMYT.
Rathinasabapathi, B., and R. Kaur. 2006. Metabolic engineering for stress tolerance In *Physiology and Molecular Biology of Stress Tolerance in Plants*, eds. K.V. Madhava, A.S. Rao Raghavendra, and K.J. Reddy, 255–299. Netherlands: Springer.
Rathinasabapathi, B., K.F. McCue, D.A. Gage, and A.D. Hanson. 1994. Metabolic engineering of glycinebetaine synthesis: plant betaine aldehyde dehydrogenases lacking typical transit peptides are targeted to tobacco chloroplasts where they confer betaine aldehyde resistance. *Planta* 193: 155–162.

Rhodes, D., and A.D. Hanson. 1993. Quaternary ammonium and tertiary sulfonium compounds in higher plants. *Annu. Rev. Plant Physiol. Plant Mol. Biol.* 44: 357–384.

Ribaut, J.M., and M. Ragot. 2007. Marker-assisted selection to improve drought adaptation in maize: The backcross approach, perspectives, limitations, and alternatives. *J. Exp. Bot.* 58: 351–360.

Ribaut, J.M., M. Bänziger, J. Betran, C. Jiang, G.O. Edmeades, K. Dreher, and D. Hoisington. 2002a. Use of molecular markers in plant breeding: Drought tolerance improvement in tropical maize. In *Quantitative Genetics, Genomics, and Plant Breeding*, ed. M.S. Kang, 85–99. Wallingford, UK: CABI.

Ribaut, J.M., C. Jiang, and D. Hoisington. 2002b. Efficiency of a gene introgression experiment by backcrossing. *Crop Sci.* 42: 557–565.

Ribaut, J.M., C. Jiang, D. Gonzalez-de-Leon, G. Edmeades, and D.A. Hoisington. 1997. Identification of quantitative trait loci under drought conditions in tropical maize. 2. Yield components and marker assisted selection strategies. *Theor. Appl. Genet.* 94: 887–896.

Ribaut, J.M., M. Bänziger, T. Setter, G. Edmeades, and D. Hoisington. 2004. Genetic dissection of drought tolerance in maize: A case study. In *Physiology and Biotechnology Integration for Plant Breeding*, eds. N. Nguyen N. and A. Blum, 571–611. New York: Marcel Dekker.

Robin, S., M.S. Pathan, B. Courtois, R. Lafitte, S. Carandang, S. Lanceras, M. Amante, H.T. Nguyen, and Z. Li. 2003. Mapping osmotic adjustment in an advanced back-cross inbred population of rice. *Theor Appl. Genet.* 107(7): 1288–1296.

Robinson, S.P., and G.P. Jones. 1986. Accumulation of glycinebetaine in chloroplasts provides osmotic adjustment during salt stress. *Aust. Jour. Plant Physiol.* 13: 659–668.

Rodrigues, S.M., M.O. Andrade, A.P.S. Gomes, F.M. DaMatta, M.C. Baracat-Pereira, and E.P.B. Fontes. 2006. *Arabidopsis* and tobacco plants ectopically expressing the soybean antiquitin-like *ALDH7* gene display enhanced tolerance to drought, salinity, and oxidative stress. *J. Exp. Bot.* 55: 301–308.

Sakamoto, A.A., N. Murata, and A. Murata. 1998. Metabolic engineering of rice leading to biosynthesis of glycinebetaine and tolerance to salt and cold. *Plant Mol. Biol.* 38: 1011–1019.

Sakamoto, A., and A.N. Murata. 2002. The role of glycinebetaine in the protection of plants from stress: clues from transgenic plants. *Plant Cell Environ.* 25: 163–171.

Sakuma, Y., K. Maruyama, Y. Osakabe, F, Qin, M. Seki, K. Shinozaki, and K. Yamaguchi-Shinozaki. 2006. Functional analysis of an *Arabidopsis* transcription factor, DREB2A, involved in drought responsive gene expression. *Plant Cell.* 18: 1292–1309.

Saranga, Y., C.X. Jiang, R.J. Wright, D. Yakir, and D.H. Paterson. 2004. Genetic dissection of cotton physiological responses to arid conditions and their interrelationships with productivity. *Plant Cell Environ.* 27(3): 263–277.

Sawahel, W.A., and A.H. Hassan. 2002. Generation of transgenic wheat plants producing high levels of the osmoprotectant proline. *Biotech. Lett.* 24: 721–725.

Serraj, R., C.T. Hash, S.M.H. Rizvi, A. Sharma, R.S. Yadav, and F.R Bidinger. 2005. Recent advances in marker-assisted selection for drought tolerance in pearl millet. *Plant Prod. Sci.* 8: 332–335.

Shen, B.R., G. Jensen. and H.J. Bohnert. 1997. Increased resistance to oxidative stress in transgenic plants by targeting mannitol biosynthesis to chloroplasts. *Plant Physiol.* 113: 1177–1183.

Shirasawa, K., T. Takabe, T. Takabe, and S. Kishitani. 2006. Accumulation of glycinebetaine in rice plants that overexpress choline monooxygenase from spinach and evaluation of their tolerance to abiotic stress. *Ann. Bot.* 98: 565–571.

Shou, H., P. Bordallo, and K. Wang. 2004. Expression of the *Nicotiana* protein kinase (*NPK1*) enhanced drought tolerance in transgenic maize. *J. Exp. Bot.* 55: 301–308.

Singh, S., J.S. Sidhu, N. Huang, Y. Vikal, Z. Li, D.S. Brar, H.S. Dhaliwal and G.S. Khush. 2001. Pyramiding three bacterial blight resistance genes (xa5, xa13 and Xa21) using marker-assisted selection into *indica* rice cultivar PR106. *Theor. Appl. Genet.*102: 1011–1015.

Sivamani, E., A. Bahieldin, J.M. Wraith, T. Al-Niemi, W.E. Dyer, T.H.D. Ho, and R. Wu. 2000. Improved biomass productivity and water use efficiency under water deficit conditions in transgenic wheat constitutively expressing the barley HVA1 gene. *Plant Sci.* 155: 1–9.

Steele, K.A., A.H. Price, H.E Shashidhar, and J.R. Witcombe. 2006. Marker-assisted selection to introgress rice QTLs controlling root traits into an Indian upland rice variety. *Theor. Appl. Genet.* 112: 208–221.

Steele K.A., D.S. Virk, R. Kumar, S.C. Prasad, and J.R Witcombe. 2007. Field evaluation of upland rice lines selected for QTLs controlling root traits. *Field Crop Res.* 101: 180–186.

Steponkus, P.L., M. Uemura, R.A. Joseph, S.J. Gilmour, and M.F. Thomashow. 1998. Mode of action, of the COR15a gene on the freezing tolerance of *Arabidopsis thaliana*. *Proc. Nat. Acad. Sci.* 95: 14570–14575.

Suzuki, N., L. Rizhsky, H. Liang, J. Shuman, V. Shulaev, and R. Mittler. 2005. Enhanced tolerance to environmental stress in transgenic plants expressing the transcriptional coactivator multiprotein bridging factor 1. *Plant Physiol.* 139: 1313–1322.

Talamè, V., M.C. Sanguineti, E. Chiapparino, H. Bahri, M.B. Salem, B.P. Forster, R.P. Ellis, et al. 2004. Identification of *Hordeum spontaneum* QTL alleles improving field performance of barley grown under rainfed conditions. *Ann. Appl. Biol.* 144: 309–319.

Tanksley, S.D. 1993. Mapping polygenes. *Annu. Rev. Genet.* 27: 205–233.

Tarczynski, M.C., R.G. Jensen, and H.J. Bohnert. 1992. Expression of a bacterial mtl D gene in transgenic tobacco leads to production and accumulation of mannitol. *Proc. Natl. Acad. Sci.* 89: 2600–2604.

Tarczynski, M.C., R.G. Jensen, and H.J. Bohnert. 1993. Stress protection of transgenic plants by production of the osmolyte mannitol. *Science* 259: 508–510.

Taylor, C.B. 1996. Proline and water deficit: ups, downs, ins and outs. *Plant Cell* 8: 1221–1224.

Thomas, J.C., M, Sepahi, B, Arendall, and H.J. Bohnert. 1995. Enhancement of seed germination in high salinity by engineering mannitol expression in *Arabidopsis thaliana*. *Plant Cell Environ.* 18: 801–806.

Tilman, D., K.G. Cassman, P.A. Matson, R. Naylor, and S. Polasky. 2002. Agricultural sustainability and intensive production practices. *Nature* 418: 671–677.

Tran, L.S.P., K. Nakashima, Y. Sakuma, S.D. Simpson, Y. Fujita, and K. Maruyama. 2004. Isolation and functional analysis of *Arabidopsis* stress-inducible NAC transcription factors that bind to a drought-responsive cis-element in the early responsive to dehydration stress 1 promoter. *Plant Cell* 16: 2481–2498.

Tuberosa, R., S. Salvi, M.C. Sanguineti, P. Landi, M. MacCaferri, and S. Conti. 2002. Mapping QTLs regulating morpho-physiological traits and yield: case studies, shortcomings and perspectives in drought-stressed maize. *Ann. Bot.* 89: 941–963.

Van Camp, W., K. Capiau, M. Van Montagu, D. Inze, and L. Slooten. 1996. Enhancement of oxidative stress tolerance in transgenic tobacco plants overproducing Fe-superoxide dismutase in chloroplasts. *Plant Physiol.* 112: 1703–1714.

Verma, V., M.J. Foulkes, A.J. Worland, R. Sylvester-Bradley, P.D.S Caligari, and J.W. Snape. 2004. Mapping quantitative trait loci for flag leaf senescence as a yield determinant in winter wheat under optimal and drought-stressed environments. *Euphytica* 135(3): 255–263.

Vinocur, B., and A. Altman. 2005. Recent advances in engineering plant tolerance to abiotic stress: achievements and limitations. *Curr. Opin. Biotechnol.* 16: 123–132.

Virmani, S.S., and M. Ilyas-Ahmed. 2007. Rice breeding for sustainable production. In *Breeding Major Food Staples*, eds. M.S. Kang and P.M. Priyadarshan, 141–191. Malden, MA: Blackwell.

Xiao, B., Y. Huang, N. Tang, and L. Xiong. 2007. Over expression of *LEA* gene in rice improves drought resistance under field conditions. *Theor. Appl. Genet.* 115: 35–46.

Xu, D., X. Duan, B. Wang, B. Hong, T.H.D. Ho, and R. Wu. 1996. Expression of late embryogenesis abundant protein gene *HVA1*, from barley confers tolerance to water deficit and salt stress in transgenic rice. *Plant Physiol.* 110: 249–57.

Xu, J.L., H.R. Lafitte, Y.M. Gao, B.Y. Fu, R. Torres. and Z.K. Li. 2005. QTLs for drought escape and tolerance identified in a set of random introgression lines of rice. *Theor. Appl. Genet.* 11(8): 1642–1650.

Xu, S.M., X.C. Wang, and J. Chen. 2007. Zinc finger protein 1 (*ThZF1*) from salt cress (*Thellungiella halophila*) is a Cys-2/His-2-type transcription factor involved in drought and salt stress. *Plant Cell Rep.* 26(4): 497–506.

Yadav, R.S., C.T. Hash, F.R. Bidinger, G.P. Cavan, and C.J. Howarth. 2002. Quantitative trait loci associated with traits determining grain and stover yield in pearl millet under terminal drought stress conditions. *Theor. Appl. Genet.* 104: 67–83.

Yadav, R.S., C.T. Hash, F.R. Bidinger, K.M. Devos, and C.J. Howarth. 2004. Genomic regions associated with grain yield and aspects of post-flowering drought tolerance in pearl millet across stress environments and testers background. *Euphytica* 136: 265–277.

Yeo, E.T., H.B. Kwon, S.E. Han, J.T. Lee, J.C. Ryu, and M.O. Byu. 2000. Genetic engineering of drought resistant potato plants by introduction of the trehalose-6-phosphate synthase TPS1 gene from *Saccharomyces cerevisiae. Mol. Cell* 30: 263–268.

Yonamine, I., K. Yoshida, K. Kido, A. Nakagawa, H. Nakayama, and A. Shinmyo. 2004. Overexpression of NtHAL3 genes confers increased levels of proline biosynthesis and the enhancement of salt tolerance in cultured tobacco cells. *J. Exp. Bot.* 55: 387–395.

Yu, Q., and Z. Rengel. 1999. Micronutrient deficiency influences plant growth and activities of superoxide dismutases in narrow-leafed lupins. *Ann. Bot.* 183: 175–182.

Zhang, Y., C. Yang, Y. Li, N. Zheng, H. Chen, Q. Zhao, T. Gao, H. Guo, and Q. Xie. 2007. SDIR1 is a ring finger E3 ligase that positively regulates stress-responsive abscisic acid signaling in *Arabidopsis. Plant Cell.* 19: 1912–1929.

Zhao, F.Y., and H. Zhang. 2006a. Salt and paraquat stress tolerance results from co-expression of the *Suaeda salsa* glutathione S-transferase and catalase in transgenic rice. *Plant Cell Tissue Organ Cult.* 86(3): 349–358.

Zhao, F.Y., and H. Zhang. 2006b. Expression of *Suaeda salsa* glutathione S-transferase in transgenic rice resulted in a different level of abiotic stress tolerance. *J. Agri. Sci.* 144: 1–8.

Zhao, F.Y., and H. Zhang. 2007. Transgenic rice breeding for abiotic stress tolerance– present and future. *Chinese J. Biotech.* 23(1): 1–6.

Zhifang, G., and W.H. Loescher. 2003. Expression of a celery mannose 6-phosphate reductase in *Arabidopsis thaliana* enhances salt tolerance and induces biosynthesis of both mannitol and a glucosyl-mannitol dimmer. *Plant Cell Environ.* 26: 275–283.

Zhu, B., J. Su, M. Chang, D.P.S. Verma, Y.L. Fan, and R. Wu. 1998. Overexpression of a Δ-1-pyrroline-5-carboxylate gene and analysis of tolerance to water- and salt-stress in transgenic rice. *Plant Sci.* 139: 41–48.

Chapter 15

Transgenic strategies for improved drought tolerance in legumes of semi-arid tropics

Pooja Bhatnagar-Mathur, J. Shridhar Rao, Vincent Vadez & Kiran K. Sharma

Genetic Transformation Laboratory, International Crops Research Institute for the Semi-Arid Tropics (ICRISAT), Patancheru, Andhra Pradesh, India

ABSTRACT

Water deficit is the most prominent abiotic stress that severely limits crop yields, thereby reducing opportunities to improve livelihoods of poor farmers in the semi-arid tropics (SAT) where most of the legumes, including groundnut and chickpea, are grown. Sustained long-term efforts in developing these legume crops with better drought tolerance through conventional breeding have been met with only limited success mainly because of an insufficient understanding of the underlying physiological mechanisms and lack of sufficient polymorphism for drought tolerance-related traits. Exhaustive efforts are being made at the International Crop Research Institute for Semi-Arid Tropics (ICRISAT) to improve crop productivity of the SAT crops by comprehensively addressing the constraints caused by water limitations. The transgenic approach has been used to speed up the process of molecular introgression of putatively beneficial genes for rapidly developing stress-tolerant legumes. Nevertheless, the task of generating transgenic cultivars requires success in the transformation process and proper incorporation of stress tolerance into plants. Hence, evaluation of the transgenic plants under stress conditions and understanding the physiological effect of the inserted genes at the whole plant level is critical. This review focuses on the recent progress achieved in using transgenic technology to improve drought tolerance, which includes evaluation of drought-stress response and protocols developed for testing transgenic plants under near-field conditions. A trait-based approach was considered, in which yield was dissected into components. Yield (Y) is defined as transpiration (T) × transpiration efficiency (TE) × harvest index (HI).

Address correspondence to Pooja Bhatnagar-Mathur at Genetic Transformation Laboratory, International Crops Research Institute for the Semi-Arid Tropics (ICRISAT), Patancheru, Andhra Pradesh 502 324, India. E-mail: p.bhatnagar@cgiar.org

Reprinted from: *Journal of Crop Improvement*, Vol. 24, Issue 1 (January 2010), pp. 92–111, DOI: 10.1080/15427520903337095

Keywords: drought, transgenic, chickpea, groundnut, transpiration efficiency, water-use efficiency

15.1 INTRODUCTION

Climate change is a major global concern that can make dryland agriculture even more risk-prone, especially in the developing world. Abiotic stresses are a primary cause of crop loss worldwide, reducing average yields by >50% in most major crops (Boyer 1982; Bray, Bailey-Serres & Weretilnyk 2000). Crops are often exposed to multiple stresses in many regions. Approximately, 19% of the world's agricultural land is subject to salt stress and 5% to drought stress (FAO, 1996).

The semi-arid tropics (SAT) cover parts of 55 developing countries and account for about 1.4 billion people of the world. The SAT has very short growing seasons, separated by very hot and dry periods. In addition, rapid and unforeseen disturbances in these environments have resulted in stressful conditions, which include paucity of water for long periods because of lack of irrigation, infrequent rains, or lowering of water table. These conditions cause drought stress. On the other hand, excess water from rain, cyclones, or frequent irrigation results in flooding, submergence or anaerobic stress. Climatologists believe that the changing global climate might produce even more severe and widespread dry conditions in these regions, with potentially serious consequences for agriculture and food availability (Wenzel & Wayne 2008). In addition, water limitation may prove to be a critical constraint to crop productivity under future scenarios (Fischer et al. 2001). Since rainfed agriculture, particularly in the developing countries of the SAT, contributes significantly to total food production, food security will be unsustainable in the absence of dramatic yield increases in the marginal environments, especially in the drought prone areas.

Conventional breeding and enhanced management practices have addressed several constraints that limit crop productivity or quality, but there are situations where the existing germplasm lacks the required traits. While plant breeding in the past has relied heavily on empirical approaches for drought tolerance in crop plants, there is a broad consensus that strategic approaches based on sound physiological and genetic understanding of yield will also be required if further yield gains are to be achieved (Jackson et al. 1996; Miflin 2000; Slafer 2003; Snape et al. 2001).

Broadly, the drought response of plants can be divided into three mechanisms: 1) drought escape, 2) drought avoidance, and 3) drought tolerance. To date, the most important contribution to drought tolerance has come from breeding for altered phenological traits like early flowering and maturity (Araus et al. 2002) that essentially involves escape from drought conditions rather than tolerance per se. Besides, plants undergo drought postponement either by conserving water or by enhanced water uptake, which is usually achieved by structural modifications such as increased cuticle thickness, deeper roots, reduced leaf area, or reduced stomatal conductance (Jones 2004). Limiting stomatal conductance can result in high water-use efficiency (WUE), which is a trait for postponement of stress. Under extreme stress, tolerance is about survival and not productivity. The drought tolerance is achieved through changes in the biochemical composition that protect the macromolecules and membranes or maintain cell turgor through osmotic adjustment.

Yield losses due to constraints like drought are highly variable in nature depending on the stress timing, intensity, and duration, which is related to location-specific environmental stress factors such as high irradiance and temperature that make breeding for drought tolerance through conventional approaches difficult. Since most crop plants have not been selected for meeting exigencies caused by these stress factors, their capacity to adjust to such conditions is usually limited. Nevertheless, efforts to breed crop species for high WUE and stomatal conductance have met with limited success. This is in part because the physiological data and information about the molecular events underlying the abiotic stress responses are scarce, besides the non-availability of molecular tools that could help in assisting the breeding activities for such complex traits. Besides, a main constraint in the advancement efforts of crop improvement for drought tolerance includes insufficient know-how of tolerance mechanisms, a poor understanding of its low inheritance, and a scarcity of efficient techniques for screening of germplasm and breeding material.

15.2 TRANSGENIC INTERVENTIONS FOR DROUGHT TOLERANCE

Nevertheless, agricultural biotechnology has the potential to address this challenge and can play a role in developing crops that use water more efficiently, thus reducing the negative consequences of drought. The use of genetic engineering technology potentially offers a more targeted gene-based approach for improving plant's adaptation to water-limiting conditions. Transgenic approaches offer a powerful means of gaining valuable information to understand the mechanisms governing stress tolerance, providing a complementary means for the genetic betterment of the genome of field crops and thus promising the alleviation of some of the major constraints to crop productivity in the developing countries (Sharma & Ortiz 2000). When a plant is subjected to abiotic stress, a number of genes are turned on, resulting in increased levels of several osmolytes and proteins, some of which may be responsible for conferring a certain degree of protection from these stresses. Therefore, it will likely be necessary to transfer several potentially useful genes into the same plant in order to obtain a high degree of tolerance to drought or salt stress. Novel genes accessed from exotic sources – plants, animals, bacteria, even viruses – can be introduced into the crop through biolistics or by using *Agrobacterium*–mediated genetic transformation (Sharma & Lavanya 2002). Further, it is possible to control the timing, tissue-specificity, and expression level of transferred genes for their optimal function.

Association of several traits with tolerance has been tested in transgenic plants. The results of transgenic modifications of biosynthetic and metabolic pathways indicate that higher stress tolerance can be achieved by engineering. However, the results of simulation modeling also suggest that changes in a given metabolic process, relatively apart from the yield architecture, may end up with little benefit for actual yield under stress (Sinclair, Purcell & Sneller 2004). Various transgenic technologies have been used to improve stress tolerance in plants (Allen 1995). In recent years, the genes responsible for low-molecular-weight metabolites have been shown to confer increased tolerance to salt or drought stress in transgenic dicot plants (mainly tobacco). Metabolic traits, especially pathways with few enzymes, have been genetically characterized and are more amenable to manipulations than structural and developmental traits. Genetically engineered plants for single-gene products include those encoding for enzymes required for the biosynthesis of osmo-protectants (Tarczynski, Jensen & Bohnert 1993; Kavikishore et al. 1995; Hayashi et al. 1997), or modifying membrane lipids (Kodama et al. 1994; Ishizaki-Nishizawa et al. 1996), late embryogenesis proteins (Xu et al. 1996), and detoxifying enzymes (McKersie et al. 1996). However, the results of the transfer of a single trait driven by a single protein are unlikely to improve a plant's tolerance beyond the short-term effects that have been reported. Hence, multiple mechanisms to engineer water stress tolerance must be utilized (Bohnert, Nelson & Jenson 1995). Indeed, drought being an extremely complex phenomenon controlled by multiple genes and regulatory pathways, drought tolerance has proved much more difficult to engineer into plants than more simply inherited traits governed by single genes. Therefore, many genes involved in stress response can be simultaneously regulated by using a single gene encoding stress-inducible transcription factor (Kasuga et al. 1999), thereby offering the possibility of enhancing tolerance towards multiple stresses including drought, salinity, and freezing. Transcription factors modulate the expression of a cascade of stress-inducible genes to impart tolerance to stressed plants (Bartels & Sunkar 2005; Chinnusamy, Jagendorf & Zhu 2005).

Several recent reviews have provided comprehensive lists of genes shown to confer drought tolerance in plants (Wang, Vinocur & Altman 2003; Zhang et al. 2004; Umezawa et al. 2006) that support the prospects of genetically engineering for drought tolerant crops. However, in most of the cases, the evidence indicating that these genes confer drought tolerance is based on laboratory-grown plants under artificial drought conditions. The translation of drought tolerance from research experiments with model species in the laboratory to crop species in the field depends on the ability to transfer the mechanism of tolerance and, on whether the mechanism of tolerance will actually lead to increased biomass or yield under drought stress in crop species. Again, here it is important to identify and understand the kind of drought tolerance that is the target. For example, genes that confer desiccation tolerance in the laboratory may not be useful for enhancing yield in the field, although they are the object of extensive studies. In order to improve the genetic make-up of the agronomically important crop plants for their integration into the breeding programs, the research focus should be on crop productivity where their successful deployment in the field will require more detailed physiological studies than are typically reported (Chaves & Oliveira, 2004).

15.3 TRANSGENIC RESEARCH FOR DROUGHT TOLERANCE AT ICRISAT

Legumes such as chickpea and groundnut constituting the important food and oilseed crops of the SAT are mostly grown in low-input, rain-fed agriculture of the world and suffer from drought due to insufficient, untimely, and erratic rainfall in these climates that become major constraints to crop productivity. Globally, on an average, about 20% of the land surface is under drought at any one time, and the proportion of the land surface in extreme drought is predicted to increase from 1% for the present day to 30% by the end of the 21st century (Burke, Brown & Christidis 2006). Thus, drought represents the major constraint to increase their yield, and drought tolerance therefore is a major aim of breeding of these legume crops. Exhaustive efforts are being made at the International Crop Research Institute for Semi-Arid Tropics (ICRISAT) to improve the crop productivity of the SAT crops by comprehensively addressing the constraints due to water limitations. This is done through cutting-edge, knowledge-based breeding practices complemented adequately by genomics and genetic transformation technologies that could lead to simpler and more effective gene-based approach for improving drought tolerance.

There has been a general consensus that increased crop performance of the legumes under drought conditions can be achieved through improvements in total water use, WUE, and harvest index (Nigam et al. 2005; Vadez et al. 2008). Amongst these, transpiration efficiency (TE), one of the components of WUE, is very critical to plant performance under water-limiting conditions, and hence new breakthroughs in water-saving strategies could be attained by improving WUE in these crop species. Sustained long-term efforts in developing groundnut and chickpea crops with better drought tolerance/WUE through conventional breeding so far have had only limited success. This is mainly due to an insufficient understanding of the physiological mechanisms underlying WUE and a lack of sufficient polymorphism for these traits. As a consequence, there is difficulty in discovering molecular markers for WUE that have the potential to ease the breeding process. Drought tolerance seems to be controlled by complex sets of traits that may have evolved as separate mechanisms in different groups of plants, thereby making the breeding of these using conventional methods difficult. Therefore, a continuing need has been felt to integrate biotechnological approaches with plant physiology and plant breeding for an efficient application of these tools for crop improvement. Hence, efforts are underway to use genetic engineering approaches for speeding up the process of improving resilience in these legumes to drought stress by specifically targeting and more quickly inserting genes known to be involved in plant response to stress.

At ICRISAT, research is being carried out to impart drought tolerance using both osmoregulatory and regulatory gene approaches. Because it is believed that osmoregulation is one of the best strategies for abiotic stress tolerance, especially if osmoregulatory genes could be triggered in response to drought, salinity, and high temperature. A widely adopted strategy has been to engineer certain osmolytes or cause overexpression of such osmolytes in plants to breed stress-tolerant crops (Ishitani et al. 1997; Holmstrom et al. 2000; Delauney & Verma 1993; Nanjo et al. 1999; Zhu et al. 1998; Yamada et al. 2005). We have introduced an osmoregulatory gene, *P5CSF129A*, encoding the mutagenized *D1-pyrroline-5-carboxylate synthetase (P5CS)* for the overproduction of proline in chickpea. The hypothesis for the present study was that the introduction of the *P5CSF129A* gene into chickpea might result in the accumulation of elevated amounts of endogenous proline, and consequently improve plant production into this important pulse crop. The transgenic chickpea plants were extensively evaluated under greenhouse conditions for various physiological, molecular, and biochemical characters under a typical dry-down setup for water deficits. The accumulation of proline in several transgenic events was more pronounced and increased significantly in the leaves when exposed to water stress coupled with a decrease in the free radicals as measured by a decrease in the malonaldehyde (MDA) levels, a lipid peroxidation product. Eleven transgenic events that accumulated high proline

(2–6 folds) were further evaluated in the greenhouse experiments based on their TE, photosynthetic activity, stomatal conductance, and root length under water stress. Almost all the transgenic events showed a decline in transpiration at lower values of the fraction of transpirable soil water (TSWV; dryer soil), and extracted more water than their untransformed parents. However, the overexpression of proline appeared to have no beneficial effect on the biomass accumulation since only a few events showed a significant increase in the biomass production toward the end of the progressive drying period. In any case, the overexpression of *P5CSF129A* gene resulted only in a modest increase in TE, thereby indicating that the enhanced proline had little bearing on the components of yield architecture, which are significant in overcoming the negative effects of drought stress in chickpea. These results agree with the previous reports in other crops (Turner & Jones 1980; Morgan 1984; Serraj & Sinclair 2002; Turner et al. 2007) and, in our above assessment, the gene affecting single protein might be less efficient in coping with water-limiting conditions.

Hence, to cater to the multigenicity of the plant response to stress, a strategy to target transcription factors that regulate the expression of several genes related to abiotic stress was considered. The purpose was to develop a large number of groundnut and chickpea transgenic plants carrying the *DREB1A* transcription factor from *Arabidopsis*, driven by a stress-inducible promoter from *rd29A* gene from *A. thaliana*. Regulatory genes or transcription factors, more specifically those belonging to the AP2/ERF family, have previously been shown to improve stress tolerance under lab conditions by regulating the coordinated expression of several stress-related genes in heterologous transgenic plants (Kasuga et al. 1999, 2004; Behnam et al. 2006; Bhatnagar-Mathur et al. 2007). Since improving WUE of a plant is a complex issue, and efforts to breed groundnut genotypes for high TE and stomatal conductance have obtained limited success, regulatory genes or transcription factors have a potential to improve stress tolerance by regulating the coordinated expression of several stress-related genes in heterologous transgenic plants. Plants overexpressing the *P5CSF129A* and *DREB1A* genes demonstrated substantial increase in TE under experimental greenhouse conditions (Bhatnagar-Mathur et al. 2007, 2009a). A few transgenic events with contrasting responses have been selected for further detailed studies on the gas exchange characteristics of leaves. Besides, the biochemical responses of plants under identical conditions of water stress have been examined critically for further understanding of the mechanisms underlying environmental stress resistance in these transgenic events (Bhatnagar-Mathur et al. 2009b).

15.3.1 Choosing a right promoter is important

Tissue specificity of transgene expression is also an important consideration while deciding on the choice of the promoter to increase the level of expression of the transgene for any transgenic technology. Thus, the strength of the promoter and the possibility of using stress-inducible, developmental-stage or tissue-specific promoters need to be considered (Bajaj et al. 1999). While for some gene products, such as LEA3, which are needed in large amounts, a very strong promoter is needed, for others like enzymes for polyamine biosynthesis, it may be better to use an inducible promoter of moderate strength. So far, most of the promoters that have been most commonly used in the development of abiotic stress tolerant plants have been constitutive in nature.

These include the *CaMV 35S*, *ubiquitin* 1, and *actin* promoters, which by and large express the downstream transgenes in all organs and at all the stages of growth and development. However, constitutive overproduction of molecules, such as trehalose (Romero et al. 1997) or polyamines (Capell et al. 1998), causes abnormalities in plants grown under normal conditions. In addition, since the production of these molecules can be metabolically expensive, the use of a stress-inducible promoter may be desirable. Various types of abiotic stresses induce a large number of well-characterized and useful promoters in plants. The transcriptional regulatory regions of the drought-induced and cold-induced genes have been analyzed to identify several cis-acting and trans-acting elements involved in abiotic stress induced gene expression (Shinwari 1999). For complex traits like drought, an ideal inducible promoter should be the one that does not show any basal level of gene expression in the absence of inducing agents, besides resulting in an expression that is reversible and dose-dependent. Most of the stress-inducible promoters contain an array of stress-specific cis-acting elements that are recognized by the requisite transcription factors. The *A. thaliana rd29A* and *rd29B* are stress responsive genes differentially induced under abiotic stress conditions. While the *rd29A* promoter includes both DRE and ABRE elements and dehydration, high salinity and low temperatures induce the gene; the *rd29B* promoter induces only ABREs; and the induction is ABA-dependent. Overexpression of *DREB1A* transcription factors gene under the control of stress-inducible promoter *rd29A* has been reported to show a better phenotypic growth of the transgenic plants than the ones obtained using the constitutive *CaMV35S* promoter (Kasuga et al. 1999; Bhatnagar-Mathur et al. 2007). Gene expression is induced by the binding of DREB1A, which is itself induced by cold and water stress, to cis-acting DRE elements in the promoters of genes such as *rd29A*, *rd17*, *cor6.6*, *cor15A*, *erd10*, and *kin1*, initiating synthesis of gene products imparting tolerance to low temperatures and water stress in plant. These basic findings on stress promoters have led to a major shift in the paradigm for genetically engineering stress tolerant crops (Katiyar-Aggarwal, Agarwal & Grover 1999). A thorough understanding of the underlying physiological processes in response to different abiotic stresses can efficiently/successfully drive the choice of a given promoter or transcription factor to be used for transformation.

15.4 DROUGHT EVALUATION OF TRANSGENICS: A REVIEW

The task of generating transgenic cultivars is not only limited to the success in the transformation process, but also proper evaluation of the stress tolerance. Understanding the physiological effect of the inserted genes at the whole-plant level remains a major challenge. Most of the procedures that have been followed to impose stress for the phenotypic evaluation of transgenic plants for their response to drought and other stresses have been questioned (Sinclair, Purcell & Sneller 2004; Bhatnagar-Mathur, Vadez & Sharma 2008). A large number of studies have been carried out to evaluate different transgenic constructs in different plant species for drought stress. Although, there has been a tendency to report in detail the expression of the transgenes as well as the level of metabolite increase due to the transgene, fewer details are given with regard to the method used to evaluate the stress response.

This lack of detail applies mostly to drought stress where a variety of methods have been applied in various laboratory settings, which include 1) osmotically adjusted media, 2) detaching leaves, and 3) withholding water from soil-grown plants. Also, the protocols used for the evaluation of transgenic plants for abiotic stresses often involved the use of young plants grown in small pots, disregarding water content in pots, usually maintained under inappropriate light and growth conditions (Tarczynski, Jensen & Bohnert 1993; Pilon-Smits et al. 1996; Xu et al. 1996; Pellegrineschi et al. 2004). Stress conditions used to evaluate the transgenic material in most of the reports so far have been too sharp (Shinwari et al. 1998), (Nanjo et al. 1999; Garg et al. 2002), meaning the plants are very unlikely to undergo them in a real field condition situation.

Most of the reported assays for drought tolerance typically involved the complete withholding of water until control plants wilted followed by re-watering and recovery (see Fujita et al. 2005; Sakuma et al. 2006) where the measure of drought tolerance was often given as percent survival. Since slight differences in the degree of wilting can translate into large differences in survival, these tend to amplify the apparent tolerance. However, this approach is useful for assessing tolerance for extremely dry conditions that may be relevant for plant survival, but less relevant for crop productivity under field conditions. These protocols are in fact used to assess transgene response and expression, which unfortunately often becomes assimilated to "tolerance." Also, the experimental means of evaluation can be misleading, resulting in inconsistent data output that leads to conclusions far from convincing and meaningful (Pilon-Smits et al. 1995, 1996, 1999; Pardo, Reddy & Yang 1998; Sivamani et al. 2000; Sun et al. 2001; Lee et al. 2003; Pellegrineschi et al. 2004).

15.4.1 Adequate protocols for transgenic evaluation for drought

Unlike what seems to be a common practice in transgenic evaluation, applying drought does not consist simply in withholding water. One of the major limitations in the evaluation and screening of drought-tolerant genotypes/transgenics has been the lack of a clear definition of the targeted drought. The lack of means to compare drought conditions across experiments, along with the differences in the timing, intensity, and duration of the moisture deficit conditions, makes the comparison of work by different groups relatively difficult. Indeed, we cannot investigate drought response of plants without understanding the different phases that a plant undergoes under drought in natural conditions. These steps have been described earlier (Ritchie 1982; Sinclair & Ludlow 1986). Briefly, in phase I, water is abundant and the plant can take up all the water required for transpiration and stomata are fully open. During that stage, the water loss is mostly determined by the environmental conditions to which the leaves are exposed. During phase II, the roots are no longer able to supply sufficient water to the shoot and stomata progressively close to adjust the water loss to the water supply, so that leaf turgor is maintained. During phase III the roots are no longer able to supply sufficient water to the shoot, and stomata progressively close to adjust to the water loss to the water supply, so that leaf turgor is maintained.

At ICRISAT, we are developing a thorough understanding of the different phases that the plant undergoes under drought, which includes a more realistic physiological response to progressive soil drying. A proper control of soil moisture depletion is done to ensure that transgenic plants are exposed to stress levels and kinetics of water deficits approaching those occurring under field conditions. This involves designing dry-down experiments, where the response of plants to drought is taken as a function of the fraction of soil-water moisture available to plant (FTSW), which allows a precise comparison of stress imposed across experiments. Here, we base our index of stress intensity on FTSW, which represents the volumetric soil water available. Based on the transpiration, the plants can be partially compensated for the water loss to apply a milder stress condition which allows plants of different sizes to be exposed to similar stress levels. This protocol has the advantage of mimicking the situation a plant would face in the field, i.e., a progressive soil drying. This concept has been successfully adapted to assess the response of transgenic plants of groundnut and chickpea with *At rd29A*-driven *DREB1A* transgene as well as with the constitutively expressed *P5CSF129A* gene under contained greenhouse conditions (Bhatnagar-Mathur et al. 2004).

Briefly, in the pot studies under contained greenhouse conditions, the pots are filled with soil taken from the field and plants are initially grown under well-watered conditions with the air temperature regulated at approximately 22 ± 2°C before initiating the experimental treatments. In all the experiments, the plants are grown under well-watered conditions until flowering prior to imposing the drought treatment. In each experiment, the pots are divided into three subsets, each set having six replicates of plants. The pretreatment set is harvested at the beginning of the experiment to record the initial plant biomass. The other two sets are used as well-watered control and stress treatment and are harvested at end of the experiment. The afternoon before the stress treatment, all pots are fully watered (saturation) and allowed to drain overnight. Next morning, the pots are covered with white plastic bags around the stem to prevent direct soil evaporation. A small tube is inserted in plastic bags to re-water the pots during the experiment. The pots are weighed thereafter, and the weight serves as the initial target pots weight. Subsequently, the pots are regularly weighed every morning at a fixed time. The daily transpiration is calculated as the pot weight difference between two successive days. To control the rapid soil drying in the drought stress treatment and to allow plants to develop stress progressively and uniformly, drought-stressed plants are allowed to lose no more than 70 g of water daily. A partial re-watering of the stressed pots is done every morning after weighing to maintain a daily net water loss of 70 g. In both the experiments, well-watered control plants of each genotype are maintained at initial target weight by adding the daily water loss back to the pots.

The above experiment is terminated when the daily transpiration rate of drought-stressed plants decreases to less than 10% that of the well-watered plants. At this endpoint, soil water is no longer available to meet the transpiration demand (Sinclair & Ludlow 1986). The transpiration data are analyzed by the procedure previously described by Ray and Sinclair (1997) and Sinclair and Ludlow (1986). To minimize the influence of large variations in daily transpiration across days, the daily transpiration rates (T) of the drought-stressed pots are normalized against the transpiration rate of the well-watered plants each day. Transpiration ratios (TR) are calculated by dividing daily transpiration of each individual plant in the drought-stressed regime by the daily mean transpiration of the well-watered control plants for each genotype. The values of TR vary among individual plants because of plant size differences among plants within the treatments. To facilitate the comparisons among plants, a second normalization (normalized transpiration ratio, NTR) is done such that the TR of each plant is divided by the TR of that plant averaged across the first three days of the drying cycle, i.e., before any water stress starts affecting T. The transpirable soil water available to the plant in each pot is calculated as the difference between the initial pot weight (field capacity) and the endpoint (no more transpirable water). The use of transpirable soil water as the basis of comparing plant response to soil drying under a range of conditions has been effectively used in a number of studies (Ray & Sinclair 1997; Serraj, Purcell & Sinclair 1999).

Sensitivity of the plant to the amount of available water in the soil profile under water deficit is used as a criterion for comparing drought intensity across experiments (Ritchie 1982). The value of FTSW threshold is calculated from the difference between the water content of the profile after a thorough wetting to saturation and the water content of the profile after healthy plants have exhausted the entire possible water source (transpirable soil water). This approach has been successfully used to monitor the response of different physiological processes to water deficit in many plant species (Sinclair & Ludlow 1986; Ray & Sinclair 1997, 1998; Serraj, Purcell & Sinclair 1999).

In our studies, the transgenic events of groundnut and chickpea were evaluated using the dry-down methodology, where a diversity of stress response patterns was observed, especially with respect to the NTR-FTSW relationship. The soil-moisture threshold where the transpiration rate begins to decline relative to control well-watered (WW) plants and the number of days needed to deplete the soil water was used to rank the genotypes using the average linkage cluster analysis. Five diverse events were selected from the different clusters for further testing. All the selected transgenic events were able to maintain a transpiration rate equivalent to the WW control in soils dry enough to reduce transpiration rate in the untransformed controls. Various transgenic events in both the crops exhibited increased TE, which is an important component of plant performance under limited moisture conditions (Passioura 1977). The TE, which is an important characteristic for plant breeding as a means of improving farm productivity under drought stress, is calculated as the ratio of biomass increase between initial and final harvest, divided by the total water transpired during that period. The most striking finding with the groundnut transgenic events was that some events showed up to 40% higher TE over the untransformed parents under drought stress. These differences can be considered very large when compared with the range of variation usually found for TE between germplasm accessions of peanut (Sheshshayee et al. 2006) and in many other crops (Krishnamurthy et al. 2007) where only a 20–30% difference in TE was observed across a mapping population developed between high and low TE parents.

15.4.2 Roots are also critical for evaluations

Much has been said about the potential of roots to improve crop yield and resilience under drought. Yet, very little breeding, specifically for root traits, has been achieved in groundnut. Although, some progress has been achieved, there are bottlenecks in assessing roots in a large number of genotypes that can be meaningful for plant breeders. This might be partly because of the time-consuming methods to measure rooting differences, limiting their use in breeding (Kashiwagi et al. 2006). Water uptake is probably crucial during key stages of plant growth and development like flowering and grain filling (Boyer & Westgate 1984), and small differences in water uptake at these stages can bring large yield benefits (Boote et al. 1982). However, so far most of the studies being carried out by various groups on root measurements do not address these critical parameters. As we progress in the direction of understanding the role of roots using a unique lysimetric system (described below), we advocate that more focus should be put on the functionality of roots rather than on their morphology, in particular with regards to direct water-uptake measurements and their related kinetics (Vadez et al. 2008).

In our transgenic events of groundnut and chickpea, *DREB1A* certainly appeared to confer a drought-avoidance mechanism by improving WUE. However, it was worthwhile testing whether *DREB1A* could also induce drought avoidance through better water capture. It is well known that under water stress, plants tend to increase their root/shoot ratio. We tested whether *DREB1A* gene could have an effect on root growth under water deficit. Interestingly, we found that *DREB1A* clearly induced a root response under water deficit conditions (Vadez et al. 2008). This response enhanced root growth under water deficit, in particular in the deep soil layers. Consequently, water uptake under water deficit was enhanced up to 20%–30% in some transgenic events compared with the WT and was well related ($r^2 = 0.91$) with the root dry weight below the 40 cm soil depth (Vadez et al. 2007). This was an interesting finding where *DREB1A* transcription factor certainly had an impressive effect on the root growth under water deficit. It was expected though, as DREB1A appears to be a major "switch" for a cascade of genes that are activated under water deficit. These results supported our argument that the capacity of roots to take up water should also be considered in the context of the capacity of shoots to limit their water loss in a comprehensive manner (Vadez et al. 2008).

15.5 THE LYSIMETRIC SYSTEM

At ICRISAT, we have designed and tested a lysimetric system in which plants are grown in long and large polyvinyl chloride (PVC) cylinders (20 cm in diameter and 1.2 m long) that can be tested either under contained greenhouse conditions or even for conducting strip trials for the selection of transgenic events (Vadez et al. 2008). Here, utmost attention is paid to develop a system that mimics the field conditions as closely as possible. The soil volume available to each individual plant is almost equivalent to the soil volume available in the field at usual planting densities (25–30 plant m^{-2}). This system allows measurement of water uptake by plants grown in a real soil profile under conditions of water deficit and to understand how roots contribute to drought tolerance and stress conditions. The soil packing is done using soil sieved in particles smaller than 1 cm. This allows controlling the bulk density to be approximately 1.4, which is the standard value for Alfisols. To ensure that moisture is available in all parts of the cylinders, 40 kg of dry soil is initially filled in each cylinder, keeping the soil level similar in each tube. A prior assessment of the water needed to fill the profile before saturation point is determined such that the water-holding capacity of the soil is approximately 20%. Therefore, 8 L of water is added to the first 40 kg of soil. An additional 10 kg of dry soil is added to each cylinder soon after the water has penetrated the profile and again 2 L water is added. All the cylinders have a very similar bulk density, close to 1.4.

At planting, the soil is wetted and seeds planted at a rate of 2 that are later thinned to 1 plant per cylinder where irrigation is done at regular intervals. This system is most suited to tailoring a number of drought regimes for a range of crops. In our preliminary experiments, we have imposed the last irrigation (to saturate the soil profile) at about one week after flowering. At that stage, low-density polyethylene beads are applied to all cylinders (600 mL per cylinder to have a bead layer of approximately 2 cm). The purpose of the beads is to limit soil evaporation by 90% (Vadez et al. 2008). Therefore, the regular weighing of the cylinders provides transpiration data with beads allowing peg penetration in the case of groundnut (Vadez et al. 2008). As this system simulates the natural soil profile of the field, it allows the penetration and growth of the roots all along the length of the tube. At the end of the experiment, the tubes can be washed for easy extraction of intact roots for further analysis through scanning for root surface area, root-length density, and root volume at varying depths that are crucial parameters to screen the genotypes for differences in root traits, correlating their water-uptake performance in general. The phenotypic evaluation of the transgenic events of groundnut and chickpea using the lysimetric system clearly indicated that the stress-inducible expression of *DREB1A* appeared to confer capture capacity in several transgenic events when compared with their untransformed parents. Moreover, several transgenic events appeared to have consistently higher TE than the WT across different water regimes. Although, these results are very encouraging, one should be cautiously certain about the fact that improved TE in these lines is not at the cost of other yield architecture components.

15.6 YIELD IS THE LITMUS TEST

ICRISAT has been involved for many years in the development of groundnut breeding lines having high WUE. These have been assessed via a major component of drought tolerance, viz., evapotranspiration efficiency (TE), because WUE has been identified as a major contributor to pod yield under intermittent water-deficit conditions (Wright et al. 1993; Wright, Nageswara Rao & Farquhar 1994). However, a yield-based approach as selection criteria to select genotypes where the lines are selected based on yield has proven to be more successful under water deficit (Nigam et al. 2005). This means that each component of the yield architecture cannot be addressed independently of the other. Ultimately, all efforts are made to study the component traits to comprehend their contribution to the yield in general according to the Passioura's model $Y = T * TE * HI$ (Passioura 1977). Following this, preliminary yield trials of the groundnut transgenic events have been conducted both under contained greenhouse and field and/or in controlled lysimetric conditions, where the effect of drought on the various potential component traits of yield architecture (Passioura 1977), such as T, TE, or harvest index (HI), were addressed comprehensively under terminal and intermittent drought conditions. Here, under terminal drought, irrigation was completely suppressed at 40 days after sowing; under intermittent drought stress, re-watering (1 L) was done at 6, 9, and 12 weeks after stress imposition. The results obtained so far have been encouraging. They indicated significantly higher yields in a few selected transgenic events when compared with their untransformed parent under drought (unpublished data). Efforts are ongoing for reconfirmation of these observations under drought stress in these transgenic events before dissecting the individual component traits of yield.

15.7 CONCLUSIONS

A very efficient delivery system is available at ICRISAT for the assessment of drought tolerance and yield in legumes, such as chickpeas and groundnuts, which would immensely influence the progress in the transgenic research for drought tolerance in the future. Results obtained with these legumes provide evidence that transgenic interventions attempted to introgress genes putatively involved in drought tolerance have a potential to further improve drought tolerance and thereby increase and stabilize yield in the SAT regions. These results are very encouraging because the range of variation observed for TE was higher than what has been found with recombinant inbred lines (RIL) populations. Further, the fact that transgenic events show such large phenotypic contrast, while being isogenic for one inserted gene, provides great opportunities to reinvestigate the mechanisms of drought tolerance in these legumes. This might in the near future lead to significant benefits in raising drought-tolerant genotypes. Further, the identification of other physiological traits linked to drought tolerance in our on-going studies with these transgenic events is expected to provide more insights into the mechanisms of drought-stress tolerance both under greenhouse and field conditions.

REFERENCES

Allen, R.D. 1995. Dissection of oxidative stress tolerance using transgenic plants. *Plant Physiol.* 107:1049–1054.

Araus, J.L., G.A. Slafer, M.P. Reynolds, and C. Royo. 2002. Plant breeding and water stress in C3 cereals: What to breed for? *Ann. Bot.* 89:925–940.

Bajaj, S., J. Targolli, L.F. Liu, T.H. Ho, and R. Wu. 1999. Transgenic approaches to increase dehydration stress tolerance in plants. *Mol. Breed.* 5:493–503.

Bartels, D. and R. Sunkar, 2005. Drought and salt tolerance in plants. *Crit. Rev. Plant Sci.* 21:1–36.

Behnam, B., A. Kikuchi, F. Celebi-Toprak, S. Yamanaka, M. Kasuga, K. Yamaguchi-Shinozaki and K.N. Watanabe. 2006. The *Arabidopsis DREB1A* gene driven by the stress-inducible *rd29A promoter* increases salt-stress tolerance in proportion to its copy number in tetrasomic tetraploid potato (*Solanum tuberosum*). *Plant Biotech.* 23:169–177.

Bhatnagar-Mathur, P., M.J. Devi, R. Serraj, K. Yamaguchi-Shinozaki, V. Vadez, and K.K. Sharma. 2004. Evaluation of transgenic groundnut lines under water limited conditions. *Int. Arch. Newsl.* 24:33–34.

Bhatnagar-Mathur, P., M.J. Devi, V. Vadez, and K.K. Sharma. 2009a. Differential antioxidative responses in transgenic peanut bear no relationship to their superior transpiration efficiency under drought stress. *J. Plant Physiol.* 166:1207–1217.

Bhatnagar-Mathur, P., M.J. Devi, V. Vadez, D.P.S. Verma, and K.K. Sharma. 2009b. Over expression of Vigna P5CSF129A gene in chickpea for enhancing drought tolerance. *Mol. Breed.* 23:591–606.

Bhatnagar-Mathur, P., D.S. Reddy, M. Lavanya, K. Yamaguchi-Shinozaki, and K.K. Sharma. 2007. Stress-inducible expression of Arabidopsis thaliana DREB1A in transgenic peanut (*Arachis hypogaea* L.) increases transpiration efficiency under water-limiting conditions. *Plant Cell Rep.* 26:2071–82.

Bhatnagar-Mathur, P., V. Vadez, and K.K. Sharma. 2008. Transgenic approaches for abiotic stress tolerance in plants: Retrospect and prospects. *Plant Cell Rep.* 27:411–424.

Bohnert,H.J., D.F. Nelson, and R.G. Jenson. 1995. Adaptation to environmental stresses. *Plant Cell* 7:1099–1111.

Boote, K.J., J.R. Stansell, A.M. Schubert, and J.F. Stone. 1982. Irrigation, water use and water relations. In *Peanut science and technology*, eds. H.E. Patee, L.T. Young, pp. 164–205. Yoakum, TX: American Peanut Research and Education Society.

Boyer, J.S. 1982. Plant productivity and environment. *Science* 218:443–448.

Boyer, J.S. and M.E. Westgate. 2004. Grain yields with limited water. In *Water-saving agriculture*. Special issue. *J. Exp. Bot.* 55(407):2385–2394.

Bray, E.A., J. Bailey-Serres and E. Weretilnyk. 2000. Responses to abiotic stresses. In *Biochemistry and molecular biology of plants,* eds, W. Gruissem, B. Buchannan, R. Jones, pp. 1158–1249. Rockville, MD: American Society of Plant Physiologists.

Burke, E.J., S.J. Brown, and N. Christidis. 2006. Modeling the recent evolution of global drought and projections for the twenty-first century with the Hadley Centre climate model. *J. Hydrometeorol.* 7:1113–1125.

Capell, T., C. Escobar, H. Lui, H. Burtin, O. Lepri, and P. Christou. 1998. Overexpression of the oat arginine decarboxylase cDNA in transgenic rice (*Oryza sativa* L.) affects normal development patterns in vitro and results in putrescine accumulation in transgenic plants. *Theor. Appl. Genet.* 97:246–254.

Chaves, M.M. and M.M. Oliveira. 2004. Mechanisms underlying plant resilience to water deficits: prospects for water-saving agriculture. *J. Exp. Bot.* 55(407):2365–2384.

Chinnusamy, V., A. Jagendorf, and J.K. Zhu. 2005. Understanding and improving salt tolerance in plants. *Crop Sci.* 45:437–48.

Delauney, A.J. and D.P.S. Verma. 1993. Proline biosynthesis and osmoregulation in plants. *Plant J.* 4:215–223.

FAO. 1996. *Quarterly Bulletin of Statistics.* Rome: FAO.

Fischer, G., M. Shah, H. van, Velthuizen, and F.O. Nachtergaele. 2001. *Global agro-ecological assessment for agriculture in the 21st century.* Laxenburg, Austria: IIASA and FAO.

Fujita, Y., M. Fujita, R. Satoh, K. Maruyama, M.M. Parvez, M. Seki, K. Hiratsu, M. Ohme-Takagi, K. Shinozaki, and K. Yamaguchi-Shinozaki. 2005. AREB1 is a transcription activator of novel ABRE-dependent ABA signaling that enhances drought stress tolerance in Arabidopsis. *Plant Cell* 17:3470–3488.

Garg, A.K., J.K. Kim, T.G. Owens, A.P. Ranwala, Y.C. Choi, L.V. Kochian, and R.J. Wu. 2002. Trehalose accumulation in rice plants confers high tolerance levels to different abiotic stresses. *Proc. Natl. Acad. Sci.* 99:15898–15903.

Hayashi, H., Alia, L. Mustardy, P. Deshnium, M. Ida, and N. Murata. 1997. Transformation of *Arabidopsis thaliana* with the cod A gene for choline oxidase: accumulation of glycinebetaine and enhanced tolerance to salt and cold stress. *The Plant J.* 12:133–142.

Holmstrom, K.O., S. Somersalo, A. Mandal, E.T. Palva, and B. Welin. 2000. Improved tolerance to salinity and low temperature in transgenic tobacco producing glycine betaine. *J Exp. Bot.* 51:177–185.

Ishitani, M., L. Xiong, B. Stevenson, and J-K. Zhu, 1997. Genetic analysis of osmotic and cold stress signal transduction in Arabidopsis: Interactions and convergence of abscisic acid-dependent and abscisic acid-independent pathways. *Plant Cell* 9:1935–1949.

Ishizaki-Nishizawa, O., T. Fujii, M. Azuma, K. Sekiguchi, N. Murata, T. Ohtani, and T. Toguri. 1996. Low-temperature resistance of higher plants is significantly enhanced by a nonspecific cyanobacterial desaturase. *Nat. Biotechnol.* 14:1003–1006.

Jackson, R.B., J. Canadell, J.R. Ehleringer, H.A. Mooney, O.E. Sala, and E.D. Schulze. 1996. A global analysis of root distributions for terrestrial biomes. *Oecologia* 108:389–411.

Jones, M.L. 2004. Changes in gene expression during senescence. In *Plant cell death processes,* ed L. Nooden, pp. 51–72. San Diego, CA: Elsevier Science.

Kashiwagi, J., L. Krishnamurthy, R. Serraj, H.D. Upadhyaya, S.H. Krishna, S. Chandra, and V. Vadez, 2006. Genetic variability of drought-avoidance root traits in the mini-core germplasm collection of chickpea (*Cicerarietinum* L.). *Euphytica* 146:213–222.

Krishnamurthy, L., V. Vadez, M.J. Devi, R. Serraj, S.N. Nigam, M.S. Sheshshayee, S. Chandra, and R. Aruna. 2007. Variation in transpiration efficiency and its related traits in a groundnut (*Arachis hypogaea* L.) mapping population. *Field Crop Res.* 103:187–197.

Kasuga, M., Q. Liu, S. Miura, K. Yamaguchi-Shinozaki, and K. Shinozaki. 1999. Improving plant drought, salt, and freezing tolerance by gene transfer of a single stress inducible transcription factor *Nat. Biotechnol.* 17:287–291.

Kasuga, M., S. Miura, K. Shinozaki, and K. Yamaguchi-Shinozaki. 2004. A combination of the *Arabidopsis* DREB1A gene and stress-inducible *rd29A* promoter improved drought- and low-temperature stress tolerance in tobacco by gene transfer. *Plant Cell Physiol.* 45:346–350.

Katiyar-Agarwal S., M. Agarwal, and A. Grover. 1999. Emerging trends in agricultural biotechnology research: use of abiotic stress induced promoter to drive expression of a stress resistance gene in the transgenic system leads to high level stress tolerance associated with minimal negative effects on growth. *Curr. Sci.* 77:1577–1579.

KaviKishor, P.B., Z. Hong, G.H. Miao, C.A.A. Hu, and D.P.S. Verma. 1995. Overexpression of Δ^1-pyrroline-5-carboxylate synthetase increases proline production and confers osmotolerance in transgenic plants. *Plant Physiol.*108:1387–1394.

Kodama, H., T. Hamada, G. Horiguchi, M. Nishimura, and K. Iba. 1994. Genetic enhancement of cold tolerance by expression of a gene for chloroplast ω-3 fatty acid desaturase in transgenic tobacco. *Plant Physiol.* 105:601–605.

Lee, S.S., H.S. Cho, G.M. Yoon, J.W. Ahn, H.H. Kim, and H.S. Pai, 2003. Interaction of NtCDPK1 calcium-dependent protein kinase with NtRpn3 regulatory subunit of the 26S proteasome in *Nicotiana tabacum. Plant J.* 33:825–840.

McKersie, B.D., S.R. Bowley, E. Harjanto, and O. Leprince. 1996. Water-deficit tolerance and field performance of transgenic alfalfa over-expressing superoxide dismutase. *Plant Physiol.* 111:1177–1181.

Miflin, B.J. 2000. Crop biotechnology. Where now? *Plant Physiol.* 123:17–28.

Morgan, J.M. 1984. Osmoregulation and water stress in higher plants. *Annu. Rev. Plant Physiol.* 35:299–348.

Nanjo,T., M. Kobayashi, Y. Yoshiba, Y. Kakubari, Yamaguchi-Shinozaki K. and K. Shinozaki, 1999. Antisense suppression of proline degradation improves tolerance to freezing and salinity in *Arabidopsis thaliana. FEBS Lett.* 461:205–210.

Nigam, S.N., S. Chandra, K. Rupa Sridevi, M. Bhukta, A.G.S. Reddy, N.R. Rachaputi, G.C. Wright, et al., 2005. Efficiency of physiological trait-based and empirical selection approaches for drought tolerance in groundnut. *Ann. Appl. Biol.* 146:433–439.

Pardo, J.M., M.P. Reddy, and S. Yang. 1998. Stress signaling through Ca^{2+}/Calmodulin dependent protein phosphatase calcineurin mediates salt adaptation in plants. *Proc. Natl. Acad. Sci.* 95:9681–9683.

Passioura, J.B. 1977. Grain yield, harvest index and water use of wheat. *J. Aust. Inst. Agri. Sci.* 43:21.

Pellegrineschi, A., M. Reynolds, M. Pacheco, R.M. Brito, R. Almeraya, K. Yamaguchi-Shinozaki and D. Hoisington. 2004. Stress-induced expression in wheat of the *Arabidopsis thaliana* DREB1A gene delays water stress symptoms under greenhouse conditions. *Genome* 47:493–500.

Pilon-Smits, E.A.H., M.J.M. Ebskamp, M.J.W. Jeuken, I.M. van der Meer, R.G.F. Visser, P.J. Weisbeek, and J.C.M. Smeekens. 1996. Microbial fructan production in transgenic potato plants and tubers. *Ind. Crops Prod.* 5:35–46.

Pilon-Smits, E.A.H., M.J.M. Ebskamp, M.J. Paul, J.W. Jeuken, Weisbeek, and S.C.M. Smeekens. 1995. Improved performance of transgenic fructan-accumulating tobacco under drought stress. *Plant Physiol.* 107:125–130.

Pilon-Smits, E.A.H., N. Terry, T. Sears, and K. van Dun. 1999. Enhanced drought resistance in fructan-producing sugar beet. *Plant Physiol. Biochem.* 37:313–317.

Ray, J.D. and T.R. Sinclair. 1997. Stomatal closure of maize hybrids in response to soil drying. *Crop Science* 37:803–807.

Ray, J.D. and T.R. Sinclair. 1998. The effect of pot size on growth and transpiration of maize and soybean during water deficit stress. *J. Exp. Bot.* 49(325):1381–1386.

Ritchie, G.A. 1982. Carbohydrate reserves and root growth potential in Douglas fir seedlings before and after cold storage. *Can. J. For. Res.* 12:905–912.

Romero, C., J.M. Belles, J.L. Vaya, R. Serrano, and F.A. Culianez-Macia. 1997. Expression of the yeast trehalose-6-phosphate synthase gene in transgenic tobacco plants: pleiotropic phenotypes include drought tolerance. *Planta* 201:293–297.

Sakuma, Y., K. Maruyama, k. Osakabe, F. Qin, M. Seki, K. Shinozaki, and K. Yamaguchi-Shinozaki. 2006a. Functional analysis of an Arabidopsis transcription factor, DREB2A, involved in drought-responsive gene expression. *Plant Cell* 18:1292–1309.

Serraj, R., L.C. Purcell, and T.R. Sinclair. 1999. Symbiotic N_2 fixation response to drought. *J. Exp. Bot.* 50:143–155.

Serraj, R. and T.R. Sinclair. 2002. Osmolyte accumulation: Can it really help increase crop yield under drought conditions? *Plant Cell Environ.* 25:333–341.

Sharma, K.K. and M. Lavanya. 2002. Recent developments in transgenics for abiotic stress in legumes of the semi-arid tropics. In *Genetic engineering of crop plants for abiotic stress,* ed. M. Ivanaga M. 61–73. JIRCAS Working Report No. 23. Tsukuba, Japan: JIRCAS.

Sharma, K.K. and R. Ortiz. 2000. Program for the application of the genetic engineering for crop improvement in the semi-arid tropics. *In Vitro Cell. Develop. Biol.* (Plant) 36:83–92.

Sheshshayee, M.S., H. Bindumadhava, N.R. Rachaputi, T.G. Prasad, M. Udayakumar, and G.C. Wright. 2006. *Leaf chlorophyll concentration relates to transpiration efficiency in peanut.* Oxford, Blackwell.

Shinwari, Z.K. 1999. Function and regulation of genes that are induced by dehydration stress. *Biosci. Agric.* 5:39–47.

Shinwari, Z.K., K. Nakashima, S. Miura, M. Kasuga, M. Seki, K. Yamaguchi-Shinozaki, and K. Shinozaki, 1998. An *Arabidopsis* gene family encoding DRE/CRT binding proteins involved in low-temperature-responsive gene expression *Biochem. Biophys. Res. Commun.* 50:161–170.

Sinclair, T.R. and M.M. Ludlow. 1986. Influence of soil water supply on the plant water balance of four tropical grain legumes. *Aust. J. Plant Physiol.* 13:329–341.

Sinclair, T.R., L.C. Purcell, and C.H. Sneller. 2004. Crop transformation and challenge to increase yield potential. *Trends Plant Sci.* 9:70–75.

Sivamani, E., A. Bahieldin, J.M. Wraith, T. Al-Niemi, W.E. Dyer, T.H.D. Ho, and R. Qu. 2000. Improved biomass productivity and water use efficiency under water deficit conditions in transgenic wheat constitutively expressing the barley HVA1 gene. *Plant Sci.* 155:1–9.

Slafer, G.A. 2003. Genetic basis of yield as viewed from a crop physiologist's perspective. *Ann. Appl. Biol.* 142(2):117–128.

Snape, J.W., K. Butterworth, E. Whitechurch, and A.J. Worland. 2001. Waiting for fine times: Genetics of flowering time in wheat. *Euphytica* 119:185–190.

Sun, W.C., Bernard, B. van de Cotte, M.V. Montagu, and N. Verbruggen. 2001. At-HSP17.6A, encoding a small heat-shock protein in *Arabidopsis*, can enhance osmotolerance upon overexpression. *Plant J.* 27:407–415.

Tarczynski, M.C., R.G. Jensen, and H.J. Bohnert. 1993. Stress protection of transgenic tobacco by production of the osmolyte mannitol. *Science* 259:508–510.
Turner, N.C., A. Shahal, J.D. Berger, S.K. Chaturvedi, R.J. French, C. Ludwig, D.M. Mannur, S.S. Singh, and H.S. Yadava. 2007. Osmotic adjustment in chickpea (*Cicer arietinum L.*) results in no yeild benefit under terminal drought. *J. Exp. Bot.* 58:187–194.
Turner, N.C. and M.M. Jones. 1980. Turgor maintenance by osmotic adjustment: a review and evaluation. In *Adaptation of plants to water and high temperature stress*, eds. N.C. Turner, and P.J. Kramer, pp. 84–104. New York: Wiley Intersience.
Umezawa, T., M. Fujita, Y. Fujita, K. Yamaguchi-Shinozaki, and K. Shinozaki. 2006. Engineering drought tolerance in plants: discovering and tailoring genes to unlock the future. *Curr. Opin. Biotechnol.* 17:113–122.
Vadez, V., S. Rao, J. Kholova, L. Krishnamurthy, J. Kashiwagi, P. Ratnakumar, K.K. Sharma, P. Bhatnagar-Mathur, and P.S. Basu. 2008. Roots research for legume tolerance to drought: Quo vadis? *J. Food Legumes* 21(2):77–85.
Vadez, V., S. Rao, K.K. Sharma, Bhatnagar-Mathur, P. and M.J. Devi. 2007. DREB1A allows for more water uptake by a large modification in the root/shoot ratio under water deficit. *Int. Ara. Newsletter.* 27:27–31.
Wang, W.B., Vinocur, and A. Altman. 2003. Plant responses to drought, salinity and extreme temperatures: towards genetic engineering for stress tolerance. *Planta* 218:1–14.
Wenzel, and Wayne. March 14 2008. H2O Optimizers, *Farm Journal.*
Wright, G.C., K.T. Hubick, G.D. Farquhar, and R.C.N. Rao. 1993. Genetic and environmental variation in transpiration efficiency and its correlation with carbon isotope discrimination and specific leaf area in peanut. *In Stable isotopes and plant carbon/water relations*, eds. J.R. Ehleringer, A.E. Hall, and G.D. Farquhar, pp. 247–267. San Diego, CA: Academic Press.
Wright, G.C., R.C. Nageswara Rao, and G.D. Farquhar. 1994. Water-use efficiency and carbon isotope discrimination in peanut under water deficit conditions. *Crop Sci.* 34:92–97.
Xu, D., X. Duan, B. Wang, B. Hong, T.H.D. Ho, and R. Wu. 1996. Expression of a late embryogenesis abundant protein gene, HVA1, from barley confers tolerance to water deficit and salt stress in transgenic rice. *Plant Physiol.* 110:249–257.
Yamada, M., H. Morishita, K. Urano, N. Shiozaki, K. Yamaguchi-Shinozaki, K. Shinozaki, and Y. Yoshiba. 2005. Effects of free proline accumulation in petunias under drought stress. *J Exp. Bot.* 56:1975–1981.
Zhang, H., Z. Huang, B. Xie, Q. Chen, X. Tian, X. Zhang, H.B. Zhang, X. Lu, D. Huang, and R. Huang. 2004. Ethylene, jasmonate, abscisic acid and NaCl responsive tomato transcription factor JERF1 modulates expression of GCC box-containing genes and salt tolerance in tobacco. *Planta* 220:262–270.
Zhu, B., J. Su, M. Chang, D.P.S. Verma, Y.L. Fan, and R. Wu. 1998. Overexpression of delta1-pyrroline-5-carboxylate synthase gene and analysis of tolerance to water and salt stress in transgenic rice. *Plant Sci.* 199:41–48.

Chapter 16

Viable alternatives to the rice-wheat cropping system in Punjab

B.S. Kang, Kuldeep Singh, Dhanwinder Singh & B.R. Garg
The Punjab Agricultural University, Ludhiana, India

R. Lal
Carbon Management and Sequestration Center, The Ohio State University, Columbus, Ohio, USA

M. Velayutham
M.S. Swaminathan Research Foundation, Chennai, India

ABSTRACT

On-farm experiments conducted on alluvial soils of Punjab with diverse crop rotations of different water requirement showed that at Nawanshahar, mean yield of wheat was similar for rice-wheat system (RWS; 4.0 Mg/ha) and maize-wheat system (MWS; 4.3 Mg/ha). Mean yield of maize was also similar in the MWS (4.4 Mg/ha) and maize-rapeseed system (MRS; 4.4 Mg/ha). Although output in gross returns (GR) and return on variable expenses (ROVE) were higher for RWS in Nawanshahar district, substituting maize for rice in the monsoon season is preferred because of savings in water. In Faridkot district, the monetary returns were higher for the cotton-wheat system (CWS), which makes judicious use of groundwater as compared to the non-traditional RWS. Soil organic carbon (SOC) concentration at Nawanshahar increased from a mean initial level of 0.44% to 0.50% when both rice and wheat residues were burned; 0.51% when rice straw was incorporated but wheat straw was removed; and 0.57% when both rice and wheat straw were incorporated. At Faridkot site, the SOC concentration increased from a mean initial level of 0.39% to 0.49% when both rice and wheat straw were incorporated. The positive effect of straw incorporation on soil quality was more pronounced in the light-textured soil (sandy loam) at Faridkot than in loamy soil at Nawanshahar. Soil bulk density at the Faridkot site decreased from 1.65 Mg/m^3 to 1.58 Mg/m^3, and water infiltration rate increased from 6.1 mm/hr to 7.6 mm/hr. Straw incorporation increased soil organic carbon pool at the Faridkot site by 1 Mg/ha in 0–15 cm depth across a two year period. Comparative economics of residue incorporation in RWS across three years showed that, although the GR (Rs/ha) was the same in straw-burned and straw-incorporated treatments, the improvements

Address correspondence to R. Lal, Carbon Management and Sequestration Center, The Ohio State University, 2021 Coffey Road, Rm 210 Kottman Hall, Columbus, Ohio 43210, USA. E-mail: lal.1@osu.edu

Reprinted from: *Journal of Crop Improvement*, Vol. 23, Issue 3 (July 2009), pp. 300–318, DOI: 10.1080/15427520902809912 in soil quality that the straw incorporation brings about for sustainable agriculture in the long-run offset any advantage of burning straw and gave higher net returns and benefit-cost ratio (BCR) in the short-term. Efficient on-farm water management using fertigation technology was demonstrated in a farmer's field for production of vegetables throughout the year. Diversified agriculture in the form of agroforestry and fish culture as alternative land-use options were also demonstrated in farmers' fields. Agroforestry interventions involving poplar trees with a range of intercrops were found to be useful cyclic interventions in the land-use and management strategy for sustainable agriculture in the intensively cultivated alluvial soils of Punjab.

Keywords: water management, Indo-Gangetic plains, cropping systems, agroforestry, inter-cropping

16.1 INTRODUCTION

The Indo-Gangetic alluvial Plains (IGP) of India extend from 21° 45′ to 31° 0′ N latitudes and 74° 15′ to 91° 30′ E longitudes, and include the states of Punjab, Haryana, Uttarpradesh, Bihar, West Bengal, Himachal Pradesh, northern parts of Rajasthan, and Tripura as well as Delhi. The IGP cover a total area of 43.7 Mha. Velayutham and colleagues (2002) have discussed the historical perspective of the pedogenesis of the soils of the IGP.

Punjab state, situated in the western part of the IGP, comprises an area of 5 million ha and contributes annually 75% of wheat (*Triticum aestivum*) and 34% of rice (*Oryza sativa*) to the national food reserves (2006–07). The rice-wheat system (RWS), being highly profitable, has been adopted even in non-traditional areas. Soil and environmental issues pertaining to the RWS were discussed by Velayutham et al. (1994; 2002). From 1960 to 2008, the area under rice cultivation increased by more than 10 times and that under wheat by 2.5 times. The RWS in Punjab is irrigated and is an intensive input-driven and a soil-exhaustive cropping system. Between 1960/61 and 2005/06, irrigated land area increased from 59% to 97%, cropping intensity from 118% to 189% (2003–04), and food grain production from 3.2 Mt to 26.81 Mt (Department of Agriculture, 2003–2009).

Irrigation has led to excessive withdrawal of groundwater in most parts of Punjab. There is a decline in water table (WT) in the central districts and a rising trend in the southwestern districts of the state. Further, there is a declining trend in soil fertility and a rising trend in salinity and degradation of irrigation water quality leading to the fatigue of the RWS (Sinha, Singh & Rai 1998). Improper management of crop residues exacerbates the problems of declining soil fertility. Burning of the rice straw, widely practiced in Punjab, leads to the loss of nutrients, the environmental problem of air pollution, and health hazards.

16.2 PHYSIOGRAPHY AND ECO-REGIONAL CHARACTERISTICS OF EXPERIMENTAL SITES

The M.S. Swaminathan Research Foundation (MSSRF), Chennai, India, The Ohio State University (OSU), USA, and the Punjab Agricultural University (PAU), India, implemented a community-based field project (MSSRF, 2006). Specific objectives of the project were: 1) demonstration and adoption of techniques of efficient management of soil, water and crop residues; 2) identification of cropping systems as viable alternatives to RWS, which alleviate the pressure on groundwater resources; 3) assessment of agronomic productivity of diversified production systems; and 4) creation of training opportunities for researchers, extension staff, and farming communities.

The field demonstrations/experiments were conducted in farmers' fields in four locations (Barbha, Sajawalpur, Rakkaran Dhahan, and Baghauraan) in Nawanshahar district, and four locations (Kothae Maur, Sadiq, Sandhwan and Nathewala) in Faridkot district (Figure 16.1). The Farm Science Centre at Langroya in Nawanshahar district and the Regional Agriculture Research Station and the Farm Science Centre at Faridkot provided technical support.

Nawanshahar district is situated in the hot, dry, sub-humid climatic zone at 31° 32′ N latitude and 75° 54′ E longitude, and forms part of northern flat plain with alluvium-derived soils. It is a part of the 9.1 agro-eco subregion, with a growing season of 120–180 days (Velayutham et al., 1999). The mean winter and summer

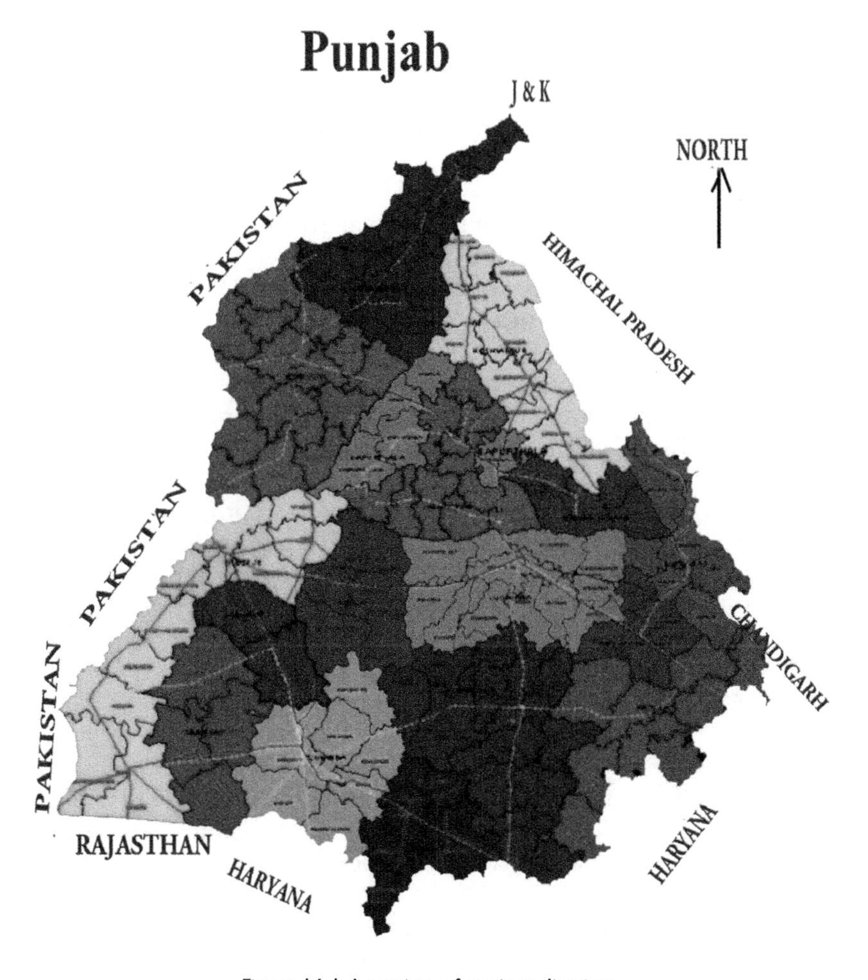

Figure 16.1 Location of project districts.

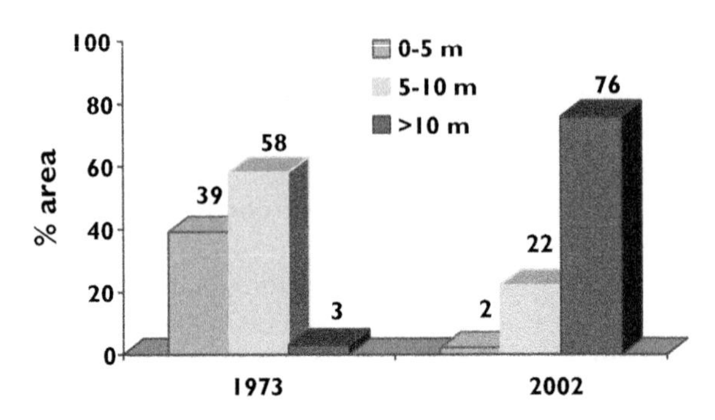

Figure 16.2 Percent area under different water table depths (m) in central zone of Punjab.

temperatures are 14.2° and 34.4°C, respectively, with a mean annual temperature of 23.5°C. The average annual rainfall is 850 mm. The soils are classified in three subgroups, viz., Typic Haplustalf, udic Ustochrept, and fluventic Ustochrept (Sidhu et al., 1995). Soils are characterized by an alkaline pH, low electric conductivity (EC), and low soil organic carbon (SOC) concentration. The RWS is the dominant cropping system in this district; 82% of the area is irrigated by tubewells (Figure 16.2), and the water table (WT) is falling rapidly.

Faridkot district is situated in the semi-arid climatic zone at 30° 40′ N latitude and 75° 45′ E longitude, and forms part of northern alluvial plain with sand dunes. It is a part of the 4.1 agro-eco subregion, with a growing season of 90–150 days. The mean winter and summer temperatures are 13.9° and 32.1°C, respectively, with a mean annual temperature of 24°C. The mean annual rainfall in the district is 427 mm (Velayutham et al., 1999). The soils of the experimental site are very deep and well drained, and developed on reworked sand dunes. These soils are classified as Ustipsamments, have alkaline pH, and are calcareous. The major source of irrigation is the groundwater (84%). Both RWS and cotton (*Gossypium hirsutum*)-wheat system (CWS) are the dominant crop rotations in the district. Soil characteristics of the two locations, one from each district, are given in Table 16.1. The chemical properties, SOC, available soil nutrients, bulk density, and infiltration rate were determined as per the procedures given in "Methods of Soil Analysis" (Klute, 2006 & Sparks et al., 1996).

16.3 PRODUCTIVITY OF ALTERNATE CROPPING SYSTEMS

Three cropping systems tested at Langroya site were RWS, maize (*Zea mays* L.)-wheat system (MWS), and maize-field mustard (*Brassica rapa* L.)-rapeseed (*Brassica napus* L.) system (MTRS). The initial soil analysis showed that the SOC concentration ranged from 0.46% to 0.60%, available P from low to medium, and available K from medium to high (Table 16.2).

Similarly, three cropping systems tested at Faridkot site were RWS, CWS, and cotton-chickpea (*Cicer arietinum*) system (CCS). The SOC concentrations at these

Table 16.1 General properties of soils at two experimental sites.

Soil characteristics	*Langroya*	*Faridkot*
Texture	Loam	Loamy Sand
pH	8.0 to 8.7	8.5 to 9.4
E.C. (ds m^{-1})	0.2 to 0.3	0.2 to 0.4
Organic Carbon (%)	0.4 to 0.6	0.2 to 0.6
Available P (kg/ha)	6.8 to 13.5	9.2 to 24.8
Available K (kg/ha)	202 to 397	405 to 562
$CaCO_3$ (%)	0.35 to 1.0	0.7 to 1.2
Extractable Calcium (μg/g)	613 to 754	557 to 681
Extractable magnesium (μg/g)	351 to 584	97 to 289
Extractable sodium (μg/g)	18 to 70	41 to 162
†DTPA-extractable Zn (μg/g)	0.4 to 0.8	0.4 to 0.6
DTPA-extractable Fe (μg/g)	3.1 to 9.8	2.0 to 5.2
DTPA-extractable Mn (μg/g)	1.7 to 4.1	0.9 to 2.5
DTPA-extractable Cu (μg/g)	0.5 to 1.4	0.3 to 1.0

†DTPA = Diethylene triamine pentaacetic acid.

Table 16.2 Antecedent properties of soils for sites selected for the alternate cropping system experiments.

District	*Location farmer's name*	*Texture*	*pH*	*E.C. (ds/m)*	*$CaCO_3$ (%)*	*SOC (%)*	*Avail. P (kg/ha)*	*Avail. (K_2O) (kg/ha)*
Langroya Nawashahr	1. Barbha	Loam	8.7	0.20	0.5	0.60	6.7	337
	2. Sajawalpur	Loam	8.4	0.20	0.5	0.51	13.5	217
	3. Rakkaran Dhahan	Loam	8.4	0.20	0.9	0.46	13.5	270
	4. KVK Farm							
Faridkot	1. Sadiq	Loamy sand	9.4	0.40	1.2	0.21	24.7	405
	2. Kothae Maur	Sandy loam	8.7	0.20	1.0	0.31	12.5	547

E.C. = electric conductivity, SOC = soil organic carbon.

sites were much lower (e.g., 0.21% to 0.31%) than those at Langroya sites, available P varied from medium to high, and available K was relatively high (Table 16.2).

All crops in diverse rotation systems were grown with the standard package of agronomic practices. The yields of crops in the different seasons (Tables 16.3, 16.4, and 16.5) show that the mean yield of wheat (2nd crop) at Langroya was similar for the RWS (4009 kg/ha) and the MWS (4362 kg/ha). Similarly, wheat yield in both systems indicates that the lead-time available for land preparation for sowing wheat, unless no-till is practiced, is more following maize than rice cultivation. The mean yield of maize was also similar in MWS (4455 kg/ha) and MTRS (4470 kg/ha).

The SOC concentration at the end of three cropping cycles did not change at either of the two sites except in the upland CCS, probably because of the beneficial

Table 16.3 Yield of winter crops (2001–02) at different sites in an alternate cropping-system trial.

	Grain yield (kg/ha)				
	Langroya			Faridkot	
Cropping system	Barbha	Sajawalpur	Rakkaran Dhahan	Sadiq	Kothae Maur
Rice-wheat	5525	4958	4588	5375	5625
Maize-wheat	5525	4958	4588	–	–
Maize-brassica	1675	1553	1738	–	–
Cotton-wheat	–	–	–	5250	5500
Desi cotton-chickpea/wheat	–	–	–	1312*	6525**

*Gram yield.
**Wheat yield.

Table 16.4 Yield of crops at different sites in alternate cropping systems (monsoon season 2002).

	Grain yield (kg/ha)					
	Langroya				Faridkot	
Cropping system	Barbha	Sajawalpur	KVK farm	Rakkaran Dhahan	Sadiq	Kothae Maur
Rice-wheat	7100	6350	5800	6610	4375	6950
Maize-wheat	4910	3910	4545	5020	–	–
Maize-mustard-brassica	5045	4272	4092	4365	–	–
Am. Cotton-wheat	–	–	–	–	1617	1770
Desi Cotton-chickpea/wheat	–	–	–	–	1012	1287

Table 16.5 Yield of crops in alternate cropping-system experiments after winter 2002–03.

	Grain yield (kg/ha)					
	Langroya				Faridkot	
Cropping system	Barbha	Sajawalpur	KVK farm	Rakkaran Dhahan	Sadiq	Kothae Maur
Rice-wheat	4413	3617	4105	3903	4467	4857
Maize-wheat	4573	3989	4685	4200	–	–
Maize-mustard-brassica	679	1024	773	737	–	–
Cotton-wheat	938	1044	825	913	–	–
Desi cotton-chickpea/wheat	–	–	–	–	4012	4555
	–	–	–	–	1062*	4617**

*Gram yield.
**Wheat yield.

effect of biological nitrogen fixation (BNF) by chickpea (Table 16.6). Thus, including a leguminous crop in the rotation cycle is important to enhance soil quality. Goswami and colleagues (2000) reported that SOC sequestration in alluvial soils of IGP increased with the application of compost/farmyard manure (FYM) and incorporation of a leguminous green-manure crop in the rotation cycle in calcareous sodic soils under RWS.

Comparative economics of diverse cropping systems was assessed for the participating farms. Wheat crop did not compete for use of scarce resources with any other crop in winter and had a good return per unit of human labor, fertilizer, pesticide use, and water (irrigation) at both the locations. The RWS was most profitable both at Langroya and Faridkot, probably because of the minimum support price (MSP) offered by the government for procurement of rice and wheat immediately after harvest. However, maize in MWS and MTRS at Langroa was also a profitable cropping system and fared better than rice in the RWS in the judicious use of scarce resources. Singh and Hossain (2003) reported the decelerating trend in total factor productivity (TFP) growth for rice in high-production RWS of north India. TFP is a variable that accounts for effects in total output not caused by inputs. In view of the receding WT, substituting maize for rice during the crop cycles in the monsoon season is a viable option that merits attention. At Faridkot, the economic returns from CWS were as good as those for RWS. However, for successful cotton production, there was heavy dependence on pesticides. In this regard, the use of Bt cotton is a viable option. The lowest economic returns were obtained from the rice-based system.

Table 16.6 Soil Organic Carbon concentration as affected by alternate cropping systems after winter 2002–03.

	SOC concentration (%)					
	Langroya				*Faridkot*	
Cropping system	*Barbha*	*Sajawalpur*	*KVK farm*	*Rakkaran Dhahan*	*Sadiq*	*Kothae Maur*
Initial	0.60	0.51	–	0.46	0.21	0.31
Rice-wheat	0.65*	0.46	0.35	0.45	0.27	0.32
	0.65**	0.50	0.40	0.50	0.23	0.47
	0.71***	0.44	0.45	0.45	0.26	0.51
Maize-wheat	0.60*	0.51	0.42	0.43	–	–
	0.62**	0.45	0.40	0.50	–	–
	0.66***	0.53	0.40	0.56	–	–
Maize-mustard-brassica	0.69*	0.47	0.40	0.51	–	–
	0.65**	0.45	0.38	0.45	–	–
	0.65***	0.51	0.38	0.48	–	–
Am. cotton-wheat	–	–	–	–	0.23	.032
	–	–	–	–	0.20	0.50
	–	–	–	–	0.18	0.44
Desi. cotton-chickpea	–	–	–	–	0.20	0.35
	–	–	–	–	0.14	0.51
	–	–	–	–	0.20	0.57

*After winter 2001–02, ** After summer 2002, *** After winter 2002–03.

A corroborative sample survey was undertaken on selected farms in the project area. The data showed that the benefit-cost ratio (BCR) was more favorable for wheat in winter and rice in summer season under the present price structure (Tables 16.7 and 16.8). Mean return was the highest for irrigation, followed by that for fertilizers for wheat, and for fertilizers, followed by irrigation for rice. Cotton competed well with rice for input use of irrigation and fertilizer in wheat and for fertilizers, followed by irrigation in rice. Cotton competed fairly well with rice with regards to use of irrigation and fertilizer.

Another economic indicator, i.e., ROVE, was found to be higher for the RWS in Nawanshahar district and for the CWS in Faridkot district compared with all other cropping systems. In Faridkot district, therefore, the adoption of CWS ensures more economical use of groundwater resource compared with the RWS. Thus, CWS has replaced RWS on large area in the major cotton growing districts of Faridkot, Muktsar, Bhatinda, Ferozepore, and Mansa during 2002–03 to 2004–05 (Table 16.8, annual Statistical Abstracts of Punjab, 2003–05). The area under cotton in the operational villages (Sadiq and Kothae Maur) increased from 80 ha in 2002–03 to 200 ha in 2004–05 in Sadiq village and from 20 ha to 144 ha during the same period at Kothae Maur. Similarly, the area under cotton in the Faridkot district increased from 16,000 ha in 2002 to 21,000 ha in 2004–05.

16.4 RESIDUE MANAGEMENT IN RICE-WHEAT CROPPING SYSTEM

With the large-scale adoption of RWS in Punjab, the management of rice straw and seedbed preparation for timely sowing of wheat consists of principal agronomic

Table 16.7 Comparative economics of crops in demonstration farms, 2003–04.

S. No.	*Economic indicator*	*Nawanshahar*			*Faridkot*		
1.	*Average returns per unit*	*Wheat*	*Rice*	*Maize*	*Wheat*	*Rice*	*Cotton*
a.	Human labor (h)	146.69	70.85	37.97	146.52	52.15	51.67
b.	NPK (kg)	155.70	195.82	115.54	146.19	149.24	214.98
c.	Pesticides (Rs.).	76.44	55.27	57.19	30.46	24.49	6.56
d.	Irrigation (ha cm)	309.60	70.8	170.57	255.52	73.29	313.18
e.	Land (ha)	30805	32060	14200	32195	32395	31005
2.	*B.C. ratio*	*2.30*	*1.78*	*1.33*	*2.51*	*1.63*	*1.72*

Table 16.8 Comparative economics of alternate cropping systems in sample farms (Rs. /ha).

District	*Rotation*	*GR*	*VE*	*ROVE*	*B-C ratio*
Nawashahr	Rice-wheat	62865	31365	31500	2.00
	Maize-wheat	45005	24093	20912	1.87
Faridkot	Rice-wheat	64590	32765	31822	1.97
	Cotton-wheat	63200	30858	32342	2.05

GR = gross returns; VE = variable expenses; ROVE = return over variable expenses; B-C ratio = benefit-cost ratio.

challenges that must be addressed. Because of its low nutritional value as fodder for livestock, rice straw is usually burned. Logistically, it is a convenient option in the combine-harvested rice fields. However, burning of rice straw has adverse impacts on soil quality, especially on SOC concentration and nutrient cycling. Rather than burning it, no-till sowing of wheat through the mulch of rice straw can be agronomically viable and economically profitable (Figure 16.3). Therefore, demonstrations on residue management were established at three locations in Langroya (Barbha, Sajawalpur, KVK farm) and at two locations (Sadiq and Kothae Maur) in the Faridkot.

A comparison of initial (Table 16.9) and final soil analyses (Table 16.10) shows management-induced changes in SOC concentration even across a short period. Residue management affected the SOC concentration after three crops at Nawanshahar (Table 16.10). The SOC concentration increased from a mean initial level of 0.44% to 0.50% when both rice and wheat straw were burned; to 0.51% when rice straw was incorporated but wheat straw was removed; and to 0.57% when both rice and wheat straw were incorporated. The increase in SOC concentration due to incorporation of both rice and wheat straw was statistically significant. Although a similar trend was observed at Faridkot, the differences in SOC concentration were statistically not significant.

The effects of these residue-management treatments on the yield of 2nd crop of wheat (3rd crop in the rotation) are given in Table 16.11. Wheat yield at Nawanshahar in plots where rice and wheat straw were burned was a little more (5.1 Mg/ha) than that where the straw was incorporated (4.4 to 4.6 Mg/ha). Even with an increase of about 13%, the difference in yield was not statistically significant. However, incorporation of both straws at Faridkot sites was more beneficial than burning. At Faridkot, the soils are light textured (sandy loam) as compared with the loamy soil at the Langroya site. Thus, yield response to residue management seemingly depends

Figure 16.3 Wheat sown under no tillage.

Table 16.9 Antecedent properties of soils for residue management in rice-wheat cropping-system experiments.

District	*Location*	*Texture*	*pH*	*E.C. (ds/m)*	*$CaCO_3$ (%)*	*SOC (%)*	*Avail. P (kg/ha)*	*Avail. K_2O (kg/ha)*
Nawashahr	Barbha	Loam	8.4	0.30	0.6	0.51	10.25	337
Langroya	KVK Farm	Loam	8.3	0.20	0.5	0.33	6.75	202
	Sajawalpur	Loam	8.4	0.20	0.3	0.48	6.75	210
Faridkot	Kothae Maur	Sandy Loam	9.0	0.20	0.7	0.45	12.5	555
	Sadiq	Sandy Loam	9.4	0.20	1.2	0.33	12.5	450

Table 16.10 Soil Organic Carbon as affected by residue management (Winter 2002–03).

	SOC concentration (%)					
	Nawashahr				*Faridkot*	
Treatment	*Barbha*	*Sajawalpur*	*KVK farm*	*Mean*	*Sadiq*	*Kothae Maur*
Initial	0.51	0.48	0.33	0.44^{b}	0.33	0.45
Burning of both rice	0.53*	0.47	0.35		0.32	0.45
and wheat straw	0.49**	0.47	0.30		0.28	0.45
	0.60***	0.51	0.40	0.50^{ab}	0.27	0.54
Incorporation of rice straw	0.54*	0.54	0.30		–	–
in wheat and wheat	0.54**	0.54	0.38		–	–
straw removed	0.58***	0.51	0.44	0.51^{ab}	–	–
Incorporation of rice	0.52*	0.51	0.39		0.41	0.56
and wheat straw + GM	0.47**	0.51	0.42		0.36	0.55
	0.61***	0.57	0.53	0.57^{a}	0.42	0.57

*After winter 2001–02, **After monsoon 2002, ***After winter 2002–03, GM = green manure.
Means within column followed by the same letter are not significantly different (L.S.D. at P 10%).

Table 16.11 Grain yield of wheat (Winter 2002–03) at different sites in residue-management experiments.

	Grain yield (kg/ha)				
	Nawanshahar			*Faridkot*	
Treatment	*Barbha*	*Sajawalpur*	*KVK farm*	*Sadiq*	*Kothae Maur*
Burning of both rice and wheat straw	5780	5243	4573	4787	5350
Rice straw incorporation and wheat straw removed	5723	4328	3923	–	–
Incorporation of rice and wheat straw + GM	5165	4348	3790	5017	5512*

*Without green manure.

Table 16.12 Effect of residue management on the yields of rice and wheat (2003–04).

	Nawashahr (Mg/ha)				*Faridkot (Mg/ha)*			
	Rice		*Wheat*		*Rice*		*Wheat*	
Treatment	*Range*	*Mean*	*Range*	*Mean*	*Range*	*Mean*	*Range*	*Mean*
Burning of rice and wheat straw	5.0–7.4	6.5	4.3–6.1	5.0	4.4–7.2	5.9	3.9–5.6	4.9
Rice incorporated & wheat straw removed	5.4–7.5	6.5	3.9–6.0	4.9	4.8–7.1	6.0	–	–
Rice & wheat straw + GM incorporated	5.9–8.2	6.8	3.8–6.0	4.8	4.8–7.6	6.3	4.0	5.0

GM = green manure.

Table 16.13 Comparative economics (per ha; 2001–04) of residue incorporation in rice-wheat system.

	Gross returns (Rs)		*Variable expenses (Rs)*		*Return over variable expenses (Rs)*		*B-C ratio*	
Treatment	*NS*	*FK*	*NS*	*FK*	*NS*	*FK*	*NS*	*FK*
GM + both rice & wheat straw incorporated	68240	62912	32194	35927	36046	27615	2.12	1.78
Rice straw incorporated, wheat straw removed	66653	63181	29000	33100	37653	30081	2.30	1.91
Rice & wheat straw burned	67754	63779	25637	33991	42117	30788	2.64	1.93

NS = Nawashahr; FK = Faridkot.
GM = green manure.

on soil texture and certain other properties (e.g., moisture retention). Experiments conducted in farmers' fields in both districts also indicated similar trends in crop yields (Table 16.12). The data were also analyzed to assess comparative economics of residue incorporation in the RWS across three years (Table 16.13). Because the GR (Rs/ha) were almost the same in straw-burned and straw-incorporated treatments, the improvements in soil quality that the straw incorporation brings about might, in the long run, offset the apparent short-term advantage of burning strawrun. This is an important management issue brought about by the changing land-use scenarios in all agro-eco-regions of the IGP (Abrol and Gupta, 1998; Velayutham et al. (1999).

16.5 CROP ROTATION AND SOIL ORGANIC CARBON SEQUESTRATION

The positive effects of incorporation of straw in Faridkot sites are also corroborated by the attendant improvements in soil physical properties (Table 16.14). There

Table 16.14 Effect of residue incorporation on Soil Organic Carbon (SOC) (%), SOC pool (Mg/ha) bulk density (Mg/m^3) and infiltration rate (mm/h) – 2003–04.

	Nawashahr				*Faridkot*			
Treatment	*SOC*	*BD*	*SOC pool*	*IR*	*SOC*	*BD*	*SOC pool*	*IR*
Burning of rice and wheat straw	0.52	1.53	11.9	3.5	0.45	1.65	11.1	6.1
Rice straw incorporated & wheat straw removed	0.53	1.50	11.9	3.5	0.42	1.60	10.1	6.7
Rice & wheat straw + GM incorporated	0.57	1.42	12.1	3.1	0.51	1.58	12.1	7.6

SOC = soil organic carbon.
BD = bulk density.
IR = infiltration rate.
SOC pool (0–15 cm depth).
GM = green manure.

was a slight increase in SOC concentration in straw-incorporated plots. With straw incorporation, the SOC pool increased by 0.2 Mg/ha in 0–15 cm depth for the Nawanshahar site. In comparison, increase in SOC pool by straw incorporation was 1.0 Mg/ha for the Faridkot site (Table 16.14). Soil bulk density decreased, and there was a perceptible increase in water infiltration rate in soils. These results demonstrate the positive effects of straw incorporation in light-textured soils on improving the soil quality (Lal, 1995).

Swarup, Manna, and Singh (2000) estimated that there was 59% depletion of SOC concentration of cultivated land compared with that in undisturbed soils in the IGP during the 1960–2000 period. Bhattacharyya and colleagues (2007) determined the modeled SOC stocks and changes in the IGP, India, from 1990 to 2030, using output of the GEFSOC modeling system. The predicted mean SOC stock change rate in agro-ecological sub-region 4.1, which includes Faridkot district, for rice-wheat-berseem (*Trifolium alexandrinum*) rotation was 0.076, 0.025, and 0.157 Tg per year in 1990, 2000, and 2003, respectively. In contrast, for the RWS, the corresponding values were –0.740, –0.636, and 0.260, respectively. This trend shows the importance of reviving berseem in the rice-wheat rotation to enhance SOC accumulation in the soil. Lal (2004) has discussed the potential of SOC sequestration in soils of India.

16.6 INTERCROPPING IN SUGARCANE AND POPLAR

The feasibility of a diversified production system, such as an intercropping system, was demonstrated in the long-duration autumn-planted sugarcane crop (*Saccharum officinarum; sinenses; barberi, etc.*). Demonstrations of intercropping a short-duration crop of brassica in sugarcane (Figure 16.4) were conducted at three sites at Nawanshahar and at two sites at Faridkot. The yield of brassica (GSL-1) at Nawanshahar ranged from 1,320 to 1,380 kg/ha, and yet intercropping did not reduce the yield of sugarcane (Table 16.15). Similar results of no reduction in sugarcane yield by intercropping were also obtained at the Faridkot sites.

Figure 16.4 Intercropping of rapeseed in sugarcane.

Table 16.15 Sugarcane yield and grain yield of brassica and green pod yield of peas as intercrop in sugarcane.

		Cane yield (Mg/ha)/Grain yield (kg/ha)				
		Nawanshahar			*Faridkot*	
Crop rotation	*Crop*	*Barbha*	*Sauna*	*KVK farm*	*Kothae Maur*	*Hariewala*
Sugarcane + Brassica	Sugarcane	60.0	60.5	70.8	79.6	68.4
	Brassica	1375	1330	1380	1300	1075
Sugarcane + Peas	Sugarcane	550	596	–	–	–
	Peas	360	150	–	–	–
Sugarcane (control)	Sugarcane (sole crop)	51.0	65.1	69.7	81.1	71.6

16.7 DRIP IRRIGATION (FERTIGATION)

In view of the rapidly receding WT, and the need for popularizing efficient on-farm water management, the project staff assisted Mr. Kuldeep Singh, a progressive farmer at Kotkapura in Faridkot district, in installing and demonstrating a drip-irrigation system and in using fertigation for growing vegetables all through the year on his farm (Figure 16.5). The soil of the farm was sandy loam (pH = 8.6, E.C. = 0.2 ds/m, $CaCO_3$ = 1.7% and SOC concentration = 0.39%). The available P status was low to medium (15.3 kg P/ha), and that of available K was high (566 kg K/ha). Irrigation with nutrients (fertigation) was used at the drip discharge rate of 2.0 l/hr for only a few min/day. The yield of tomato (*Lycopersicon esculentum* L.) was 52.5 Mg/ha and that of bittergourd (*Momordica charantia* L.) was 8.75 Mg/ha. The quality of produce in both crops was better and fetched higher prices in the market. This farm is now a focal point for dissemination of the fertigation technology to the farming community in the region (MSSRF, 2006).

Figure 16.5 Drip irrigation in vegetables.

Figure 16.6 Turmeric and ginger as intercrops in poplar.

16.8 DIVERSIFICATION OF AGRICULTURE

16.8.1 Agroforestry

Only 5.6% of the land area in Punjab is under forests, compared with >84% under cultivation. It is thus logical to popularize agroforestry systems. Poplar (*Populus deltoids*) is among the fast-growing winter trees, which can grow well in association with field crops (Figure 16.6). Being a deciduous tree, a large amount of organic matter and nutrients can be recycled through leaf litter added to the soil.

Two demonstration sites were established to assess the feasibility of intercropping with poplar. Maize as a monsoon crop and wheat and mung bean (*Vigna radiata* L.) as winter crops were intercropped with poplar. There was no deleterious shade effect of poplar trees on the yield of any of these crops (Table 16.16). The leaf samples of poplar trees were collected during the autumn and analyzed for nutrient contents. Nitrogen

Table 16.16 Yield of intercrops in poplar (*Populus deltoids*).

	Grain yield (kg/ha)	
	Baghauran, Langroya	*Khara, Faridkot*
Crop rotation	*Wheat/Maize*	*Wheat/Moong*
Poplar + Wheat/Maize	4463/1150	–
Poplar + Wheat/Moong	–	4625/217

Figure 16.7 Ponded agricultural field for fish culture.

concentration ranged from 1.7% to 1.9%, P from 0.08% to 0.09%, K from 0.26% to 0.28%, S from 1.2% to 1.4%, Ca from 2% to 2.5%, and Zn from 4.2 to 4.7 ppm. Thus, a large amount of nutrients is recycled. In addition, poplar generates another income stream from the same unit area under the agroforestry system. The agroforestry system, when practiced as a cyclic agroforestry intervention, also leads to SOC sequestration in these calcareous, sodic soils of the IGP (Gupta and Rao, 1994).

16.9 FISH CULTURE DEMONSTRATION

In an attempt to diversify agriculture in Punjab, a poorly drained agricultural field with impounded water was used for aquaculture (Figure 16.7). On-farm demonstrations on the farm of Mr. Gursharan Singh were established on a 0.4 ha (1 acre) pond at Nathawala Nawan in Faridkot district. The fish feed consisted of: farmyard manure (0.48 Mg), poultry manure (0.24 Mg), single superphosphate (15 kg), deoiled rice bran (8.8 kg), deoiled mustard cake (*Brassica juncea*; 8.8 kg), soya meal (6 kg), mineral mixture (240 g), and salt (80 g). Four species of fish were grown in the pond.

Table 16.17 Fish culture demonstration.

Farmer: Mr. Gurcharan Singh	*Village: Nathawala Nawan*	*District: Faridkot*		
		Yield (Mg/ha)		
Species	*Weight at maturity*	*2002*	*2003*	*2004*
Catla catla (Catla)	1.0 kg	2.34	2.75	2.49
Labeo rohita (Rohu)	600–800 g			
Cirrhina mrigala (Mirgal)	700 g			
Cyrpinus carpio (Common carp)	600–800 g			

Fish production (Table 16.17) proved successful in three years and was an economically viable intervention in regions where water is available in adequate quantities. This is an important diversification to the traditional cropping system.

16.10 CONCLUSIONS

Soil fertility and on-farm water management are crucial to the sustainability of intensive agricultural production systems of Punjab. On-farm demonstrations indicated several viable alternatives to the traditional rice-wheat system. Substituting rice with maize/sugarcane in some areas at Nawanshahar proved to be a profitable cropping system. Thus, there occurred a decline in the area under rice by about 10% in Nawanshahar district between 2000 and 2003, and an increase in area under maize by about 13%.

In regions where water table is rising, as in the canal-irrigated areas in Faridkot district, replacement of rice with cotton is a viable option. The area under cotton in the project villages increased from 80 ha in 2002–03 to 200 ha in 2004–05 in Sadiq village, and from 20 ha to 144 ha at Kothae Maur. The area under cotton increased from 16,000 ha in 2002–03 to 30,000 ha in 2004–05 in the project district, Faridkot.

With forest cover of only 5%–6% of the total land area, agroforestry interventions are an integral part of a long-term land-cover and land-use-management strategy. Incorporating poplar in the rotation with intercrops is a suitable agroforestry system.

ACKNOWLEDGEMENTS

This project was implemented with a generous grant provided by the Sir Dorabji Tata Trust and Tata Education Trust of the Tata House, Mumbai. We are grateful to them for providing the financial and monitoring support and to Dr. M.S. Swaminathan for his guidance during the execution of the project.

REFERENCES

Abrol, I.P., and R.K. Gupta. 1998. Indo-Gangetic Plains–Issues of changing land use. *LUCC Newsletter,* March 1998 (3):8–9.

Bhattacharyya, T., D.K. Pal, M. Easter, N.H. Batjes, E. Milne, K.S. Gajbhiye, and P. Chandran et al. 2007. Modelled soil organic carbon stocks and changes in the Indo-Gangetic Plains, India, from 1980 to 2030. *Agri. Eocsyst. Environ.* 122:84–94.

Department of Agriculture. 2003–2009. Agriculture at a Glance. Economic Advisor to Punjab Government, Statistical Abstracts. Chandigarh, Punjab, India.

Goswami, N.N., D.K. Pal, G. Narayanasamy, and T. Bhathacharyya. 2000. Soil organic matter: Management issues. In *Proc. Intl. conference on management of natural resources for sustainable agriculture towards 21st century,* eds. J.S.P. Yadav and G. Narayanasamy, 87–96. New Delhi: ICAR.

Grossman, R.B., and T.G. Reinsch. 2002. Bulk density and linear extensibility. In *Methods of soil analysis, part 4,* eds. J.H. Dane and G.C. Topp, 201–225. Madison WI: Soil Science Society of America.

Gupta, R.K., and D.L.N. Rao. 1994. Potential of wastelands for sequestering carbon by reforestation. *Current Sci.* 66:373–380.

Singh, J., and M. Hossain. 2003. Component analysis of total factor productivity in high potential rice-wheat systems in the Indian Punjab. In *Proc. Intl. conference on impact of agricultural research and development.* Why has this impact assessment research not made more of a difference? ed. D.J. Watson. Mexico City: CIMMYT.

Klute, A. (Ed) 2006. Methods of Soil Analysis. Part 1. Physical and Mineralogical Methods. Soil Sci. Soc. Am. Monograph #a, Madison ,WI, 1188pp.

Lal, R. 1995. The role of residue management in sustainable agricultural systems. *J. Sustainable. Agric.* 5:71–78.

Lal, R. 2004. Soil carbon sequestration in India. *Climatic Change.* 65:277–296.

MSSRF. 2006. Bulletin on sustainable management of natural resources for food security and environmental quality. Project achievements 2001–05. M.S. Swaminathan Research Foundation, Chennai, Tamil Nadu, India.

Nelson, D.W., and L.E. Sommers. 1996. Total carbon, organic carbon, and organic matter. In *Methods of soil analysis: Chemical methods, part 3,* ed. D.L. Sparks, 961–1010. Madison, WI: Soil Sci. Soc. of America.

Reynolds, W.D., D.E. Ehrick, and E.G. Youngs. 2002. Single-ring and double-concentric-ring infiltrometers. In *methods of soil analysis, part 4,* eds. J.H. Dane and G.E. Topp, 821–826. Madison, WI: Soil Sci. Soc. of America.

Sidhu, G.S., C.S. Walia, L. Tarsem, K.P.C. Rena, and J. Sehgal. 1995. Soils of Punjab for optimizing land use. National Bureau of Soil Survey and Land Use Planning, Pub. No. 45. Nagpur, India: ICAR.

Sinha, S.K., G.B. Singh, and M. Rai. 1998. Decline in crop productivity in Haryana and Punjab: Myth or reality? New Delhi, India: Indian Council of Agricultural Research.

Sparks, D.L. (Ed) 1996. Methods of Soil Analysis: chemical Methods, Part 3. Soil Sci. Soc. Am., Madison, WI, 1390pp.

Swarup, A., M.C. Manna, and G.R. Singh. 2000. Impact of land use and management practices on organic carbon dynamics in soils of India. In *Global climate change and tropical ecosystems,* eds. R. Lal, J.M. Kimble, and B.A. Stewart, 261–281. Boca Raton, FL: CRC/ Lewis Publishers.

Velayutham, M. 1994. Sustainable productivity under rice-wheat cropping system – issues and imperatives for research. *Bull. Indian Soc. Soil Sci.* 18:1–6.

Velayutham, M., D.K. Mandal, M. Champa, and J. Sehgal. 1999. Agro-ecological sub-regions of India for planning and development. National Bureau of Soil Survey and Land Use Planning Pub. No. 25, Nagpur, India: ICAR.

Velayutham, M., G.S. Sidhu, S.P. Singh, D. Sarker, A.K. Holder, and T. Bhattacharyya. 2000. Soil series of the Indo-Gangetic Plains. National Bureau of Soil Survey and Land Use Planning, Nagpur, MP, India.

Chapter 17

Weed management in aerobic rice in northwestern Indo-Gangetic plains

G. Mahajan
Punjab Agricultural University, Ludhiana, India

B.S. Chauhan & D.E. Johnson
International Rice Research Institute (IRRI), Los Baños, Philippines

ABSTRACT

Aerobic rice systems, wherein the crop is established via direct seeding in non-puddled, non-flooded fields, are among the most promising approaches for saving water and labor. However, aerobic systems are subject to much higher weed pressure than conventionally puddled transplanted rice (CPTR). Experiments were conducted for two years to develop effective and economical methods for managing weeds in aerobic rice grown by direct seeding rather than by conventional transplanting method. The proportion of mean grass-weed dry matter was 28.3% more in aerobic direct-seeded rice (ADSR) as compared to CPTR. Both weed density and dry weight were negatively correlated with rice grain yield. ADSR treatment produced yield similar to CPTR treatment when weeds were controlled effectively. Post-emergence application of bispyribac Na 25 g/ha and penoxsulam 25 g/ha could effectively control all the weeds in ADSR. Irrigation water productivity remained statistically the same in both ADSR and CPTR under the weed-free situation or when bispyribac Na herbicide was applied as post-emergence because of effective weed control in ADSR. The variation in net profitability between the ADSR and CPTR decreased with herbicide treatments, viz., bispyribac Na, followed by penoxsulam and sequential application of pretilachlor and metsulfuron.

Keywords: Bispyribac Na, penoxsulam, aerobic rice, *Oryza sativa* L., establishment methods, economic weed management

Address correspondence to G. Mahajan, Punjab Agricultural University, Ludhiana, Punjab 141 004, India. E-mail: mahajangulshan@rediffmail.com

Reprinted from: *Journal of Crop Improvement*, Vol. 23, Issue 4 (October 2009), pp. 366 – 382, DOI: 10.1080/15427520902970458

17.1 INTRODUCTION

The area and productivity of rice (*Oryza sativa* L.) in the trans–Indo-Gangetic Plains (IGP) increased dramatically between the 1960s and 1990s. Even being an agro-ecologically unsuitable crop for this region, the crop flourished in this region because of improved varieties, assured irrigation, stability, matching machinery, assured market and socio-political system (flat/free electricity). However, during the past decade, yields have stagnated and because of higher input expenses and as a result, the profitability of farmers decreased. Further, poor-quality irrigation systems and greater reliance on groundwater have led to a water table decline of 0.1 to 1.0 m yr^{-1} in parts of the IGP, resulting in a scarcity of and higher cost of pumping water (Harrington et al., 1993; Gill, 1994; Sharma et al., 1994; Sondhi et al., 1994). In Punjab, where irrigated rice cropping has been practiced on alluvial coarse-textured soils since the early 1970s, there has been an alarming scarcity of groundwater resources. In central Punjab, where groundwater is good, the areas with water table below 10 m depth increased from 3% in 1973 to 76% in 2002 (Hira et al., 2004), thereby threatening the sustainability of rice culture. The sustainability of rice of the trans-IGP and the ability to increase production in pace with population growth are major concerns as the contribution from this region to the central pool is the largest and plays a vital role in the food security of India.

Among various methods of rice planting, transplanting of young seedlings in puddled field is the most popular method of rice cultivation in this region. More recently, this method of rice cultivation has been criticized because of high requirements of water and labor. Transplanting in puddled soils (wet tillage) with continuous flooding is the most common method of rice crop establishment. Transplanted rice requires a large amount of water and labor. During peak periods of transplanting, labor also becomes very scarce. Puddling also affects soil health because of the dispersion of soil particles and soil becoming compact, which makes tillage operations difficult and increases requirement of energy in succeeding crops, such as wheat (Singh et al., 2002). Fujisaka, Harrington, and Hobbs (1994), on the basis of a diagnostic survey conducted in several rice-wheat areas in South Asia, observed low wheat yields in a rice-wheat system, mainly because of deterioration of soil structure and the development of subsurface hardpans. Hobbs and colleagues (2002) described the emerging issues of sustainability of rice-wheat systems and stressed the need to improve water-use efficiency, soil structure, and weed management against the backdrop of increasing scarcity of labor and water. An alternative to puddling and transplanting of rice could be aerobic direct seeding because it requires less water, labor, and capital input. The direct-seeded crop also matures 7 to 10 days earlier than the transplanted crop; thus, allowing timely planting of the succeeding wheat crop (Giri, 1998; Singh et al., 2006). Irrigated "aerobic rice" is a new system being developed for lowland areas with water shortage and for favorable upland areas with access to supplementary irrigation (Tuong & Bouman, 2003; Belder et al., 2005). Aerobic rice systems, wherein the crop is established via direct seeding in non-puddled, non-flooded fields, are among the most promising approaches for saving water (Wang et al., 2002; Tuong & Bouman, 2003; Bhushan et al., 2007). Aerobic rice systems can reduce water application by 44% relative to conventionally transplanted systems by reducing percolation, seepage, and evaporative losses, while maintaining yield at an acceptable level (6 Mg ha^{-1}) (Wang et al., 2002; Bouman et al., 2005). However, aerobic systems are subject to much higher weed pressure than conventional puddled transplanting systems (Rao et al., 2007; Balasubramanian & Hill, 2002) in which weeds are suppressed by standing water and by transplanted rice seedlings, which have a size advantage over germinating weed seedlings (Moody, 1983). On the other hand, aerobic soil dry-tillage and alternate wetting and drying conditions are conducive to the germination and growth of weeds, causing grain yield losses of 50%–91% (Elliot et al., 1984; Fujisaka, Harrington & Hobbs, 1994; Rao et al., 2007). Thus, weeds are one of the severest constraints to aerobic rice production, and timely weed management is crucial to increasing the productivity of aerobic rice (Rao et al., 2007).

Singh and colleagues (2002) reported good success with dry-seeded rice production technology in large-scale farmer participatory trials in the Terai area of Uttaranchal, India, when the stale seedbed technique was combined with the application of pre-emergence herbicide, pendimethalin, within two days of seeding (DOS). Thus, timely weed control is crucial to increasing rice productivity. Herbicides are considered an alternative/supplement to hand weeding. The development of new, improved herbicides for dry-seeded rice is also needed (Gupta et al., 2003; Singh et al., 2006). Several pre-emergence herbicides, including butachlor, thiobencarb, pendimethalin, oxadiazon, oxyfluorfen, and nitrofen alone or supplemented with hand weeding, have been reported to provide a fair degree of weed control (Estorninos & Moody, 1988; Castin & Moody, 1985; Janiya & Moody, 1988; Moorthy & Manna, 1993; Pellerin & Webster, 2004). Some difficulties are, however, associated with pre-emergence herbicides, such as their limited application duration (0–5 DOS) and requirement of adequate soil moisture at the time of their application. In such situations, post-emergence herbicides are superior. Hence, it is necessary to evaluate different pre- and post-emergence herbicides that are formulated from time to time to provide wider options to farmers for weed control in rice.

17.2 MATERIALS AND METHODS

17.2.1 Experimental site

The field experiment was conducted for 2 years (2006 and 2007) at the experimental farm of the Rice Section, Department of Plant Breeding and Genetics, Punjab Agricultural University, Ludhiana, India (30°56/N and 75°52/E), situated at an elevation of 247 m above mean sea level. The climate of Ludhiana is broadly classified as semiarid subtropical, characterized by very hot summers and cold winters. The hottest months are May and June, when the maximum temperature reaches 45°–46°C, whereas during December and January, the coldest months of the year, the minimum temperature often goes below 3°C. The average annual rainfall is 750 mm, 75%–80% of which is received through the northwest monsoon during July–September. The experimental soil was loamy sand in texture. The surface soil (0–15 cm) had a total organic nitrogen content of 220 mg kg^{-1} and phosphorus 8.1 mg kg^{-1}; with a pH of 8.1 and electrical conductivity (EC) of 0.19 ds m^{-1}.

17.2.2 Experimental design and treatments

The experiment was arranged in a factorial randomized-block design with three replications. The two treatments (see Table 17.1) assigned to the main plot were: (a) aerobic direct-seeded rice (ADSR) on flat land, and (b) conventionally puddled transplanted rice (CPTR). The rice was direct seeded on flat land at 50 kg ha^{-1} at 20 cm row spacing using *kera* method. For puddled transplanted crop, the nursery was sown on the same day as seeding of aerobic direct-seeded crop was done and transplanted in the puddled field 25 days after sowing of direct seeded crop. The size of subplots was 2 m × 4 m. Six weed-control treatments (Table 17.1) were assigned to the subplots: (a) weedy; (b) weed-free; (c) pendimethalin (0.75 kg/ha 2 DOS), followed by one hand-weeding (1 HW) at 30 DOS/DOT (days of sowing/days of transplanting); (d) pretilachlor at 0.4 kg/ha, followed by metsulfuron (15 g/ha 30 DOS/DOT); (e) bispyribac Na at 25 g/ha; (f) penoxsulam at 25 g/ha at 12 DOS/DOT: and (g) penoxsulam at 25 g/ha at 3 DOS/DOT during 2006. Due to poor efficacy of penoxsulam at 25 g/ha at 3 DOS/DOT, it was replaced with bispyribac Na at 25 g/ha at 20 DOS/DOT during 2007. Bisyribac Na is a newly introduced herbicide and was not available during 2006. Herbicides were applied using a knapsack sprayer with a flat fan nozzle and water as a carrier at 300 l ha^{-1}. For the weed-free subplot treatment, six hand-weedings were done to maintain fields weed-free. In the weedy control, no weeding was done.

17.2.3 Experimental details and measurements

Rice (cv. Superbasmati, a medium-long-duration basmati or scented variety) was planted in June and harvested in November in each year. For ADSR, land was prepared with two plowings with a disc harrow and one planking. In ADSR treatment, seed at 50 kg ha^{-1} was direct-seeded in rows with 20 cm spacing between rows. Irrigation was applied after seeding. Soil moisture was maintained up to field capacity and later irrigation was given at 4 to 5 day intervals. The land preparation for the CPTR main-plot treatment consisted of one dry plowing, followed by irrigation and two harrowings to puddle the soil under wet conditions. Two rice seedlings per hill were transplanted at 20 cm × 15 cm spacing after puddling. Nitrogen was applied at 120 kg ha^{-1} in three equal splits as basal, at 20 DOT/45 DOS, and at 40 DOT/65 DOS. Weed density and weed dry weight were measured at 35, 75 DOS/DOT, and at harvest. Weed density was recorded with the help of a quadrate (0.5 m × 0.5 m) placed randomly at two spots in each plot. Weeds were cut at ground level, washed with tap water, sun dried, oven dried at 70°C for 48 h, and then weighed. The data on actual number of weeds were subjected to square-root transformation for statistical analyses. Grain yield was taken from the net plot of 1.2 × 3 = 3.6 m^2 area and expressed in g ha^{-1} at 14% moisture. The statistical analysis of the data was done using CPCS-1 software. Unless indicated otherwise, differences were considered significant at $p = 0.05$.

17.3 RESULTS AND DISCUSSION

17.3.1 Yield response and weeds

The major weeds associated with rice include grasses (*Digitaria sanguinalis, Echinochloa colona, Echinochloa crus-galli*), sedges (*Cyperus iria* and *Cyperus rotundus*), and broadleaf weeds (*Sphenoclea spp., Euphorbia hirta, Eclipta prostrata, Trianthema*

Table 17.1 Impact of varying crop establishment methods and weed control treatments on total weed dry matter (gm^{-2}).

Weed control methods	*ADSR 35 DOS*	*CPTR 35 DOT*	*Mean*	*ADSR 75 DOS*	*CPTR 75 DOT*	*Mean*	*ADSR At harvest*	*CPTR At harvest*	*Mean*
2006									
Pendi + 1 HW	19.89	10.21	15.04	245.30	56.99	*151.15*	262.77	38.54	*150.66*
Pretila + Metsulfu	17.44	6.32	11.88	120.44	22.03	*71.24*	145.11	24.45	*84.78*
Penoxsulam 12 DOT	20.80	13.22	17.01	159.64	61.69	*110.67*	136.04	45.39	*90.72*
Penoxsulam 3 DOT	26.89	13.85	20.37	214.69	70.39	*142.54*	200.88	58.11	*129.49*
Weedy	48.55	30.12	39.33	448.10	226.62	337.36	435.61	226.53	*331.06*
Mean	*26.71*	*14.74*		*237.63*	*87.54*		*236.08*	*78.60*	
LSD (0.05)	*CE: 3.46; WC: 5.46; CExWC: NS*			*CE: 3.46; WC: 5.46; CExWC: NS*			*CE: 26.53; WC: 41.95 ; CExWC: 59.32*		
2007									
Pendi + 1 HW	19.42	28.13	*23.77*	326.48	80.34	*203.40*	319.42	82.15	*200.78*
Pretila + Metsulfu	40.94	6.81	*23.88*	185.10	21.97	*103.54*	179.42	20.83	*100.12*
Penoxsulam 12 DOT	36.76	6.74	*21.75*	200.32	45.77	*123.05*	183.37	43.48	*113.42*
Bispyribac Na	28.03	9.50	*18.77*	91.43	39.54	*65.48*	117.47	38.46	*77.96*
Weedy	142.13	40.46	*91.30*	596.02	292.50	*444.26*	563.46	271.78	*417.62*
Mean	*53.46*	*18.33*		*279.87*	*96.02*		*272.6*	*91.34*	
LSD (0.05)	*CE: 10.80; WC: 17.07; CExWC: 24.14*			*CE: 20.26; WC: 32.03; CExWC: 45.30*			*CE: 21.80; WC: 34.46; CExWC: 48.74*		

Pendi = pendimethalin; Pretila = pretilachlor; Metsulfu = metsulfuron; ADSR = aerobic direct seeded rice; CPTR = conventional puddle transplanted rice; CE = crop establishment methods; WC = weed control treatments; DOS = days of sowing; DOT = days of transplanting.

portulacastrum, Ammania spp. and *Ludwigia spp.).* The weeds *Euphorbia hirta, Eclipta prostrata, Digitaria sanguinalis,* and *Trianthema portulacastrum* were more prominent in aerobic direct-seeded crop.

The weed density and weed dry matter were higher in 2007 than in 2006. The total weed dry weight (Table 17.1) and total weed density (Table 17.2) were lower for CPTR at all stages of crop growth in both years. This was attributable to the head-start advantage of the CPTR crop, which offered competition in favor of the crop and ultimately helped in smothering the weed flora. The aerobic direct-seeded crop faced severe weed pressure, particularly at the early stage and as a result, crop-weed competition was more in ADSR than in CPTR. These results are in conformity with Singh and colleagues (2008) and Moody (1996). During both the years, the proportion of grass dry weight was higher (52% of total weed dry weight in 2006 and 61% in 2007) than for other weeds in both the systems of rice establishment. The proportion of mean grass weed dry matter (Figure 17.1) was more in ADSR (63.5% of total weed biomass) than in CPTR (49.5% of total weed biomass).

In both years, crop-establishment methods and weed-control treatments significantly influenced dry matter accumulation by weeds (Table 17.1). The application of pendimethalin + 1 HW proved inferior among all the weed-control treatments; however, this combination caused a reduction in dry matter accumulation by weeds to the extent of 54.5% in 2006 and 51.9% in 2007. The application of penoxsulam at 25 g/ha at 12 DOT proved superior against weeds in both the years and caused a reduction in dry matter accumulation by the weeds to the extent of 72.5% in 2006 and 72.8% in 2007. The application of penoxsulam provided better control of weeds when applied at 12 DOS/DOT than at 3 DOS/DOT. Most of the weeds emerged after the application of penoxsulam at 3 DOS/DOT. In terms of penoxsulam, which is a foliage-translocated herbicide, some weeds escaped when it was applied at an early stage. Penoxsulam kill the weeds by inhibiting the acetolactase synthase enzyme, which ultimately reduce the transport of photosynthate from source leaves to roots, resulting in root growth inhibition (Devine, 1989; Devine, Bestman & Vanden Born, 1990; Shaner, 1991). Because most of the weeds emerged late, they escaped the mechanism of transport of photosynthate from source leaves to roots. During 2007, the new herbicide bispyribac Na provided an excellent control of weeds and caused a reduction in dry matter of weeds to the tune of 81.3%, 61.7%, 22.1%, and 31.2% over the weedy check, pendimethalin + 1 HW, pretilachlor + metsulfuron, and penoxsulam, respectively. Bispyribac Na is a selective herbicide effective for the control of grasses, sedges, and broadleaf weeds in rice and is effective as a soil or foliar treatment (Schmidt et al., 1999). It is a member of the pyrimidinyloxybenzoic chemical family (Darren & Stephen, 2006) and inhibits the enzyme acetohydroxy acid synthase, also known as acetolactate synthase (ALS), in susceptible plants. The interaction effect of crop-establishment method and weed-control treatment on dry matter accumulation by weeds was significant in both years of the study. The results (Table 17.1) revealed that in both years, a sequential application of pretilachlor + metsulfuron and penoxsulam at 12 DOS/DOT caused a significant reduction in dry matter of weeds when compared with the weedy check in CPTR; however, all the herbicide treatments provided significantly superior weed control in CPTR compared with ADSR. Interestingly, bispyribac Na in ADSR was statistically no different in dry matter production of weeds when compared with pendimethalin + 1 HW treatment in the CPTR.

Table 17.2 Impact of varying crop establishment methods and weed control treatments on total weed density (No. m^{-2}).

Weed control methods	*ADSR 35 DOS*	*CPTR 35 DOT*	*Mean*	*ADSR 75 DOS*	*CPTR 75 DOT*	*Mean*	*ADSR At harvest*	*CPTR At harvest*	*Mean*
2006									
Pendi + 1 HW	296	110	*203*	261	54	*157*	288	57	*172*
Pretila + Metsulfu	173	65	*119*	149	27	*88*	178	26	*102*
Penoxsulam 12 DOT	221	118	*169*	173	61	*117*	162	39	*100*
Penoxsulam 3 DOT	256	113	*184*	205	68	*136*	207	73	*140*
Weedy	489	289	*389*	442	224	*333*	427	210	*318*
Mean	*287*	*139*		*246*	*87*		*252*	*81*	
LSD (0.05)	*CE: 34.0; WC: 53.7; CExWC: NS*			*CE: 30.0; WC: 47; CExWC: NS*			*CE: 15.2; WC: 24.0 ; CExWC: 34.0*		
2007									
Pendi + 1 HW	221	81	*151*	363	76	*219*	356	75	*215*
Pretila + Metsulfu	279	36	*157*	259	36	*147*	260	41	*151*
Penoxsulam 12 DOT	268	61	*164*	267	45	*156*	254	45	*150*
Bispyribac Na	162	45	*103*	107	40	*73*	102	35	*68*
Weedy	631	336	*484*	589	261	*425*	571	249	*410*
Mean	*312*	*112*		*317*	*91*		*309*		
LSD (0.05)	*CE: 32.09; WC: 50.75; CExWC: 71.77*			*CE: 29.55; WC: 46.74; CExWC: 66.1*			*CE: 28.9; WC: 45.7; CExWC: 64.6*		

Pendi = pendimethalin; Pretila = pretilachlor; Metsulfu = metsulfuron; ADSR = aerobic direct seeded rice; CPTR = conventional puddle transplanted rice; CE = crop establishment methods; WC = weed control treatments; DOS = days of sowing; DOT = days of transplanting.

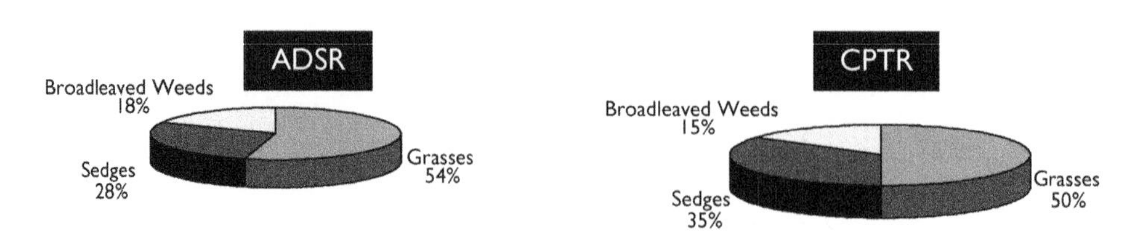

Proportion of weed species in 2006

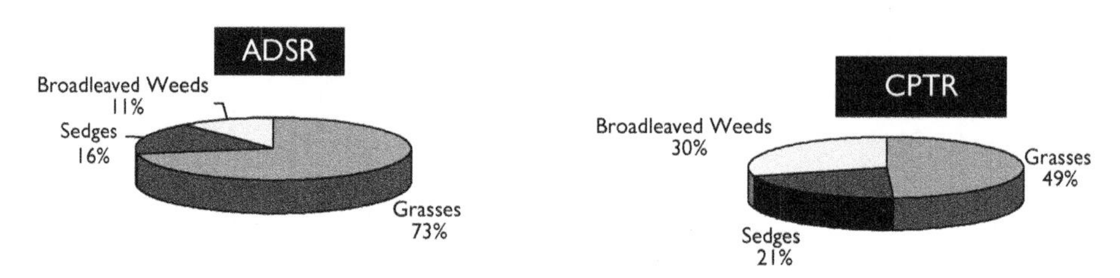

Proportion of weed species in 2007

Figure 17.1 Proportion of weed species during different years.
ADSR = aerobic direct seeded rice; CPTR = conventional puddle transplanted rice.

During 2006, the lowest weed density was recorded for penoxsulam at 25 g/ha applied at 12 DOS/DOT, followed by a sequential application of pretilachlor and metsulfuron (Table 17.2). However, an application of bispyribac Na registered lowest weed density among all the treatments during 2007. The weed density following the application of bispyribac Na in ADSR was statistically no different than that in CPTR when supplemented with pendimethalin at 0.75 kg/ha + 1 HW.

All the herbicide treatments caused significant reduction in dry matter of grass weeds as compared with weedy treatment (Table 17.3). The treatment pendimethalin + 1 HW proved inferior in decreasing dry matter of grass weeds among all the weed control treatments. This was attributable to repeated flushes and poor control of *Echinochloa crus-galli* and *Digitaria sanguinalis* in the pendimethlin + 1 HW treatment. The superior control of sedges and broadleaf weeds was achieved with a sequential application of pretilachlor and metsufuron. The application of bispyribac Na at 25 g/ha also provided excellent control of sedges in 2007 (Table 17.3). However, all the tested herbicide treatments provided excellent control of broadleaf weeds when compared with the weedy check.

The highest rice yield was recorded with the CPTR method of establishment in both years (Table 17.4). Grain yield in CPTR increased to the tune of 28.7% and 34.4% over ADSR in 2006 and 2007, respectively. Many researchers have reported higher yields in CPTR than in ADSR (Brar, Mahajan & Sardana, 2002; Singh et al., 2002, Singh et al., 2008). This was because of greater competition offered by the crop in CPTR, as reflected in the number of panicles/sq m, which increased by 36.7% and 24.8% over ADSR during 2006 and 2007, respectively. Rice yields for all the tested weed control treatments were higher than for the weedy check. During 2006, yield with a sequential application of pretilachlor + metsulfuron, penoxsulam 3 DOS/DOT,

Table 17.3 Impact of varying crop establishment methods and weed control treatments on weed dry matter (g/m^2) of individual weeds species at harvest.

Weed control methods	*ADSR Grass weeds*	*CPTR Grass weeds*	*Mean*	*ADSR Sedges*	*CPTR Sedges*	*Mean*	*ADSR BLW*	*CPTR BLW*	*Mean*
2006									
Pendi +1 HW	138.4	14.6	*76.5*	82.4	17.5	*50.0*	41.9	8.8	*25.4*
Pretila + Metsulfu	99.7	12.5	*56.1*	26.4	8.3	*17.3*	19.0	3.6	*11.3*
Penoxsulam 12 DOT	78.7	22.1	*50.4*	44.0	12.3	*28.1*	13.4	11.1	*12.2*
Penoxsulam 3 DOT	95.3	23.3	*59.3*	75.9	24.4	*50.2*	29.7	10.3	*20.0*
Weedy	223.5	125.1	*174.3*	104.6	74.4	*89.5*	107.5	27.0	*67.2*
Mean	*127.1*	*39.5*		*66.7*	*27.4*		*42.3*	*12.2*	
LSD (0.05)	*CE: 22.4; WC: 35.4 ; CExWC: NS*			*CE: 12; WC: 18.9 ; CExWC: NS*			*CE: 17.6; WC: 27.9; CExWC: NS*		
2007									
Pendi +1 HW	263.6	29.5	*146.5*	27.6	13.9	*20.8*	28.2	38.8	*33.5*
Pretila + Metsulfu	127.4	8.97	*68.2*	24.7	11.4	*18.0*	27.1	3.8	*15.5*
Penoxsulam 12 DOT	120.1	15.8	*67.9*	45.0	9.60	*27.3*	18.3	18.1	*18.2*
Bispyribac Na	126.3	10.8	*68.5*	20.3	14.6	*17.5*	21.5	12.8	*17.2*
Weedy	389.2	162.0	*275.6*	111.4	46.9	*79.1*	62.8	62.8	*62.8*
Mean	*205.3*	*45.4*		*45.8*	*19.3*		*31.6*	*27.3*	
LSD (0.05)	*CE: 22.8; WC: 36.1; CExWC: 51.1*			*CE: 16; WC: 25.2; CExWC: NS*			*CE: NS; WC: 30.7; CExWC: NS*		

Pendi = pendimethalin; Pretila = pretilachlor; Metsulfu = metsulfuron; ADSR = aerobic direct seeded rice; CPTR = conventional puddle transplanted rice; CE = crop establishment methods; WC = weed control treatments; DOS = days of sowing; DOT = days of transplanting; BLW: broadleaved weeds.

Table 17.4 Impact of varying crop establishment methods and weed control treatments on panicles m^{-2} and grain yield (q/ha).

	Panicles m^{-2}			*Grain yield*			*(%) Increase in yield over weedy check*		
Weed control methods	*ADSR*	*CPTR*	*Mean*	*ADSR*	*CPTR*	*Mean*	*ADSR*	*CPTR*	*Mean*
2006									
Pendi + 1 HW	288	308	*298*	11.86	17.82	*14.84*	118.4	26.29	*72.34*
Pretila + Metsulfu	291	378	*334*	21.06	26.53	*23.79*	287.8	88.02	*187.9*
Penoxsulam 12 DOT	274	333	*303*	24.51	27.89	*26.20*	351.4	97.66	*224.5*
Penoxsulam 3 DOT	253	328	*291*	20.96	26.17	*23.56*	286.0	85.47	*185.7*
Weed free	321	395	*358*	29.16	33.02	*31.09*	437.0	134.0	*285.5*
Weedy	173	266	*220*	5.43	14.11	*9.77*			
Mean	*267*	*335*		*18.83*	*24.25*		*296.1*	*86.28*	
LSD (0.05)	*CE: 21.6; WC: 37.4; CExWC: NS*			*CE: 2.11; WC: 3.66; CExWC: NS*					
2007									
Pendi + 1 HW	253	275	*264*	15.43	23.26	*19.34*	136.3	60.8	*98.55*
Pretila + Metsulfu	282	358	*320*	21.98	31.88	*26.93*	236.6	120.5	*178.55*
Penoxsulam 12 DOT	267	331	*299*	20.76	28.58	*24.67*	217.9	97.6	*157.75*
Bispyribac Na	253	324	*289*	25.69	30.02	*27.85*	293.4	107.6	*200.5*
Weed free	322	379	*351*	28.27	31.37	*29.82*	332.9	116.9	*224.9*
Weedy	168	266	*217*	6.53	14.46	*10.49*			
Mean	*258*	*322*		*19.78*	*26.59*		*243.4*	*100.68*	
LSD (0.05)	*CE: 23.1; WC: 40.1; CExWC: NS*			*CE: 1.81; WC: 3.13; CExWC: NS*					

Pendi = pendimethalin; Pretila = pretilachlor; Metsulfu = metsulfuron; ADSR = aerobic direct seeded rice; CPTR = conventional puddle transplanted rice; CE = crop establishment methods; WC = weed control treatments; DOS = days of sowing; DOT = days of transplanting and penoxsulam 12 DOT was statistically similar; however, considering grain yield, pendimethalin + 1 HW proved inferior among all the treatments. The highest mean grain yield was registered with weed-free treatment (31.09 q/ha) during 2006 and this treatment was significantly superior among all the weed control treatment. Similarly, during 2007, highest yield (29.82 q/ha) was registered with weed free treatment. Interestingly, an application of bispyribac Na and a sequential application of pretilachlor and metsulfuron registered no statistically significant differences in grain yield when compared with the weed-free treatment in 2007 because of a broad-spectrum weed control.

17.3.2 Irrigation water productivity

Crop establishment methods had no influence on irrigation water productivity during 2006 (Table 17.5); however; all the weed-control treatments proved significantly superior to weedy check in relation to irrigation water productivity. Weed-free treatment registered highest water productivity (27.9 kg/ha-cm), followed by a post-emergence application of penoxsulam (12 DOT). The interactive effect of weed-control treatments and crop-establishment methods in relation to water productivity was significant. A cursory examination of data in Table 17.5 revealed that weedy check under ADSR registered lowest water productivity (5.69 kg/ha-cm). The crop water productivity remained statistically the same in both ADSR and CPTR, except under the weed-free treatment and post-emergence application of penoxsulam (12 DOT), where it was significantly higher in ADSR than in CPTR. During 2007, because of greater pressure of weeds in ADSR, water productivity was significantly higher in CPTR than in ADSR. Irrigation water productivity remained statistically the same in both ADSR and CPTR under the weed-free treatment or with a post-emergence application of bispyribac Na herbicide. Application of penoxsulam (12 DOT) in ADSR resulted in lower water productivity than in CPTR because of less control of weeds in ADSR and, as a result, crop and water productivity decreased.

17.3.3 Economic analysis

Net returns from ADSR under weedy situation were negative (loss), which revealed that weed control is an important component of profitability of the farmers in ADSR. During 2006, highest net profit was realized from weed-free treatment; and among the herbicide treatments, penoxsulam at 25 g/ha at 12 DOT was superior in retaining higher net returns, followed by a sequential application of pretilachlor and metsulfuron (Table 17.6). During 2007, bispyribac Na registered highest net returns in ADSR, followed by a sequential application of pretilachlor and metsulfuron. However, in CPTR during 2007, highest net returns were realized from a sequential application of pretilachlor and metsulfuron, closely followed by bispyribac Na. During both the years, the treatment with pendimethalin + 1 HW gave lower net income compared with all other herbicide treatments in both the crop-establishment methods. Irrespective of weed-control treatment, mean net profit was 39.5% higher in CPTR than in ADSR. Therefore, it was inferred from the study that although the expenses were reduced in the ADSR, net profitability did not increase as compared with CPTR because of reduced yield in ADSR. The variation in net profitability between the ADSR

Table 17.5 Impact of varying crop establishment methods and weed control treatments on water productivity of basmati rice (kg/ha-cm).

Year 2006	*Pendi + 1 HW*	*Pretila + Metsulfu*	*Penoxsulam 12 DOT*	*Penoxsulam 3 DOT*	*Weed free*	*Weedy*	*Mean*
ADSR	12.43	22.08	25.69	21.97	30.56	5.69	*19.73*
CPTR	13.68	20.37	21.41	20.09	25.35	10.83	*18.62*
Mean	*13.05*	*21.22*	*23.55*	*21.03*	*27.95*	*8.26*	
LSD (0.05)	*CE: NS; WC: 3.25; CExWC: 4.59*						
Year 2007	*Pendi + 1 HW*	*Pretila + Metsulfu*	*Penoxsulam 12 DOT*	*Bispyribac Na*	*Weed free*	*Weedy*	*Mean*
ADSR	14.62	20.83	19.68	24.35	26.80	6.18	*18.74*
CPTR	22.05	27.82	26.77	28.46	29.73	13.70	*24.73*
Mean	*18.33*	*24.32*	*23.22*	*26.40*	*28.26*	*9.94*	
LSD (0.05)	*CE: 2.05; WC: 3.55; CExWC: NS*						

Water applied during 2007: ADSR: 105.5 cm. CPTR: 140.75.
Water applied during 2006: ADSR: 95.4 cm. CPTR: 130.25.
Pendi = pendimethalin; Pretila = pretilachlor; Metsulfu = metsulfuron; ADSR = aerobic direct seeded rice; CPTR = conventional puddle transplanted rice; CE = crop establishment methods; WC = weed control treatments; DOS = days of sowing; DOT = days of transplanting.

Table 17.6 Cost and return analysis of different weed control treatments under different crop establishment methods.

	Total expenses (USD)		*Gross income (USD)*		*Net profit (USD)*	
Weed control methods	*ADSR*	*CPTR*	*ADSR*	*CPTR*	*ADSR*	*CPTR*
2006						
Pendi + 1 HW	422.5	512.5	533.7	801.9	111.2	289.4
Pretila + Metsulfu	426	516	947.7	1193.85	521.7	677.5
Penoxsulam 12 DOT	425	515	1102.95	1255.05	677.95	740.05
Penoxsulam 3 DOT	425	515	943.2	1177.65	518.2	662.65
Weed free	540	630	1312.2	1485.9	772.2	855.9
Weedy	390	480	244.35	634.95	−145.65	154.95
2007						
Pendi + 1 HW	430.5	520.5	870.3	1046.7	439.8	526.2
Pretila + Metsulfu	426	516	989.1	1434.6	563.1	918.6
Penoxsulam 12 DOT	433	523	934.2	1286.1	501.2	763.1
Bispyribac Na	436	526	1156.05	1350.9	720.05	824.9
Weed free	548	638	1272.15	1411.6	724.15	773.6
Weedy	398	488	293.85	650.7	−104.15	162.7

1 USD = Rs. 40 (Value adjusted to nearest USD); Hand Weeding 2.5 USD day^{-1}; Pretilachlor = 16 USD ha^{-1}; Metsufuron: 10 USD ha^{-1}; Pendimethalin: 7.5 USD ha^{-1}; Penoxsulam: 35 USD ha^{-1}; Bispyribac Na: 38 USD ha^{-1}.

Pendi = pendimethalin; Pretila = pretilachlor; Metsulfu = metsulfuron; ADSR = aerobic direct seeded rice; CPTR = conventional puddle transplanted rice; CE = crop establishment methods; WC = weed control treatments; DOS = days of sowing; DOT = days of transplanting.

and CPTR decreased with bispyribac Na, followed by penoxsulam and a sequential application of pretilachlor and metsulfuron.

17.4 CONCLUSION

Conventional practices of puddled transplanting in rice require a large amount of water and labor. The emerging shortages and increasing costs of water and labor will therefore force a change in the way farmers grow rice crop. It is concluded that the ADSR could be as effective as conventionally puddled transplanted rice in attaining high rice grain yield and net returns when weeds were kept under control using these identified effective weed management treatments. The water-saving feature of direct seeding is largely attributed to the avoidance of puddling used in transplanted rice. However, savings in irrigation largely depended on the occurrence and distribution of rainfall during the crop period and infestation of weed flora during the crop-growing period. Therefore, more efforts will be needed to evaluate and improve the technologies for weed control on a site- and season-specific basis. Shifting from conventional tillage practice to ADSR may cause changes in soil properties, microflora, microfauna, weed flora, and additional use of herbicides affecting long-term crop productivity and input-use efficiency. It may also pose the problem of herbicide resistance. Therefore, long-term changes in the crop performance, herbicide efficacy, input efficiencies, and weed flora should be monitored to achieve a paradigm shift in farmers' practices.

REFERENCES

Balasubramanian, V. and J.E. Hill. 2002. Direct seeding of rice in Asia: Emerging issues and strategic research needs for the 21st century. In *Direct seeding: Research strategies and opportunities,* eds. S. Pandey, M. Mortimer, L. Wade, T.P. Tuong, K. Lopez, and B. Hardy, 15–39. Los Baños, Philippines: International Rice Research Institute.

Belder, P., B.A.M. Bouman, J.H.J. Spiertz, S. Peng, A.R. Castaneda, and R.M. Visperas. 2005. Crop performance, nitrogen and water use in flooded and aerobic rice. *Plant Soil* 273:167–182.

Bhushan, L., J.K. Ladha, R.K. Gupta, S. Singh, A. Tirol-Padre, Y.S. Saharawat, M. Gathala, and H. Pathak. 2007. Saving of water and labor in rice–wheat system with no-tillage and direct seeding technologies. *Agron. J.* 99:1288–1296.

Bouman, B.A.M., S. Peng, A.R. Castaneda, and R.M. Visperas. 2005. Yield and water use of tropical aerobic rice systems. *Agric. Water Manage.* 74:87–105.

Brar, L.S., G. Mahajan, and V. Sardana. 2002. Weed management practices for basmati rice planted by different methods. Paper published in Proceedings of Second International Conference on Sustainable Agriculture for Food, Energy and Industry. Vol (2) held at Beijing, China, pp. 1672–1677.

Castin, E.M. and K. Moody. 1985. Weed control in dry-seeded wetland rice as affected by time and method of tillage. In Proceedings of the 10th Asian Pacific Weed Science Society Conference, Chiangmai, Thailand, pp. 645–661.

Darren, W.L. and E.H. Stephen. 2006. Foliar and root absorption and translocation of bispyribac-sodium in cool-season turfgrass. *Weed Technol.* 20:1015–1022.

Devine, M.D. 1989. Phloem translocation of herbicides. *Rev. Weed Sci.* 4:191–225.

Devine, M.D., H.D. Bestman, and W.H. Vanden Born. 1990. Physiological basis for the different phloem mobilities of chlorsulfuron and clopyralid. *Weed Sci.* 38:1–9.

Elliot, P.C., D.C. Navarez, D.B. Estario, and K. Moody. 1984. Determining suitable weed control practices for dry-seeded rice. *Philipp. J. Weed Sci.* 11:70–82.

Estorninos, Jr., L.E. and K. Moody. 1988. Evaluation of herbicides for weed control in dry-seeded wetland rice (*Oryza sativa*). *Philipp. J. Weed Sci.* 15:50–58.

Fujisaka, S., L. Harrington, and P.R. Hobbs. 1994. Rice–wheat in South Asia: systems and long-term priorities established through diagnostic research. *Agric. Syst.* 46:169–187.

Gill, K.S. 1994. Sustainability issues related to rice–wheat production system. pp. 30–61. In R.S. Paroda et al. (ed.) Sustainability of rice–wheat production systems in Asia. FAO, Bangkok, Thailand.

Giri, G.S. 1998. Effect of rice and wheat establishment techniques on wheat grain yield. In Proceedings of Rice–Wheat Research End-of-Project Workshop, 1–3 October 1997, Hotel Shangri-La, Kathmandu, Nepal. Jointly published by CIMYT, NARC, Rice-Wheat System Research Consortium, eds. P.R. Hobbs and R. Bhandari, 65–68.

Gupta, R.K., R.K. Naresh, P.R. Hobbs, Z. Jiaguo, and J.K. Ladha. 2003. Sustainability of post-green revolution agriculture: The rice–wheat cropping systems of the Indo-Gangetic Plains and China. In *Improving the productivity and sustainability of rice–wheat systems: Issues and impacts,* eds. J.K. Ladha, J.E. Hill, J.M. Duxbury, R.K. Gupta, and R.J. Buresh, 1–25. ASA Special Publication no. 65. Madison, WI: ASA Inc.

Harrington, L.W., S. Fujisaka, M.L. Morris, P.R. Hobbs, H.C. Sharma, R.P. Singh, M.K. Choudhary, and S.D. Dhiman. 1993. Wheat and rice in Karnal and Kurukshetra districts, Haryana, India: Farmers' practices, problems and an agenda for action. ICAR, HAU, CIMMYT, Mexico, and IRRI, Los Baños, the Philippines.

Hira, G.S., S.K. Jalota, and V.K. Arora. 2004. Efficient management of water resources for sustainable cropping in Punjab. Technical Bulletin. Department of Soils, Punjab Agricultural University, India, 20 p.

Hobbs, P.R., Y. Singh, G.S. Giri, J.G. Lauren, and J.M. Duxbery. 2002. Direct-seeding and reduced tillage options in the rice–wheat systems of the Indo-Gangetic plains of South Asia. In *Direct-seeding in Asian rice systems: Strategic research issues and opportunities,* eds. S. Pandey, M. Mortimer, L. Wade, T.P. Tuong, and B. Hardy, 201–215. Los Baños, Philippines: International Rice Research Institute.

Janiya, J.D. and K. Moody. 1988. Effect of time of planting, crop establishment method, and weed control method on weed growth and rice yield. *Philipp. J. Weed Sci.* 15:6–17.

Moody, K. 1983. The status of weed control in Asia. *FAO Plant Prot. Bull.* 30:119–123.

Moody, K. 1996. Weed community dynamics in rice fields. In *Herbicides in Asian rice: Transitions in weed management.* ed. R. Naylor, 27–35. Palo Alto, CA: Institute for International Studies, Stanford University.

Moorthy, B.T.S. and G.B. Manna. 1993. Studies on weed control in direct seeded upland rainfed rice. *Indian J. Agric. Res.* 27:175–180.

Pellerin, K.J. and E.P. Webster. 2004. Imazethapyr at different rates and timings in drill and water seeded imidazolinone-tolerant rice. *Weed Technol.* 18:223–227.

Rao, A.N., D.E. Johnson, B. Sivaprasad, J.K. Ladha, and A.M. Mortimer. 2007. Weed management in direct-seeded rice. *Adv. Agron.* 93:153–255.

Schmidt, L.A., R.E. Talbert, F.L. Baldwin, E.F.S. Rutledge, and S.S. Wheeler. 1999. Performance of V-10029 (bispyribac) in rice weed control programs. *Proc. South Weed Sci. Soc.* 52:49.

Shaner, D.L. 1991. Physiological effects of the imidazolinone herbicides. In *The imidazolinone herbicides,* eds. D.L. Shaner and S.L. O'Conner, 129–138. Boca Raton, FL: CRC.

Sharma, H.C., S.D. Dhiman, and V.P. Singh. 1994. Rice–wheat cropping system in Haryana: Potential, possibilities and limitations. pp. 27–39. In S.D. Dhiman et al. (ed.) Proc. of the

Symp. on Sustainability of Rice–Wheat System in India. CCS Haryana Agric. Univ., Regional Research Station, Karnal, India.

Singh, Y., G. Singh, V.P. Singh, R.K. Singh, P. Singh, R.S.I. Srivastava, A. Saxena, M. Mortimer, D.E. Johnson, and J.L. White. 2002. Effect of different establishment methods on rice–wheat and the implication of weed management in Indo-Gangetic plains. In Proceedings of the International Workshop on Herbicide Resistance Management & Zero Tillage in Rice–Wheat Cropping System, 4–6 March 2002, pp. 188–192. Hisar, India: Department of Agronomy, CCS Haryana Agricultural University.

Singh, S., L. Bhushan, J.K. Ladha, R.K. Gupta, A.N. Rao, and B. Sivaprasad. 2006. Weed management in dry-seeded rice (*Oryza sativa*) cultivated on furrow irrigated raised bed planting system. *Crop Prot.* 25:487–495.

Singh, S., J.K. Ladha, R.K. Gupta, L. Bhushan, and A.N. Rao. 2008. Weed management in aerobic rice systems under varying establishment methods. *Crop Prot.* 27:660–669.

Sondhi, S.K., M.P. Kaushal, and P. Singh. 1994. Irrigation management strategies for rice–wheat cropping system. p. 95–104. In S.D. Dhiman et al. (ed) Proc. of the Symp. on Sustainability of Rice–Wheat Systems in India, CCS Haryana Agricultural Univ., Regional Research Station, Karnal, India:

Tuong, T.P. and B.A.M. Bouman. 2003. Rice production in water scarce environments. In *Water productivity in agriculture: Limits and opportunities for improvement,* eds. J.W. Kijne, R. Barker, and D. Molden, 53–67. IRRI, Phillippines: CABI Publishing.

Wang, H.Q., B.A.M. Bouman, D.L. Zhao, C. Wang, and P.F. Moya. 2002. Aerobic rice in northern China: Opportunities and challenges. In *Proceedings of the International Workshop on Water-wise Rice Production, 8–11 April 2002, Los Baños, Philippines,* eds. B.A.M. Bouman, H. Hengsdijk, B. Hardy, P.S. Bindraban, T.P. Tuong, and J.K. Ladha, 143–154. Los Baños, Philippines: International Rice Research Institute.

Subject index

35S promoter 241, 243–4
9:7 ratio 243

A. thaliana 242, 266–7
ABA-dependent 267
ABA-inducible promoter complex 232
ABA-responsive element 232
Abelmoschus esculentus 119, 158
abiotic constraints 182
abiotic stress 229–31, 233–8, 241–6, 261–3, 265–8
  tolerance 229, 233–8, 242–3, 265
abscisic acid 245, 248
acetohydroxy acid synthase 302
acetolactate synthase 302
acid soil 212, 203
acidification 114, 145
acidity 210–11
actin promoters 267
adaptability 53, 216
  of alfalfa 53
adaptive management 133, 139–40
adaptive strategies 34
  early-maturing varieties 34
  irrigation methods 34
  new cropping systems 34
  time of sowing 34
adoption of water-saving practices 51
adsorption of chemicals 16
ADSR. *See* aerobic direct-seeded rice
advanced waterconservation technologies 135
adverse pH 230
adverse temperature 230
aerobic direct seeding 298
aerobic direct-seeded rice 297
aerobic rice 297
aerobic soil dry-tillage 299
aerobic systems 297–8
aerobic-rice production 37
afforestation 58, 66–7, 214–5
AFLP. *See* amplified fragment length polymorphism
agrichemicals 143, 162–3
agri-clinics 219
agricultural and grazing runoff 65
agricultural biotechnology 263
agricultural chemicals 14, 16–7, 152
agricultural ecosystems 133
agricultural electricity use 24
agricultural growth rate 72
agricultural intensification 29
agricultural production 30
  systems 9, 17
agricultural productivity 31, 39, 96, 182, 184, 195, 199
agricultural sustainability policies 21
agricultural water demands 135
agricultural water productivity 25
agricultural water use relative to total withdrawal 6
agricultural watersheds 3, 184
agriculture intensification 21
agriculture water issues 55
agri-horticulture 217
Agrobacterium 229, 232, 239, 241–3, 263
agrochemicals 145
agro-climatic zones 216
agro-climatologic parameters 220
agro-eco subregion 281–2
agro-ecologically unsuitable crop 298
agroecosystem 33, 38, 182
agroforestry 67, 111, 215, 217, 280, 292–4
agroforestry interventions 280

agronomic management practices 184
agronomic practices 95, 101, 187, 283
agronomic productivity 34, 109, 116, 281
agronomic stagnation causes 30
AICRPDA. *See* All India Coordinated Research Project for Dryland Agriculture
air and water quality 133, 138
air pollution 230, 280
alarm bell 72
Albezzia lebbeck 222
alfalfa 51–5, 163, 241
Alfisols 109–11, 119, 124, 129, 190–3, 196, 208, 210, 222–3, 271
algae 149
algal blooms 149
alkaline pH 282
alkalinity 211
All India Coordinated Research Project for Dryland Agriculture 171
Allium cepa 119
alluvial aquifers 143
alluvial coarse-textured soils 298
alluvial plain 282
alluvial soils 279–80, 285
alluvium 73, 94, 110, 281
ALS. *See* acetolactate synthase
alternate cropping system 29
productivity 282
alternate land-use system 220
alternate wetting and drying 299
alternative land-use options 280
alternatives to traditional rice 294
aluminum hydroxides 114
amla 110, 124
Ammania spp. 302
ammonia toxicity problem 64–5
Amoco planted sweetgrass 64
amplified fragment-length polymorphism 246, 248
ANACO19 245
ANACO55 245
ANACO72 245
anaerobic stress 262
animal dung 118, 151
for cooking 33
animal fodder 211
animal productivity 209
animal waste 39
anjan grass 222
annual potential evapotranspiration 170
annual water-table fall 77
anther-silk interval 250
anthropogenic contamination 144
anthropogenic factors 145
antioxidant 242
antioxidant genes 242
AP2/ERF family 266
APX. *See* ascorbate peroxidase
aqua-agriculture 102, 106
aquaculture 93, 102, 106, 111, 293
aquatic ecosystems 137
aquatic life 65, 138
aquatic resource-protection programs 61
aquifer storage 135
aquifers 3, 14, 24, 134, 143, 145, 148
Arabidopsis 229, 232, 240–1, 244–5, 266
arable land 30, 44, 110, 208, 213–4
area irrigated with tubewells in Punjab 14
area under rice 82, 280, 294
arid and semiarid climates 34
arid and semiarid regions water table problem 14
arid horticulture 110, 124, 129
arid mountains 210
arid regions 170
Aridisols 110, 210
aromatic plants 217
arsenic 143–5, 147–8, 162
concentration 143, 148
Artiplex 232
AsA 242
ascorbate peroxidase 241
ascorbate peroxidase gene 242
ASI. *See* anther-silk interval
assured employment 207, 218
and wage system 207
assured irrigation 298
in Punjab 24
assured jobs 218–9
assured market 298
assured marketing 217, 219
assured marketing linkage 217
ATAF family 245
atmospheric humidity 170
Australian Centre for International Agricultural Research 222
autumn-planted sugarcane 290
available S 199, 222
available soil nutrients 282
available water capacity 94, 96, 110
Avena sativa 243
average linkage cluster analysis 270
avoidance 201, 230, 248, 262, 271, 310
AWC. *See* available water capacity
azospirillum 118
Azotobacter 104
Azucena alleles 248

backcross breeding 246
backcross inbred population 248
backcrossing 231, 250
BADH. *See* betainealdehyde dehydrogenase
Bagpodi 98
balanced application of nutrients 197
balanced fertilization 207, 212–3
balanced fertilizer use 209
balanced nutrition 189, 197
balanced-nutrient management 162
balancing and increasing nutrient reserves 35
banding of herbicides 15
effect on water quality 15
bar (herbicide resistance) gene 243
barley 221, 230, 232, 242–4, 249–51
basalt 94
base map for watershed 220

baseline soil properties 113
base-rich rocks 94
basin-management approach 65
basket-making industry 64
basmati 300, 308
BBF (broad bed furrow system) landform 187
system 184, 190–1, 194–6
BCR. *See* benefit-cost ratio
bed-planting method 87
belowground biomass 215
bench terracing 67
beneficial soil microorganisms 214
benefit-cost ratio 93, 280, 286
Bengal famine 73
berseem 290
best crops 216
best management practices 15–6, 58, 61, 93, 95, 106, 118, 129, 139
best plant types 207, 215
best-bet management 201
practices 197, 202
Beta vulgaris 239
betaine 232, 239
betainealdehyde 232, 239
better plant types 216
Bhakhra main canal 74
bimodal rainfall 109, 111, 115–6
biochemical markers 248
biochemical oxygen demand 26, 154, 162
biochemical responses of plants 266
bioclimatic zones 170
biodiesel plants 218
biodiversity, loss of 182
biofertilizer application 211
biofertilizers 106, 207, 214
and microorganisms 214
bio-fuel plantations 67
biogeochemical 133, 136–7
biolistics 263
biological control 16, 127
for soil reclamation 16
biological nitrogen fixation 285
biological productivity 137
biological soil quality 214
biological yield 52
biomass production 33, 214–5, 239, 266
productivity 133, 242
biosolids 29, 36, 39, 214
biotechnological and agronomic strategies 227
biotechnological approaches 229–30, 265
biotechnology 229–30, 251, 263
and drought 229
bioterrorism 134
bird perches 201
bispyribac Na 297, 302, 304, 307, 310
Bist Doab canal network 74
bittergourd 291
black soils 94, 110, 186
blackgram 109, 115
blue baby syndrome 149
blue water 133–5
BMPs. *See* best management practices
BNF. *See* biological nitrogen fixation
BOD. *See* biochemical oxygen demand
boron 143, 145, 147–8, 189
Boulder creek 64
bovine milk 152–3
Brassica campestris 31, 243
brassica in sugarcane 290
Brassica juncea 158–9, 232, 293
Brassica napus 243, 282
Brassica oleracea 158
Brassica rapa 282
broad bed 106
and furrow system 93
broadleaf weeds 300, 302, 304
broad-spectrum weed control 307
Bt cotton 285
Budda Nullah 155–9
buffel 222
buffer strips 17, 61, 65
bulk density 271–2, 279, 282, 290
bunding 101, 121, 171, 196, 218
treatment 121
burning of rice straw 287
butachlor 299

C3 crops 34
temperature regime 34
cabbage 158, 243
cadmium 144, 155
Cajanus cajan 93, 115, 222
calcareous 282, 285, 293
CaMV 35S 243–4, 267
promoter 239, 267
canal irrigation 33, 73, 81, 85–6, 95, 144
canal network 72–6, 86
canal water 24, 39, 86, 149, 153, 162–3
courses 81, 86
canal-irrigated areas 14, 294
canal-irrigation systems 13
cancer 163
capacity building 60, 219
CAPS. *See* cleavable amplified polymorphic sequences
Capsicum annum 124
capture capacity 272
carbon accounting 215
carbon balance in soil 215
carbon emission 17–18
and global warming 17
carbon pools 212
carbon sequestration 133, 137–8, 207, 214–15, 290
carbon sink capacity 215
carbon stocks 215
carbon turnover 212
carbon-sequestering systems 212
case studies from Punjab 143
castor 199, 221–2
CAT. *See* catalase
catalase 241–2
cation exchange capacity 94, 113
cattle population 171
cauliflower 158
causes of pollution 139
CCS. *See* cotton-chickpea system
Cd toxicity 163
CDH 239
cDNA 239
celery 240
cell lines 232
cell turgor 262
cell wall peroxidase 241
cell water content 216
Cenchrus ciliaris 222
Central Great Plains 44
Central Highlands of Gujarat 208
central pool of India 71–2

centrifugal pumps 78
technology 75
cereal grain consumption 31
C-footprint 39
chain survey 220
check dams 67, 87
chemical deterioration 210–1
chemical leaching to groundwater, reduction in 16
chemical oxygen demand 154
chemical rates and methods of applications 15
chemical weed control 45, 54
chickpea 31, 93, 98–9, 102, 105–6, 182–3, 185, 187, 189–90, 192, 196–7, 202–3, 261–2, 264–6, 269–73, 282, 285
chillies 84, 192
Chinese cabbage 243
chisel plow 12, 16, 173
chloroplasts 239–40, 241–3
choline 232, 239
choline monooxygenase 232, 239
chromium 144, 155
Chromusterts 95
Cicer arietinum 31, 93, 282
cis-acting elements 267
clay content 94–6
minerals 95
clayey vertic Ustochrepts 95
cleavable amplified polymorphic sequences 246
climate change 3–4, 6, 21, 32, 34, 134–5, 262
challenges 6
impact of 4, 32, 34
threats from 21
and water availability 4
climate variability 25
climate-change impact on sustainability of water resources 22
climatic aridity index 170
climatic variability 22, 184
cloned gene 231
closed conduits 190
clump planting 178
cluster bean 117
CMO. *See* choline monooxygenase
coarse cereals 129, 171, 208, 221
coastal area 62
coconut husk 110, 125
cold storage 207, 217
cold-induced genes 267
colluvium 94
Colorado watershed project 64
co-mapping 249
combine-harvested rice 287
common pasturelands 68
community movement 217
community watersheds 181–2, 195
community-based field project 281
community-based soil- and water-conservation 195
compaction 210
component of watershed management 133, 137
compost 39, 109–11, 118–22, 127, 129, 223, 285
composted crop residues 187
conjunctive use of canal water with saline water 39
conjunctive use of rainfall and limited irrigation water 192
consequences of dumping 163
conservation agricultural practices 207, 215
conservation agricultural-management practices 211
conservation agriculture 211
conservation furrows 196, 222
conservation tillage 11, 16, 45, 51, 174–5, 211–2, 215
defined 174
conserving limited water for crop production 43
conserving soil in the root zone 35
conserving water in the root zone 29, 35, 39
constitutive overproduction 267
constraints in mechanization 210
contamination of groundwater with agricultural chemicals 14
contamination of water resources 17
contingency crop planning 220
continuous corn practice 17
continuous cropping system 47, 49
continuous flooding 101, 298
contour 67, 109, 114, 171, 196
contour bunding 109, 196
contour farming 196
contouring 13
contract farming 219
controlling soil degradation 211
controlling soil erosion 13
conventional breeding 229–31, 261–2, 265
conventional tillage 12, 36, 177, 221–3, 310
conventionally puddled transplanted rice 297
conventional till 176
Cooper river corridor project 64
cooperative farming 219
cor15A 267
COR15a 243
cor6.6 267
coriander 159, 202
Coriandrum sativum 159
corn residue 222
corn-soybean cropping patterns 10
corn-soybean rotation 12, 17
co-segregation 248
cost of pumping water 298
cost recovery for electricity 25
cost-effective management strategies 139
costs of inputs 209
co-transformation 231
cotton 81–2, 84–5, 109, 118, 144, 163, 171, 173, 183, 208, 221, 229, 279, 282, 285–6, 294
cotton-chickpea system 282
cotton-wheat system 282
cover crops 215
cowpea 110, 115–6, 119–21, 123, 125, 163
CPTR. *See* conventionally puddled transplanted rice
cracked furrows 190
cracking patterns 95

credit availability 207, 209
credit marketing 220
cress 245
critical erosion period 175
crop diversification 84, 111, 117, 145, 160, 202
crop failures 49, 73, 208
crop improvement 184, 230, 246, 263
crop insurance 219
crop intensification 201
crop manipulations 177
crop planning 35, 220
crop productivity 101, 162, 181–4, 189, 194, 196, 201, 203, 208–9, 230, 261–5, 268, 310
crop residue 16, 29, 36, 39, 43, 46–7, 49, 54, 109, 111, 118, 120, 125–6, 173–5, 187, 211, 214, 222, 280–1
  management 29, 44, 174–6
  mulch 34, 40, 120
  removal 44, 214
  use 29
crop rotations 17, 43, 51, 152, 279, 282
crop water productivity 307
  use 33
cropland area decrease 9
cropping intensification 54
cropping intensity 46, 49–50, 202, 216, 280
cropping pattern 24, 76, 81
cropping system production potential 52
cropping systems 17, 29, 38, 43–4, 46–51, 54–5, 102, 109–10, 114–17, 126, 129, 152–3, 162, 187–9, 193, 196–7, 201, 207–8, 212–6, 218, 220, 280–2, 284–6
  and reduction in chemical leaching 17
crop-weed competition 302
Crotalaria juncea 118
cultivated furrows 190
custom-hiring services 217
customized fertilizer application 213
cuticle thickness 262
CWS. *See* cotton-wheat system
Cyamopsis tetragonolobo 117
cyanide 155
cyanobacteria 149
cyclic agroforestry intervention 293
cyclic interventions 280
cyclones 262
Cyperus iria 300
Cyperus rotundus 300
cytosol of potato 241

Dalbergia sisso 222
data inventory for the watershed 139
database on rainfed agriculture 220
dead furrows 196
deciduous tree 292
decision support system for fertilizer recommendation 214
declining availability of water per capita 22
declining water tables 44, 55
decrease in erosion potential 46
deep plowing 171
deep tillage 117, 221
deep Vertisols 216
deeper roots 262
deep-rooted crops 162
deep-tubewell waters 149, 162
deforestation and exploitation of grasslands 13
degenerated bones 163
degradation of natural resources 182
degradation of soil quality 29, 39, 45
  impact on productivity 31
  and water 31
degradation of soil structure 45
degraded lands 60–1, 66, 215
degraded soils 111, 134, 214
dehydration-responsive element binding 229, 243–4
delayed transplanting of rice 82, 145, 160
deltas 60
dental fluorosis 147
deoiled mustard 293
deoiled rice 293
depletion and pollution of water resources 17
depletion of natural resources 21
depletion of water resources 160
desalination 135
desert development program 66
desert grass species 249
deserts 70
desiccation tolerance 264
desiltation of village tanks 67
desilting 114
detaching leaves 268
detachment of the hoof 146, 161
detoxifying enzymes 229, 241, 264
developing countries challenges 7
DH. *See* doubled haploid
di-ammonium phosphate 116
diesel generator sets 78
diethylene triamine pentaacetic acid 157
Digitaria sanguinalis 300, 302, 304
direct seeded crop 298, 300, 302
direct seeding 297–8, 310
direct soil evaporation 33, 269
direct sowing of wheat 40
direct water use 33
direct-seeded rice 300
disc plowing 109, 114–5, 129
discharge consortium 65
dissection of quantitative traits 246
distraction from agriculture 210
distribution of quality seeds 106
distribution of rainfall 207, 209, 216
District Rural Development Agencies 67
diverse cropping system adoption 54
diversification 84, 110, 129, 202, 204, 209, 294
  of agriculture 292
  of cropping 129

of land use management 110
diversified agriculture 280
diversified production system 281, 290
diversifying enterprises 207, 217
DL-788–2 98
DNA-chip analysis 245
Dolariya 98
domestic sewage water 143
domestic zone sewage 154
doomsday 80, 88
double cropping 109, 114–5, 118, 129, 193, 216
double transgenic 242
double-cropping systems 117, 129
doubled haploid 246, 248
drainage lines 68
DREB. *See* dehydration-responsive element binding
DREB1/CBF cold-responsive pathways 244
DREB1/CBF3 244
DREB1A 244–6, 266–7, 269, 271–2
DREB1A protein 244
DREB2A 244
drill sowing 171
drinking water 3, 7, 10–11, 59, 61, 63, 66, 87, 134, 138, 149, 152, 160
  protection 61
  supply 138
  treatment facilities 7
drip irrigation 6, 8, 109, 111, 114, 124, 129, 191, 216, 221, 291
  benefits 8
  vs. flood irrigation 8
  water savings 8
  and water-use reduction 8
drought 22, 33–5, 43, 47–50, 52–3, 55, 66, 71–3, 80, 110, 117, 129, 171, 173, 175, 182, 202–3, 208, 216, 219, 229–30, 232, 241–51, 261–73
  avoidance 262
  of the century 71
  conditions 47, 49, 230, 242, 262, 265, 268, 272
  cycle 47
  escape 216
  evaluation of transgenics 267
  intensity 270
  monitoring and relief 219
  prone area program 66
  resistance 216
  severity 47–8
  stress 34–5, 110, 117, 129, 229, 244–6, 248–9, 251, 262–8, 270, 273
  stress-induced genes 229
  tolerance 43, 55, 216, 229–30, 245–51, 261–5, 268, 273
  transgenic interventions 263
  years 73, 80, 173, 175
drought-induced dormancy 53
drought-induced genes 267
drought-resistant crops 171
drought-stress response 261
drought-tolerant genetically modified 229
dry farming 29, 40, 94, 112, 118, 208
dry matter accumulation 302
dry subhumid regions 170
dry wells 33
dry-farming technologies 109, 114, 129
dryland agriculture 170, 172, 262
dryland alfisols, sustainable management of 109
dryland and irrigated cropping systems 44
dryland areas 169, 171–2, 174, 182
dryland cropping 46, 50
dryland farming 171–2
dryland horticulture 220
dryland regions 44, 110, 172, 175, 211
drylands 35, 169–71, 211, 222
drylands, defined 170
dry-seeded rice 299
dry-up philosophy 50
DTPA. *See* diethylene triamine pentaacetic acid
DTPA-Zn 223
dugout tanks 184
Dust Bowl 45, 48, 172–3
dust mulch 172
dyes 155
dying 162

E. coli 239–40
early transplantation of rice 80
early vegetative growth 169
EC. *See* electric conductivity
Echinochloa colona 300
Echinochloa crus-galli 300, 304
Eclipta alba 302
Eclipta prostrata 300
eco-development projects 66
ecological balance 59
ecological indicators 137
ecological sustainability of earth 9
economic analysis 307
economic development 9–10, 65, 135
economic incentives 84
economic indicator 286
economic sustainability 43
economic thresholds 199
economic weed management 297
economic yields 173
ecosystem productivity 134
ecosystem services 33
ecosystem stability 138
edaphic factors 207, 211, 215
education guarantees 220
effective growing season 216
effective land use 209
efficiency indices 120–2
efficiency of irrigation 15, 84
efficient cropping systems 215
efficient crops 207, 215
efficient irrigation methods 29, 34
efficient irrigation technologies 85
efficient management of natural resources 194–5
efficient use of natural resources 181, 194
efficient use of rainwater 209
efficient water use policy reform 25
efficient watershed management 140

effluents, disposal of 156
Egyptian clover 146, 163
electrical conductivity
  94, 145, 147, 157, 282
electroplating 154–5, 162
Eleusine coracana 117, 221
emission of greenhouse
  gases 214
employment support 218
endogenous BADH 239
energy cost of pumping
  groundwater 77
energy efficiency 120
energy plantation 215
energy subsidy 25
energy-based inputs 31
energy-conservation
  measures 67
energy-input cost
  reduction 16
energy-intensive
  technologies 135
energy-security needs 3
enhancing agronomic
  productivity, strategies
  for 34
enhancing crop productivity
  in rainfed regions 182
enhancing soil structure 35
enriched compost
  118–21, 127
enriched pressmud 115
ensuring food security 208
Entisols 110, 208, 210
entrapping of carbon 215
environmental degradation
  of soil 15
environmental improvement
  136
environmental pollution 144
environmental
  protection 135
environmental quality 9,
  111, 137
environmental quality of
  natural resource, impact
  of population on 9
environmental stress
  resistance 266
environmental stresses 242
erd10 267
erosion of agricultural and
  forest lands 13
erratic rainfall 182, 192,
  209, 264
escape 216, 230, 249, 262
establishment methods 297,
  301–3, 305, 307–9
estuaries 133–4, 136–8, 140
ET. *See* evapotranspiration
e-technologies 140
ETP. *See* annual potential
  evapotranspiration
eucalyptus 161
Euphorbia hirta 300, 302
eutrophication 29, 149
evaporation 8, 34–5, 40,
  54, 82, 110, 120, 171,
  178, 272
  and percolation losses 8
  rate 54
evaporative
  environments 148
evaporative losses 298
evapotranspiration 33, 46,
  52, 80–1, 134, 169–70,
  176, 178, 272
excess water 196–7, 262
excessive flood irrigations
  cause of salinity
  problem 14
excessive irrigation
  in Punjab 12
excessive use of fertilizers
  and pesticides and non-
  point-source pollution 21
excessive withdrawal of
  groundwater 280
exotic sources 263
experimental-cum-
  demonstration farm 113
export 221
extensive root system 151,
  215–16
extractable sulfur 189

fall plowing 171
falling water table 33, 82, 84
fallow 43–7, 49–50, 54,
  95, 102, 106, 112, 169,
  173–6, 185, 201, 203
  efficiency 45
  management 46
  period 45, 47, 50, 169,
    173–6
  water-storage efficiency 45
family health cards 219
famine history 71–3
farm ponds 67
Farm Science Centre 281
farm water-use efficiencies 24
farmer nutrient inputs 199
farmer participatory
  approach 195
farmer participatory
  operational research 105
farmer participatory trials
  197, 299
farmer-based conservation
  measures 195
farmers' average yield 183
farmers' livelihoods 194
farmers' practice 93, 98,
  101, 121, 196
farming systems
  136, 195, 217
farming-systems
  approach 207
farmyard manure
  36, 98, 106, 163, 171,
  185, 211, 285
fast-growing winter trees 292
feedlots 150–1, 162–3
feed-tree species 217
ferredoxin-dependent
  enzyme 232
fertigation 221, 280, 291
fertilizer use 21, 207, 209,
  211–12
fertilizer-use efficiency 29
field bunding 171
field-water management 35
finger millet 117, 171, 197,
  199, 221–2
fish culture 102–3, 280, 293
fish feed 102, 293
fish kills 149
fish production 294
fisheries 59, 138
flat bed 124
flat fan nozzle 300
flat small grain residue 175
flatbed system 84, 93, 98
flat-on-grade system 196
flood irrigation 6, 8, 33
flood methods of irrigation,
  wastefulness of 8
flood plain 73, 86
Flood Prone Rivers
  Program 66
flooded rice alternatives
  37, 40
  cultivation 34
flooding 13, 34, 230, 262
flood-plain areas 73, 140
flood-prone areas 86
floods and droughts
  frequency 22

floral sterility in rice caused by high temperature 34
flow-weighed NO3–N 10
fluorescence imaging 230
fluoride 143, 145–7, 161
fluorosis 146–7, 161
fluventic Ustochrept 282
fodder 33, 38, 81, 84, 208, 218, 287
  scarcity 208
  species 218
food and nutrition security 117
food basket of the country 144
food chain 143, 145, 152–3, 160–1
food grain procurement program 24
food grain production 10, 24, 72, 152, 208, 280
  need to double 24
food insecurity 24, 181
food production 4–6, 10, 17, 33, 39, 134, 144, 229–30, 262
  sustainable system 5
food reserves in Punjab 73
food security 3–4, 6, 10, 18, 21–2, 24, 71–2, 76, 82, 111, 134, 144, 182, 202, 208, 262, 298
  components 30
  issues 3
  and melting of glaciers 5
  needs 6
  at risk 7
  threats to 21
  world population 4
food self-sufficiency 24
forage crops 217
foreground selection 251
forest and woodland 214
forestland in Nepal 13
fraction of soil-water moisture available to plant 269
free or highly subsidized electricity 21–2
free supply of inputs 195
free-radical scavenger 242
freezing 232, 243–4, 264
freezing stress 244
frequency of flooding 101
frequent droughts 182, 207, 209
fresh cow dung 102
fresh groundwater reserves 73
freshwater 4, 6, 22, 40, 144, 230
  agriculture major user of 22
  limited availability of 7
  pollution of supplies 3
  use 230
fructose-6–P 240
FTSW. *See* fraction of soil-water moisture available to plant
functionality of roots 271
furrow irrigation 191
FYM. *See* farmyard manure

Ganga-Yamuna River Basin 208
Ganges 5, 26
garlic 202
gas exchange characteristics 266
gastric cancer 149
G-E interaction 248
GEFSOC modeling system 290
gel-shift assay 245
gene expression patterns 230
gene pyramiding 231, 242
gene stacking 231
genetic engineering technology 263
genetic gain 251
genetic map 248
genetic resistance 230
genetic transformation 239, 263, 265
genetic variance for yield 230
genetically modified organisms 231
genetic-engineering techniques 231
gene-transfer methods 231
genome-wide tools 230
genotype-phenotype gap 230
geogenic and anthropogenic sources 143
  contaminants 144
  factors 145
  pollutants 145, 161
  sources 145, 160
geographical information system 58, 68
German Agency for Technical Cooperation 58, 60
germplasm accessions of peanut 270
GIS. *See* geographical information system
glacier melting 4
glacier-fed rivers 5
Gliricidia 187, 222–3
  loppings 222–3
global agriculture 134
global food security 5, 30
  water sustainability for 9
global land and water resources 134
global temperature impact on crop growth 34
global warming 3–5, 17, 214
  consequence of 5
  and sea level rise 5
  and water sustainability 4
global water consumption 134
global water use for agriculture 5–6
  for industry 5–6
  for municipal purposes 5–6
global watersheds 134
glufosinate ammonium herbicide resistance 243
glutamate 231
glutamyl phosphate 231
glutathione reductase 241
glutathione S-transferase 242
glycinebetaine 229, 232, 239–40
glyphosate-based herbicides 45
Gm TP55 antiquitin homologue 242
GMOs. *See* genetically modified organisms
goat 13, 208
Godavari and Krishna river basins 26
Gossypium 118, 221, 282
GR. *See* glutathione reductase
graded broad-beds 185
grain reserves 30
grain sorghum 169, 171, 174–6, 178
grain to straw ratio 216
grain yields of rice and wheat 31
grain-filling growth stages 177
grain-filling period 169, 178

gram cultivation 76
granary of India 72–3
grass dry weight 302
grass weeds 304
grassed waterways 101, 184, 201
grasses 217–8, 222, 300, 302
grass-saver tiles 87
grassy plains 62
grassy weed control 54
gravitational drip irrigation system 124
Great Plains 43–5, 47, 54, 169, 171–7
green biomass 118
green chillies 124
green fodder 209
green gram 109–10, 115–6, 199, 223
green manuring 109, 118, 151, 211–2
green peas 201
green revolution 5, 71, 73, 75–6, 87, 110, 171, 210, 216
  impact of 76
green water 134
greenhouse gases 133, 137, 214
grey areas 210
grey water 39, *See also* urban water
groundnut 31–2, 34, 109–10, 115–7, 119–20, 126, 129, 152–3, 182–3, 196–7, 199–200, 221, 229, 261–2, 264–6, 269–72
groundwater 7–8, 10–2, 14–7, 21–6, 34, 37–38, 40, 43–4, 50, 62, 67, 71–82, 84, 86–7, 95, 114, 133, 135, 137, 140, 143–52, 156, 160–3, 182, 187–9, 197–8, 204, 279, 281–2, 286, 298
  contamination 12, 16, 152
  nitrate 151
  overexploitation 13
  pollution 16, 145, 160, 163
  quality 72, 77, 79–80, 145
  recharge 37, 40, 67, 71, 182, 197
  resources 7, 72, 82, 160, 163, 281
  table 8, 16, 24, 72, 74–81, 87, 95, 144, 160
  use 14, 24
  withdrawal 37, 84
growing-season precipitation 169, 173, 176–7
growth stage-specific irrigation timing 50
GST. See glutathione S-transferase
guidelines for *hariyali* 59, 66–7
gully control measures 220
gypsum use on sodic soils 16, 161

H. spontaneum 250
H. vulgare 250
habitat degradation 62
habitat restoration 65
hair loss 146, 161
hand weeding 299
Happy Seeder 83–5
hard setting 210
hard-rock regions 70
hard-to-decompose root system 215
Hardwickia binata 222
hardy vegetation 62
Harike headworks 74
harvest index 249, 261, 265, 272
harvest of rainwater 59, 217
Haussainewala headworks 74, 86
Haveli system 93, 95, 102
HCH. *See* hexachlorocyclohexane
HCN in water 162
health hazards 153, 163, 280
heat stress 244
heavy dependence on pesticides 285
heavy metals 145, 157, 214
heavy textured soil 95
hematite 148
hepatoxins 149
herbicide efficacy 310
herbicide loss to groundwater 12
herbicide resistance 243, 310
herbicide treatments 297, 302, 304, 307
herbicide-resistant transgenic plants 251
herbicides 10, 12, 45, 169, 173, 175–6, 299, 310
heterologous transgenic plants 266
hexachlorocyclohexane 153
high irradiance 263
high-income modules 207
high-value crops 9, 171, 190–1, 193, 197, 202, 204
high-value enterprises 216
high-value vegetable crops 129
high-yielding crop varieties 194
high-yielding cultivars 93, 105, 197, 171, 209
high-yielding hybrids 171
hilly areas 70
Himalayan glaciers 4–5
Himalayas 34, 73, 85–6
histochemical studies 243
holistic farming systems approach 195
Hordeum vulgare 221, 242, 249
horsegram 109, 118
horticultural crops 187, 193
human and animal pathogens 137
human development 69
human greed 17
human population in rainfed regions 209
human resource development 207, 219
hurricanes 134
HVA1 gene 242
HVA1 protein 242
HVA1 stress tolerance 243
HVA1-like promoter 243
HVA22 gene 232
HvCBF4 244–5
Hydel channel 74
hydraulic conductivity 190, 222–3
hydrogen peroxide 241–2
hydrologic cycle 34
hydrologic natural resources 60
hydrological regimes 59
hydroponic culture 248
hydropower 58

hydroxyl radicals 240–1
hyper-accumulation capability 159

ice caps 210
ICRISAT. *See* International Crops Research Institute for the Semi-Arid Tropics
IGB. *See* Indo-Gangetic Basin
IGNRM. *See* integrated genetic and natural resource management
IGNRM approach of ICRISAT 184
IGP. *See* Indo-Gangetic Plains
imbalanced fertilization 207
imbalanced fertilizer use 211
impounded water 293
impounding water 196
improper use of water resources 39
improved crop varieties 10, 111, 197, 204
improved farm implements 220
improved irrigation-efficiency practices 51
improvement of soil quality 46
improving activity and species diversity of soil biota 35
improving soil health 162
improving the productivity of rainfed farms 207
improving water-infiltration 35
in situ moisture conservation 196–7, 216
in situ soil and water conservation 195
in situ soil moisture conservation 106, 123
in situ soil-conservation measures 87
inadequate use of fertilizers 44
incentive-based policies 4
Inceptisols 110, 186–7, 189, 210
incidence of insects and pests 81
income generation 215
income-generating activities 111, 126
increased potential economic return 46
indebtedness 71, 77–8, 88
India's population 31, 72, 144, 208
  vs. water and land resources 21
Indian clover 159
Indian Council of Agricultural Research 222
Indian Institute of Soil Science 201
Indian Plate 73
Indira Gandhi Canal 14
Indo-Gangetic Basin 29, 31, 37, 39, 110
Indo-Gangetic Plains 34, 280, 285, 289–90, 293, 297–8
Indo-German collaborative projects 60
Indus river basin 5, 73
industrial effluents 143, 153–4, 160, 162–3, 218
  pollution 160
  use 29, 44, 135
industrialization 8, 29, 135, 144
infant methemoglobinemia. *See* blue baby syndrome
infiltration 16, 35, 43, 47, 54, 95, 117, 190, 210, 221, 279, 282, 290
information and computing technologies 69
information hubs 219
information-sharing mechanism 221
infrastructure
  constraints 207
  cost of tubewells 78
  lack of water storage 7
ingested nitrate 149
INM. *See* integrated-nutrient management
innovative water management strategies 3
inorganic fertilizers 9, 26, 121, 211
  and excessive use of pesticides 26
input-use efficiency 29, 213, 310
input-use pattern 220
insecticide sprays 81
installation of check dams 35
institutional strategies 57–8
integrated afforestation & eco-development scheme 58, 66
integrated genetic and natural resource management 182, 184
integrated nutrient and pest management practices 215
integrated nutrient management 29, 36–7, 93–5, 119–21, 129, 187, 198, 204, 211, 213, 223
integrated nutrient sources 105
integrated pest management 116, 199, 204, 220
integrated soil fertility management 218
integrated wasteland development program 66
integrated watershed management 181–2, 184, 194, 220
intensified cropping systems 47
intensive agricultural production systems 9–10, 13, 16, 294
intensive agriculture 9, 12
intensive cropping 46–7, 110
intensive crop-production systems 12
intensive grain-production systems 12
intensive irrigation 9, 14
intensive tillage 48
intercrop systems 188–9, 192
intercropping 101, 109–10, 114–18, 124, 129, 216, 222, 280, 290, 292
intercropping in sugarcane and poplar 290
intercropping system 115
intercropping with pigeon pea 115
intercrops 115, 280, 292–4
intercultivations 171
inter-generic hybridizations 230
intermittent drought 272
International Crops Research Institute for the Semi-Arid Tropics 171, 181
international health standards 25
International Rice Research Institute (IRRI) 76, 297
inter-specific hybridization 230

intervention system and management 120
interventions in the watersheds 59
introgression 248–50, 261
inversion tillage 210
IPM. *See* integrated pest management
irrational pricing of agricultural inputs 25
irrational pricing policies 21
irrigated agriculture 5, 44, 50, 55, 95
  sustainability 44
irrigated area of India 10
irrigated area worldwide 5
irrigated crop production 33, 43
irrigated rice-wheat system 29
irrigating lawns 161
irrigation directorate 74
irrigation efficiency 13, 15–16, 135
irrigation management 16, 37, 101–2
irrigation methods 16, 34, 84, 111, 191
irrigation practices, mismanagement of 14
irrigation scheduling 84
irrigation systems 7
irrigation water 5, 8, 14, 17, 23, 50, 54, 71, 79, 84–5, 88, 146–7, 149, 152, 189–93, 240, 280, 307
  productivity 297, 307
  use 85
irrigation with saline water 29
irrigation with saline water and urban water 38
Irugur series 111

Jatropha 218
Jawaharlal Nehru Agricultural University 95
joint forest management 66

Kandi area 86–7
kera method 300
kharif sorghum 183
kin1 267
kinetics of water deficits 269
kisan credit cards 219–20
knowledge centers 60
Kodo millet 117

labile P 152
lablab 109, 117
labor scarcity 298
Lactuca sativa 243
lady's finger. *See* okra
land and water management 195
land capability classification 220
land care 207–8, 210–11
land configurations 106
land cover 211–2
  management 211
land degradation 10, 182, 184, 187, 208, 211
  salt accumulation 9
land equivalent ratio 109
land leveling 220
land reclamation 66
land use 63, 93, 110, 134, 140, 151
  activity 135
  patterns 58
  practices 134
  production systems 66
landless farmers 68
landless laborers 218–9
Larix leptoeuropaea 232
laser land leveler 84–5
late embryogenesis abundant 242
late embryogenesis proteins 264
Laterites 110
LEA. *See* late embryogenesis abundant
LEA3 266
leaching losses to groundwater 15
leaching of chemicals to groundwater 15
leaching-requirement level 38
leaf turgor 268
leather complex 156
leather industry 162
legume-based rotations 110
legumes 111, 183, 187, 199, 202, 212, 218, 222, 261, 265, 273
Lens culinaris 221
lentil 221
LEPA. *See* low-energy precision application
lepidopteron pests 201
LER. *See* land equivalent ratio

lettuce 243
Leucaena leucocephala 222
ley farming 215
LID. *See* limited irrigation dryland system
lift irrigation 24, 217, 221
light interception 177
lime 102
limited application duration 299
limited irrigation 54
  alfalfa 54
  crop production systems 50
  cropping systems 55
  dryland system 192
  practices 43, 52, 54
  strategies 54
  water 44, 192
  water availability 44
  yields 52
limited water management practices 55
  supply 50, 216
linkage drag 251
lipid peroxidation product 265
lipid peroxidation-derived reactive aldehydes 242
livestock 22, 111, 149, 208, 217, 287
location-specific technologies 66
long-duration varieties 109
longleaf pine 64
  reforestation program 64
long-term memory of pollutant accumulation 138
long-term perennials 215
long-term sustainability 17, 44
loss of body condition 146, 161
loss of forest lands 12
loss of hair 146, 161
loss of nutrients 280
low productivity 198, 202, 209–10, 239
low temperature, tolerance to 267
low water-use crop 50
low-density polyethylene 272
low-energy precision application 16
low-molecular-weight metabolites 263

lucerne mitochondria 241
Ludwigia spp. 302
Lycopersicon esculentum 291
lysimetric system 271–2

M.S. Swaminathan Research Foundation (MSSRF) 93, 95, 109, 111, 279–80
M6PR gene 240
MABC. *See* marker-assisted backcross
macromolecules 232, 262
macropores 12, 152
Macrotyloma 118
maize 12, 24, 31–2, 37, 81–2, 84–5, 95, 146, 151, 158–9, 161, 163, 185, 192, 196, 199–200, 210, 229–30, 232, 240, 242, 246, 250–1, 279, 282–3, 285, 292, 294
maize-rapeseed system 279
maize-wheat production systems 12
maize-wheat system 279
malonaldehyde 265
Malwa Plateau 95
management of crop water stress 50
management of groundwater 71–2
management of land and water on watershed basis 216
Mangifera indica 125
mango 110, 125
Manilkara 124
manipulating plant geometry 169
manipulating tillage 169
mannitol 229, 240
mannose-6–P isomerase 240
manure 9–11, 39, 118, 151, 163, 216, 285, 293
marginal lands, cost of bringing under irrigation 7
marker-assisted backcross 250
marker-assisted introgression 249
marker-assisted recurrent selection 250
marker-assisted selection (MAS) 246
MARS. *See* marker-assisted recurrent selection
MAS. *See* marker-assisted selection
maturity 185, 248, 262
maximization of water use 46
MDA. *See* malonaldehyde
MDAR. *See* mono-dehydro-ascorbate reductase
mean weight diameter 223
mechanisms of drought stress 273
medical insurance 219
medicinal and aromatic plants 217
ME-lea N4 gene 243
melting 3–5
melting of glaciers 6
membrane lipids 264
membrane-stabilizing effect 243
Merrimack river initiative 64
metabolic traits 263
metabolite increase 267
metal toxicity 230
metallic products 155
methemoglobinemia 149
method of irrigation 16, 38, 40
Methods of Soil Analysis 282
methyl viologen-induced oxidative stress 240
metsulfuron 297, 300, 302, 304, 307, 310
Mexican wheat varieties 76
micro-aggregates 211
microbial activity 16
microbial biomass 110, 137, 222–3
microbial diversity 214
microfauna 310
micro-finance-assisted micro-enterprise 127
microflora 310
microirrigation 111, 216–7
micronutrients 33, 189, 198–201
microorganisms 207, 214, 239
microsatellites 246
micro-watershed 195
migration to cities 210, 223
millet 31–2, 46, 50, 171–2, 197, 215, 249
mine spoils, rehabilitation of 218
mineral deficiency 230
mineral mixture 293
mineral weathering 62
minimum support price (MSP) 24, 285
minimum tillage 45, 212
minor millets 117, 129
mitigating water pollution 160
mixed farming 215
mixed-cropping systems 212
mixing of water of different salt concentrations 38
Mn deficiency-mediated oxidative stress 241
mobile soil-testing laboratory 126
model species 264
modern agronomic practices 209
moisture carrying capacity 201
moisture retention 289
moisture scarcity 209
moisture-use efficiency 203, 217
molecular introgression 261
molecular markers 229–30, 246, 250, 265
molecular techniques 246
molecular-marker technology 230, 246
Mollisols 110, 210
molybdenum 203
Momordica charantia 291
monocropping system 93
mono-dehydro-ascorbate reductase 242
monsoon failures and floods, social impact of 22
monsoon rains 7, 13, 87, 119
monsoon season 14, 34, 85–7, 93, 95, 98, 106, 111, 173, 208, 215, 279, 284–5
monsoonal flooding 134
moong 84
morphological changes 241
Mriga 102
MRS. *See* maize-rapeseed system
MSP. *See* minimum support price
MSSRF. *See* M.S. Swaminathan Research Foundation

mtlD 240
MTRS. *See* mustard-rapeseed system
mulberry 161
mulch farming 215
mulch of rice 287
mulching 84, 110, 116, 119–22, 124–5, 172, 222
multi-component strategies 210
multigenicity of the plant response to stress 266
multi-nutrient deficiencies 211
multiple well points 83
multipurpose trees 218
mung bean 192, 199, 292
mustard 146, 159, 161, 183, 282, 293
mustard-rapeseed system 282
mutagenesis 230
MWS. *See* maize-wheat system

N concentration in groundwater 151
N concentration in well-water 150
N in groundwater 150–151
N. tabacum 242
NAC gene family 245
NAC transcription factor genes 229
nail drop 146, 161
nalla bunds 67
Nangal headworks 74
narrow beds 84
narrow ridge and furrow 190–1
national average yield gap 183
national environmental policy, goal of 138
National Rural Employment Guarantee Act (NREGA) 218
national watershed development project 66
native rhizobial population 202
natural resource management 184
natural resources, mismanagement of 17
natural systems 62, 134
natural vegetation 218
near-famine situation 208
near-isogenic lines 248
necrosis of tip of tail 146, 161
negative water balance 80
net photosynthesis 242
net profitability 307
neurotoxins 149
NGOs. *See* non-governmental organizations
nickel 144, 155
night soil 151
NILs. *See* near-isogenic lines
nitrate 10–2, 17, 64, 87, 143–4, 149–51, 162
  leaching 11, 145
  and pesticide losses to surface water 17
  and pesticide pollution of groundwater 10
  and phosphate leaching 143, 162
nitrification 149
nitrite 149
nitrofen 299
nitrogen 11–2, 17, 31, 87, 149, 285, 293, 299
no tillage 176, 287
nodulation 121, 203
non-agricultural uses 29
non-classified soils 210
non-conventional energy-saving devices 67
non-drought period 49
non-forest wastelands 66
non-irrigated environment, crop rotations for 43
non-point source pollution 17, 21, 26, 62, 133, 135, 137–8
normalized transpiration 270
no-till 12–3, 16, 29, 43, 45–6, 49, 54, 169, 174–8, 283, 287
NREGA. *See* National Rural Employment Guarantee Act
NT. *See* no-till
NTR. *See* normalized transpiration ratio
NUE. *See* nutrient use efficiency
Nullah-irrigated soils 157
nutrient budgets 140
nutrient cycling 287
nutrient deficiencies 208, 211–2
nutrient dissolution 214
nutrient management 16, 36–8, 100, 106, 121–3, 135, 138, 140, 152, 187, 198, 201, 212, 214, 249
nutrients in aquatic systems 138
nutritional constraints in soils 213

OA. *See* osmotic adjustment
oats 146, 161, 163, 243
Occoquan reservoir watershed 65
off-farm income 126
off-season employment 218
off-season generation of biomass 212
off-season tillage 221
Ohio State University 29, 93, 95, 109, 111, 279–80
oil cake 102
oil seeds 81, 109, 171, 183, 208, 215, 221, 264
okra 109, 119, 124
old-age pensions 220
Olsen-P 152–3
on-farm community watersheds 195
on-farm evaluation of watershed technologies 193–4
on-farm trials 199, 203
on-farm water management 38, 291, 294
on-farm watersheds 195, 203
onion 109, 119
on-station research 195, 201
on-station watersheds 184, 193
OPEC. *See* Organization of Petroleum Exporting Countries
open channels for water transport 8
optimum fertilization 215
organic carbon 16, 31, 39, 48, 96, 111, 152–3, 157, 186–9, 210–1, 214, 221–3, 279, 282, 290
organic farming 207, 218–9
organic manuring 215
organic matter 13, 87, 187, 211–2, 292
organic nitrogen 299

organic treatments 223
organic wastes 111
Organization of Petroleum Exporting Countries 175
Oryza sativa 31, 98, 151, 221, 280, 297–8
OsLEA 3-1 gene 243
osmolyte 229, 231–2, 240, 263, 265
osmoprotectant 232, 264
osmoprotection 240
osmoregulation 265
osmoregulatory genes 265
osmotic adjustment 246, 262
 mediator 232
osmotic stress 232, 240
osmotically adjusted media 268
OsNAC6 245
otsA and *otsB* 241
overexploitation and pollution of groundwater 21
overexploitation of free energy 25
overexploitation of water 13, 24–5, 33, 145
over-pumping of aquifers 8
oxadiazon 299
oxidation of choline 239
oxidation of organic matter 211
oxidation reduction 148
oxidative stress 241–2
oxi-hydroxides of iron 148
Oxisols 110, 210
oxyfluorfen 299

P5CS. *See* pyrroline-5-carboxylate synthase
P5CSF129A 265–6, 269
package of practices 100, 220
paddy straw 83–4
Palmer drought severity index 47–8
pan evaporation 82
Panchayati Raj Institutions 58–9, 66
panicle harvest index 249
panicle number 248
paraquat-induced oxidative stress 242
participatory approach 66, 195
participatory operational research model 129
participatory research 94, 110–11
participatory rural appraisal approach 111
particle gun 229, 232, 240, 242, 244
Paspalum 117
Passioura's model 272
pasture development 67, 215, 218, 220
pathogenic microorganisms 160
PAU 201 rice 83
paucity of water 262
PE. *See* production efficiency
pearl millet 146, 163, 171, 182–3, 192, 209, 221, 230, 249, 251
PEG. *See* polyethylene glycol
peg formation 117
peg penetration 272
Pekinensis 243
Pellusterts 95
pendimethalin 299–300, 302, 304, 307
Penman 170
Pennisetum glaucum 249
penoxsulam 297, 300, 302, 304, 307, 310
per capita availability of land 10
per capita availability of water 22
per capita water storage 22
percolating soils 162
percolation 8, 14, 26, 34–5, 67, 187, 217, 298
perennial flow channels 73
perennial growth habit 43, 55
perennials of industrial importance 215
permanent crops 214
permanent pasture 214
permeable soils 196
pesticide losses to shallow groundwater systems 11
pesticide pollution of groundwater 10, 12
PET 170, 172–3
phenology 246
phenological traits 262
phenotypic evaluation 267, 272
 of transgenic plants 267
phenotypic selection 251
pheromone traps 199
phosphate 104, 110, 143–4, 149, 162, 231, 240–1
phosphate fixation capacity 110
phosphobacteria 118
photo-inhibition 239
photoperiod-insensitive variety 248
photoperiod-sensitive variety 248
photosynthesis 33–4, 177, 242, 245
photosynthetic activity 266
Phyllanthus emblic 124
physical deterioration 210
physiographic regions 57
phytoremediation 159, 161
pigeonpea 93, 101, 106, 109–10, 115–19, 121–4, 129, 182–3, 185, 187, 189, 192, 194, 222
pitcher-pot irrigation 110, 125
placement of chemicals 15
plant architecture 178
plant available nutrients 113
plant available soil water 169, 173, 176
plant biomass 269
plant environmental stress 230
plant extractable water 186
plant genetic engineering 229–31
plant geometry 169
plant growth and development 271
plant height 243, 248
plant population 177–8
plant productivity 137
plant residues 138, 169
plant responses to drought 229
plant water relations 248
plant water use 36
plant water-stress indicators 246
plow layer 157
pod borer 116
pod filling 117
point source of nitrates 151
point-source controls 65
policies for water management 217
policy changes 207, 210, 219

pollutant loads 133, 138–9
pollutant trading schemes 61, 65
polluted soils 158–9
polluted water 157–8, 163
pollution control 61
pollution of rivers, lakes, and groundwater 25
polyamines 266–7
polyethylene glycol (PEG) 242, 248
polyvinyl chloride (PVC) 124, 271
Pongamia 218
poor quality water, reuse of 3
poor soil fertility 210
poor soil health 207, 209
poplar 161, 280, 292–4
population pressure 17
Populus deltoids 292–3
positive C budget 39
post-emergence herbicides 299
post-harvest processing 217, 221
post-harvest soil-fertility 118
post-harvest value-addition 210
post-rainy season crops 192–3, 197
potable water 160
potassium 151, 153
potato 84, 151, 202, 229, 241
potential yield 183, 210, 229–30
poultry manure 151, 293
poverty 21, 59–60, 171, 182, 184, 210, 217
poverty alleviation 60
PRA technique 220
precipitation-storage efficiency 45–7, 54
precipitation-use efficiency 43, 174
precision agricultural approach 207, 217
percision agricultural operations 217
precision farming principles 213
pre-emergence herbicides 299
premature graying of hair 163
pre-monsoon planting of flooded rice 33
pre-monsoon tillage 221
pre-sowing irrigation 102, 193
pressmud compost 129
pressurized irrigation 189
pretilachlor 297, 300, 302, 304, 307, 310
PRI. *See* Panchayati Raj Institutions
principles of watershed management 61
private sector engagement 25
problematic soils 212
processed agricultural wastes 109, 114
process-oriented research 140
procurement price 71, 85, 88, 209–10, 219
production deficit 182
production diversification 111
production efficiency 120
productivity enhancement 184, 204
profile-stored soil moisture 95
profitability 209, 212, 217, 297–8, 307
prolific root system 216
proline 229, 231–2, 265–6
  synthesis pathway 231
promoter elements 232
promoter, choosing a 266
Prosopis spp. 112
protecting environment 162
protein kinases 243
proteinases 243
proteomics 230
public environmental policies 17
puddled transplanted rice 297, 299, 310
pulses 76, 81, 109, 111, 129, 171–2, 208, 215, 221
pump-type tubewells 75
Punjab Agricultural University 71, 81, 143, 151, 229, 279–80, 297, 299
Punjab agriculture 72, 77
Punjab farmers' indebtedness 78
Punjab irrigation system 72
Punjab State Electricity Board 77
Pusa 44 rice 83
PVC. *See* polyvinyl chloride
pyrimidinyloxybenzoic chemical family 302
pyrroline-5-carboxylate reductase 232
pyrroline-5-carboxylate synthetase 231, 265

QTL. *See* quantitative trait loci
quality seeds 93, 105, 216
quantitative trait loci (QTL) 230

R & D organization 220
rabi sorghum 183
radish 117
rainfall runoff 197
rainfall-use efficiency 120, 181, 187, 196, 199, 204
rainfed agricultural commodities 209, 219
rainfed agriculture 7, 10, 35, 71–3, 110, 129, 167, 182, 184, 207–11, 215, 217, 219–21, 223, 262, 264
  agro-ecology 214
  areas 66, 182, 184, 198, 208–12, 215, 221
  conditions 72, 116, 208, 212
  crops 24, 30, 182–3, 207, 209–10, 212, 217–21
  database 220
  ecosystem 209
  environments 203
  farmers 219
  farms 209
  horticulture 124, 217
  land holdings 218
  regions 182, 208–9, 211–12, 215–16
  rice 182–3
  soils 112
  system 197–8
  wastelands 218
  wheat 182–3
rainfed-area rural community 218
rainwater conservation 35, 93, 98, 218
rainwater harvesting 87, 120, 161, 218
rainwater management 95, 105
rainwater runoff 60
raised sunken bed system 93–4, 98–9, 106

random amplification of polymorphic DNA 246
RAPD. *See* random amplification of polymorphic DNA
rapeseed 31–2, 158–9, 163, 183, 282, 291
Raphanus sativus 117
ravine reclamation 218
raw pressmud 109, 118
rd 29A:DREB1A 244
rd17 267
rd29A 266–7, 269
rd29B 267
rd29A promoter 267
rd29B promoter 267
reactive oxygen species 241
realized yields 210
recharge of groundwater 81–2, 85–7
reclamation of salt-affected soils 76
recombinant inbred lines 273
recombinant-DNA technology 231
recommended management practices 37, 214
recovery technology 135
recreational use 138
recurrent parent genome 251
recycling 29, 36, 39–40, 109, 114, 161, 197, 211–12, 215, 218
red soils 109–11, 117, 129, 196
redox proteins 229, 241
reduced consumptive water use 50
reduced input costs 50
reduced irrigation water quantities 50
reduced leaf area 262
reduced plant populations 169
reduced stomatal conductance 262
reduced tillage 174, 176, 215, 221, 223
regulatory genes 229, 243, 266
  proteins 243
regulatory gene pathways 264
rehabilitation of rainfed wastelands 207
relative growth rate (RGR) 248
relative humidity 81–2
reluctance to move 146, 161
remote sensing 35
renewable fresh-water resources of India 30
renewable surface water resource from rains 7
renewable water resources 6–7
renewable water supplies 7
renewable water withdrawals, user of 6
renovation of ponds 87
reservoir storage of surface water 135
residue incorporation, economics of 279, 289
residue management 29, 36, 174–5, 212, 287–9
residue protection of the soil surface 46
residue recycling 211–12, 215
residue removal or burning 39
residue retention 36, 39, 49
resilience to water stress 53
resource use efficiency 121, 184, 186, 195
resource-poor farmers 212
respiration rate 34
restoring degraded soils 35
restricted layer 87
restriction fragment-length polymorphisms 246
retention of crop residue 54
return on variable expenses 279
reverse flow mechanism 79–80
RFLPs. *See* restriction fragment length polymorphisms
RGR. *See* relative growth rate
rhizobium 104, 118
Rht-B1 249
rice 24, 29–31, 33–4, 36–40, 71–3, 75–6, 78, 80–6, 98–9, 101–2, 106, 144–6, 151–2, 160–2, 182–3, 202–3, 210, 221, 229–30, 232, 239, 241–6, 248, 251, 279–80, 283, 285–90, 293–4, 297–300, 302, 304, 308, 310
  bowl of India 24
  straw burning 33
  transplantation 71, 80, 83–5
rice-wheat cropping system, alternatives to 279
rice-wheat cropping system, residue management in 286
rice-wheat system 39, 279–80, 294, 298
Ricinus communis 221
ridge planting 101
ridge-furrow system 93, 100, 106, 121, 124
ridge-till 12, 16, 174–5
RIL. *See* recombinant inbred lines
riparian vegetative buffer strips 61
riparian zone 62
rising of water tables in canal-irrigated areas 14
risk-bearing capacity 207, 209
river basins 5, 60, 63, 134
River Inter-Linking project 22
  future of 22
River Valley Program 58, 66
river water systems, sustainability of 4
rivers of the world 5
river-valley projects 73–4
RMPs. *See* recommended management practices
rock phosphate 109, 118, 129
rockout 210
Rohu 102
rooftop rainwater-harvesting technology 87, 217
root biomass 211–2, 232
root dry weight 271
root growth 121, 221, 271, 302
root length 243, 248, 266
root QTL9 248
root response 271
root rot 202
root surface area 272
root system 216
root traits, breeding for 270
root volume 272

root-length density 272
root-restrictive ground layer 110
roots 59, 158, 173, 187, 212, 242, 246, 268, 270–2, 302
Ropar headworks 74
ROS. *See* reactive oxygen species
rotation cycle 215, 285
rotations with low water-use crops 50
ROVE. *See* return on variable expenses
row spacing 84, 300
RUE. *See* rainfall-use efficiency
runoff problems 171
runoff processes 211
rural development programs 220
rural livelihoods 22, 182
rural poor 25, 182
RVP. *See* River Valley Program
RWS. *See* rice-wheat system

Saccharum 118, 232, 290
safflower 31–2, 192
saline and sodic soils 212
saline drainage water for irrigation 38
saline groundwater 73, 79–80, 83
saline land utilization 218
saline water for irrigation 38
salinity 14, 16, 73, 145, 210–11, 230, 232, 240–2, 244–5, 264–5, 267, 280
  and alkalinity in Gangetic Plains 14
salinity in productive soils 14
salinization 13–14, 24, 211, 230
  and alkalization 14
salt concentration 79
salt flats 210
salt stress 232, 239–40, 242–3, 245, 249, 262–3
  tolerance 232, 245
sample survey 286
sand dunes 218, 282
sand dune stabilization 67, 218
sandy loam 279, 287, 291
sandy soils 34, 152, 190
sapota 110, 124–6
SAT. *See* semi-arid tropics
saturation point 271
scaled-back point-source 65
scarcity of groundwater 298
scarcity of moisture 211
scavenging heavy metals 214
scented variety 300
scheduled castes 66
scheduled tribes 66
scope of watershed planning effort 139
scouting through pheromone traps 199
scum 138
Se accumulation in soil 146
Se toxicity 146, 161
sea levels rise 3
secondary carbonates 214
secondary nutrients 198
securing India's water future, challenges 21
sedges 300, 302, 304
sedimentation problems 17
seed availability 216
seed banks 216
seed delivery system 105
seed farmers 93, 105
seed priming with sodium molybdate 203
seed production 105, 221
seed replacement 93, 105
seed village 93, 105–6, 216
seedbed preparation 36, 95, 286
seed-cum-fertilizer drills 210
seeding plants in clumps 169
seeding rates 171
seepage 75, 80, 189–90, 298
selective herbicide 302
seleniferous region 146, 161
selenium 143–6, 161
selenium toxicity 143
self help group 126–7, 129
semi-arid areas 146, 170
semi-arid climatic zone 282
semi-arid environment 43–5, 178
semi-arid regions 14, 33, 54, 129, 169, 172–3
  groundwater table fall 14
  subtropical 299
  tropics 124, 181–2, 208, 261–2
senescence 149, 187, 248
sequence-tagged sites 246
sequential crop 115–6
sewage effluents 144, 157
sewage irrigation 157, 159, 163
sewage sludge 153–4, 162
sewage water 87, 143, 153–7, 162–3
Shahpur Kandi canal 74
shallow cultivation 190
shallow hand-pumps 143, 156, 163
shallow polluted water 163
shared use of river waters, disputes on 25
shedding of horn 146, 161
sheep 13, 208
sheet erosion 95
shelter belt 67, 218
shift in cropping pattern 81
Shivalik hills 86
short stem 216
short-duration varieties 109
short-rotation forestry 215
short-season crops 216
shrinking land and water sources 202
side slope 46
Sierozems 110
signal transduction 243
silvi-pasture 217, 222
simple sequence repeats 246
simulated potential yields 183
simulated rainfed potential yield 183
single strand conformation polymorphisms 246
single-gene breeding 231
single-gene products 264
single-nucleotide polymorphisms 246
sink capacity 215
sinks for carbon 215
Sirhind canal 74, 86
site-specific land use 140
skeleton fluorosis 147
skip rows 178
skip-row configurations 177
slicken sides 96
sludge 39, 138, 153–4, 157, 162
small farm size 207
small implements 221
small-farming communities 210

small-scale rainwater-harvesting systems 25
smectites 96
smokeless stove 129
SNAC1 245
snow-white chlorosis 146
SNPs. *See* single-nucleotide polymorphisms
SOC. *See* soil organic carbon
social factors 207
social mobilization 126, 129
social security 4
social welfare 69
socio-economic aspects 59
socio-economic base 209
socio-economic conditions 57, 135, 207, 220
socio-economic shifts 134
socio-economic well-being 134
socio-political system 298
SOD. *See* superoxide dismutase
sodic and saline soils, reclamation of 16
sodic soils 285, 293
sodicity 145, 210
sodium carbonate 145
soft loans to small and marginal rainfed farmers 219
soil acidification 145
soil acidity 211
soil amendments 38, 137
soil and crop management systems 43
soil and crop-residue management 44
soil and water conservation 95, 119, 129, 195
soil and water management 167
  goal of 33
soil and water quality 10, 131, 133, 135–6, 139
soil and water resources 29, 31, 95
soil C sequestration 214
soil carbon levels
  in agroforestry 215
soil conservation 58, 68, 94, 119, 171
soil degradation 33–4, 110, 134, 182, 209, 211
soil erosion 9, 13, 15–6, 43–5, 49, 61, 93, 112, 135, 138, 174, 182, 185–7, 196, 211, 214, 221
  causes of 13
  reduction in 16
  as source of water pollution 13
soil evaporation 272
soil fertility enhancement 106
soil fertility management 118, 198, 218
soil fertility, mining of 214
soil health 126–7, 214, 219, 298
soil health cards 126–7, 129, 214, 219
soil hydrologic budget 33
soil inorganic pool 214
soil loss 186–8, 221–2
soil macroaggregate formation 47
soil microorganisms 214
soil moisture conservation 121
soil moisture depletion 269
soil organic carbon 31, 188
soil organic matter 43, 110, 138, 173–4, 210
  loss of 45
soil pH 157
soil quality 31, 35–6, 110, 129, 133, 136–8, 185–6, 207–8, 210–2, 214–5, 218, 222–3, 280, 285, 287, 289
soil quality improvement 207, 210
soil salinity 33, 73, 80
soil structure 36, 54, 298
soil test network laboratories 214
soil test-based fertilizer recommendation 211
soil testing 93, 106, 126, 129
soil water conservation 216–17
soil water management 29
soil-conservation treatments 121
soil-disturbing activities 174
soil-exhaustive cropping system 280
soil-fertility management 93, 199
soil-management treatments 122
soil-moisture availability 208
soil-nutrient bioavailability 137
soil-plant-animal-human food chain 153, 160
soil-quality assessment 211
soil-quality decline 137
soil-quality improvement 207, 211
soil-quality index 212
soil-quality indicators 137–8
soil-surface management 38
soil-water availability 171
soil-water management 29, 31, 36
  strategies of 29
soil-water reserves 117, 120
soil-water storage 33, 176
soil-wetting pattern 190
Solanum 151
solar radiation 120, 170
solar radiation-use efficiency 120
soluble salts 75, 147
SOM. *See* soil organic matter
sorghum 31–2, 82, 84, 171–4, 176–7, 182–3, 185, 191–2, 194, 209, 221–3, 232
  yield 222
sources and loads of pollutants 139
sources of pollution in watersheds 138
sowing date 217
sowing of rainfed crops 221
soya meal 293
soybean 10–2, 17, 31–2, 93, 95, 98–106, 182–3, 186–7, 189, 196, 201, 221, 232, 242
  yield 101
soybean-based cropping systems 201
SOYGRO model 201
spatial distribution of rainfall 23
spatial variability 22
  in rainfall 22
spatial variation 183
spatial variation in soil properties 212
special subsidies 219
specific water use 204, 248
Sphenoclea spp. 300
spinach 158–9, 163, 232, 239

Spinacia oleracea 158, 239
split application of fertilizer N 152, 162
split-N or multiple applications 15
spring wheat 173
sprinkler irrigation 191, 216
SRUE. *See* solar radiation-use efficiency
SSCP. *See* single strand conformation polymorphisms
SSR. *See* simple sequence repeat
stable isotopes 230
stale seedbed technique 299
standard of living and renewable water resources 7
state-based agro-industries 217
stiff gait 146, 161
stomata 268
  closure in maize caused by drought 34
stomatal conductance 262–3, 266
stomatal pores 245
stopping free power to tube-wells 85
stored soil moisture 222
stored soil water 50, 169, 175
strategic planning, need for 210
strategy to adapt to climate change 34
straw incorporation 279–80, 289–90
straw yield 248
stress imposition 273
stress inducible *rd 29A* promoter 241, 244
stress timing 263
stress tolerance 52, 232, 241–2, 244, 249, 261, 263, 266–7, 273
strress tolerant crops 267
stress-induced LEA proteins 229, 242
stress-inducible expression 241, 243, 272
stress-inducible expression of *ZmDREB2A* 244
stress-inducible genes 244, 264
stress-inducible proline 232
stres-inducible promoter 266–7
stress-inducible *rd 29A* promoter 241, 244
stress-inducible transcription factor 264
stress-inducible transgene expression 241
stress-related genes 245, 266
stress-responsive NAC 1 245
stress-tolerant crops, breed for 265
stress-tolerant legumes 261
strip cropping 13, 61
strip-plot design 119
structural genes 229, 231
STS. *See* sequence-tagged sites
stubble mulch tillage 45, 173–4, 176
stubble retention 212
stylo 222
Stylosanthes hamata 222
Suaeda salsa 242
subabul 222
sub-humid climatic zone 281
submergence 262
submersible pumps 78
sub-molecular basis of plant responses to stress 241
sub-montane soils 110
Sub-Saharan Africa 30, 135
subsidies on agricultural inputs 220
subsidized electric power to tubewells 71, 77–8, 81–2, 85, 88
subsidized inputs and assured income 24
subsidized purchase of implements 217
subsidy 21, 25, 71, 85, 88
subsurface drain water 11–2
subsurface drainage 11, 14, 16
  lack of 14
subsurface hardpans 298
successful watershed-management 63–4
sugar beet 232, 239
sugarcane 109, 118, 127, 146, 161, 229, 232, 245–6, 290–1, 294
suicides 210
summer crops 173
summer fallow 172, 174
summer green gram 84
summer moong 84
summer plowing 117
summit 46
sunflower 50–1, 146, 222
sunlight 44, 178
sunnhemp 109, 118, 129
superbasmati 300
superoxide dismutase 241
superphosphate 102, 293
supplemental irrigation 35–7, 40, 115, 189, 191–3, 197, 202, 204
  crop responses to 192
  efficient application of 190
  water 190–1, 193
supply chain of knowledge 26
supra-optimal 110
surface and ground water pollution 29
surface irrigation 14–5, 111, 124, 189–91
surface retention of residues 36
surface runoff 10, 95, 110, 184, 187–8
surface water bodies 17
surface water-based irrigation 21–2
surface-residue application 212
surge flow system 191
sustainability
  of ecosystems 13
sustainability of rainfed-farming systems 209
sustainable agricultural systems 54
sustainable agriculture 58, 129, 230, 280
sustainable approach to managing pests 199
sustainable biological productivity 137
sustainable economic agricultural productivity 139
sustainable food production 229–30
sustaonable food production system 4
sustainable institutional mechanism 70
sustainable livelihoods 60
sustainable management of soil and water resources 29, 31

sustainable management of dryland alfisols 109
sustainable management of water resources 182
sustainable soil management strategies 91
sustainable sources of income 59
sustainable use of water resources 33
sustainable water management strategies 27
sustainable water system challenges 3
sustainable watershed management 133, 136
SWAMI 34
sweep plow 173–4
sweep tillage 175
sweet poison 78, 88
sweetgrass 64
system productivity 204

Talkatora Lake 12
Talwara headworks 74
Tamilnadu Highlands 208
target population of environments 248
target-area approach 66
Tar-Pamlico River Basin 65
Tawa Command 95, 96
TE. *See* transpiration efficiency
technology development and evaluation 184
technology exchange model 182
technology transfer 139, 195, 221
temporal variation 183
temporary dormancy 216
tensiometer 84, 85
terminal drought 203, 272
terminal stress 249
terracing 13, 218
terracing grass waterways 13
terrestrial ecosystem 215
terrestrial soil carbon sequestration 214
textiles 155
TFP. *See* total factor productivity
the Beas 73
the Ravi 73
the Sutlej 73–4, 86
Thellungiella halophila 245
thermal imaging 230
thermal power 135
thermo tolerance 244
thiobencarb 299
threshold levels for Cd and Ni toxicity 163
threshold value 176
ThZF1 245
Tibetan Plateau 5
tidal wetlands 62
tied ridges 110, 119–22, 124, 129
tillage method and crop management 38
tillers 178, 243
timely agricultural operations 207, 217
timing of chemical application 15
timing of irrigation 189
tissue moisture stress 216
tissue-culture techniques 231
tissue-specific promoters 266
TMDL. *See* total maximum daily load
tobacco 229, 232, 239–42, 263
toe slope 46
tolerance mechanisms 230, 263
tomato 110, 124, 192, 229, 291
top-down technical approach 66
topography 73, 86
toposequence 186
total factor productivity 285
total maximum daily load 138
total weed biomass 302
toxic concentrations of ammonia 65
toxic elements 143, 154–6, 159, 162–3
toxic metals 144, 156–60
toxic organic compounds 137
toxic substances 138
TPE. *See* target population of environments
TPP. *See* trehalose-6-phosphate phosphatase
TPS. *See* trehalose phosphate synthase
TPS1 and *TPS2*, 241
TPS1–TPS2 fusion gene construct 241
TR. *See* transpiration ratio
trace metals 137
tractor-drawn small and medium implements 217
traditional farming system 185
traditional system and management 120
training of farmers 219
trans-acting elements 267
transcription factor 229, 243–5, 264, 266–7, 271
transformation process 261, 267
transgene expression 266
transgene response 268
transgenes 246, 267
transgenesis 231
transgenic approach 261, 263
transgenic breeding 231
transgenic constructs 267
transgenic cultivars 261, 267
transgenic dicot plants 263
transgenic evaluation for drought 268
transgenic events 244, 265–6, 270–3
transgenic interventions 263, 273
transgenic peanut 244
transgenic plants 229–30, 232, 239–44, 251, 261, 263, 266–9
transgenic research 264, 273
transgenic rice 245
transgenic technologies 261, 263, 266
transgenic tobacco 239
transgenic wheat 242
transgenics 229–31, 240–1, 268
transgenics, drought evaluation of 267
trans-Indo-Gangetic Plains 298
transpiration 9, 216, 261–2, 265–6, 268–70, 272
transpiration efficiency 244
transpiration ratio 269
transplanted rice 161, 298, 300, 310
transplanting in puddled soils 298

transplanting method 297
trapezoidal embankments 196
treated municipal water for irrigation 8
treated wastewaters for forestry plantations 9
treated wastewaters for non-edible crops 9
treated wastewaters for public lawns 9
tree-based systems 215
tree-green leaf manuring 211–2
tree-growth models 215
trehalose 229, 240–1, 267
trehalose-6-phosphate phosphatase 241
trenching 218
Trianthema 300, 302
Trichoderma viride 118
Trichogramma parasitoid 127–8
Trifolium alexandrinum 159, 290
trihalomethane 149
Triticum aestivum 31, 44, 93, 221, 280
Tropicultor 194
tubewell infrastructure cost 85
tubewell irrigation 33–4, 76, 78, 80–1, 158
tubewell-irrigated soils 157–9
two-row cultivar 250
Typic Haplustalf 282

U.S. Environmental Protection Agency 11, 58, 61, 63–4, 138
U.S. National Research Council committee 136
ubi1 promoter 242
ubiquitin 1, 240, 267
udic Ustochrept 282
uidA
GUS 243
uncertain rainfall pattern 211
uncultivated furrows 190
undernourished population 6
undisturbed soils 290
undulating farmlands 191
UNCOD. *See* United Nations Conference on Desertification
unequal access to water 33
United Nations Conference on Desertification 169
unpalatability of drinking water 149
unregulated groundwater pumping 21–2
unsustainable irrigation system 14
untreated municipal and industrial wastewater 25
upland crops 30, 62, 95, 98, 208
Upper Bari Doab canal 74–5
urban pollution 12
urban wastewater 40
urban water 25, 39, 55
urbanization 8, 29, 86, 135, 144
uridine diphosphoglucose 241
USDA Natural Resource Conservation Service 175
use of organics and inorganics 212
USEPA. *See* U.S. Environmental Protection Agency
using an efficient irrigation system 35
ustic soil moisture 95
Ustipsamments 282
Ustorthents 95
utilization of CO2 from the atmosphere 215

value-addition modules 207, 217
varagu 117
Vayalogam series 111
vegetables 8–9, 81, 111, 152, 159, 163, 190–1, 193, 197, 202, 280, 291–2
vegetative barriers 218
vegetable filter strips 17
vegetable growth 178
vernalization genes 249
Vertic Inceptisols 186–7, 189
vertical mulching 222
Vertisol catchment 185
Vertisol technology package 194
Vertisol watersheds 184
Vertisols 93–6, 101–2, 105, 110, 171, 181, 184–6, 190, 192–3, 195–6, 201, 208, 210, 216, 222
Vigna aconitifolia 232
Vigna mungo 115
Vigna radiata 115, 292
Vigna unguiculata 115
village ponds 67, 87, 112, 114
village watersheds 194
Vindhyan and Narmada Valley 95
v-shaped sweeps 173

WAE. *See* water-application efficiency
wage system 218
washing of floors and vehicles 161
waste storage 138
wasteland development 66, 124
wasteland management 219
wastelands, rehabilitation of 61
wastewater 8–9, 64–5, 145, 155, 159
  production in India 8
  treatment 8
water and agricultural sustainability challenges 1
water application efficiency 190, 192
water application methods 189
water availability, major sources of 4
water borne diseases 3
water budget 33
water capture 271
water chlorination 149
water conservation 7–8, 18, 26, 36, 58, 66, 69, 95, 101, 109–1, 114, 172–3, 182, 184, 195, 204, 211, 216–7
  via improved drainage system 15
  via improved irrigation system 15
  options 50
  practices 43, 129
water conserving practices 54
water contamination 8, 10, 16

water deficit 184, 241–2, 244, 249, 261, 265, 269–72
conditions 250, 272
stress 243
tolerance 242
water depletion and pollution 160
water distribution 7, 134, 190–1
lack of distribution systems 7
water efficient crop rotations 51
water erosion 13, 73, 174, 210–11
water harvesting 29, 35, 40, 103, 109, 111, 114–15, 119, 129, 182, 197, 216–17
plans 5
in ponds 204
projects 13
and storage 35
strategies 7
structures 67, 197–8, 220
sustainability of 5
water holding capacity 96, 186, 191, 222, 271
water infiltration 95, 290
water insecurity 26
water intensive crop (paddy) 24
water limited agroecosystems 55
water limited cropping systems 44
water logging 201, 210
water loss 190, 268–9, 271
water management 3, 27, 35, 37–8, 55, 69, 72, 82, 101, 110, 112, 167, 189, 195–6, 217, 231, 280, 291, 294
policies 18
practices 65
systems 134
Water Management Training Institutes 60
water pollution 13, 58, 143, 145, 160, 162
Water Pollution Control Commission 64
water pollution potential 162
water pollution standards 58
water potential 240
water productivity 33, 35, 43, 52, 54, 191, 297, 307–8
water projects, investment in 4
water quality 3, 9, 11, 17, 62–3, 80, 133–4, 136–8, 145, 152, 280
affected by animal manure 9
affected by chemicals 9
decline 137
enhancement 16
problems 24
in rivers 26
in the watersheds 62
water recharge 111, 145
water relation studies 212
water reservoir/sources 4, 184
water resources, degradation of 15
depletion 24
management 3
problems 135
in Punjab 77
quality 136
undervaluation of 30
water retentive capacity 208
water saving 145
crops 182
feature 310
cropping systems 50
practices 145, 160
technologies 84–5
water scarcity 182, 198
water security 9
water shortages 134
water storage 5, 173, 218
water stress 22, 52–3, 171, 178, 216, 240, 242, 244, 248, 251, 264–7, 270–1
water sustainability 3–4, 9, 15, 18, 21
best practices for 15
challenges to 21
for global food security 9
water table depth
in Punjab 33
fall 14, 33, 44, 71, 76–8, 80, 82–4, 86, 88, 94, 262, 280, 282, 294, 298
water uptake 271–72
water use by weeds 33
water use efficiency (WUE) 6–7, 26, 29, 33, 35, 44, 51, 53–4, 169, 177, 181, 184, 195, 242, 248, 262, 298
water use for irrigation 21
water use in 20, 25–30
water use in India for agriculture 7
water use-grain yield relation 176
water withdrawal rates 6
waterlogged soils 212
waterlogging 8, 14, 73, 75, 83, 95, 201
watershed approach 60–1, 63, 69
watershed based interventions 194
watershed defined 60
watershed development project 59–60, 66
watershed development strategies 208
watershed hydrological units 184
watershed landscape 62
watershed management 35, 57–63, 67–70, 133, 136–40, 181–2, 184, 194–5, 203, 220
frameworks 63
models 139
planning 138–9, 220
programs 57–8, 64–5, 202
watershed project implementation 70
watershed research 194
watershed scale 184, 212
watershed soil erosion 61
watershed technology 133, 184
watershed toolbox 64
watersheds and water quality 135
water-storage ponds 218
water-stress management 229
water-stress tolerance mechanisms 52
water-stress-tolerant transgenic plants 240
water-table depth 77
wave-shaped broad beds 191
WCMF system. *See* wheat-corn-maize-fallow

Web-based technologies 68
weed biomass 302
weed control 45, 176, 212, 221, 297, 299–305, 307–10
weed density 297, 300, 302–4
weed dry matter 297, 301–2, 305
weed dry weight 300, 302
weed flora 302, 310
weed management 297–9, 310
 in aerobic rice 297
weed pressure 297, 299, 302
weed-free treatment 307
weedy check 302, 304, 307
well buffer zones 17
well watered 270
west-central Great Plains 44
wet tillage 298
wetlands 60–2, 70, 133–5, 137–8, 140
wheat 12, 29–31, 33–4, 36–9, 43–6, 49–52, 54, 71–3, 76, 80–1, 83–5, 87, 93, 98–100, 102–6, 144, 146, 151–2, 161, 171–4, 182–3, 196, 210, 221, 229–30, 232, 240, 242, 244, 248, 251, 279, 280, 282–3, 285–90, 292, 294, 298
 import 72
 production 172
 straw for fodder use 33
wheat-corn-millet-fallow 46
wheat-fallow 173
 cropping system 43
whole-genome microarrays 244
wildlife habitat protection 61
wind 13, 45–6, 48–9, 170, 173–5, 178, 210–11, 218
wind erosion 13, 211, 218
wind-break plantations 218
winter crop 173
winter wheat 44–5, 49–50, 173–6
withholding water from soil-grown plants 268
woolen industries 155, 162
world food grain production 5
world food production by 2030 4
world peace 4
world population 4–5, 10, 17, 134, 144
world-market standards 209
worm holes 12, *See also* macrophores
WP. *See* water productivity
WT. *See* water table
WUE. *See* water use efficiency
WW. *See* well watered

Yamuna river basin 26
Yangtze 5
Yellow river 5
yield architecture 263, 266, 272
yield gap 183
yield response to residue management 287

Zea mays 31, 46, 95, 151, 158, 221, 282
zero-energy 110, 124
 and drip-irrigation system 110
ZFPs. *See* zinc finger proteins
zinc 104, 118, 155, 189, 245
zinc finger proteins 229, 245
zinc sulfate 104, 118
ZmDREB2A 244